T0360408

Enantioselective Titanium-Catalysed Transformations

CATALYTIC SCIENCE SERIES

Series Editor: Graham J. Hutchings *(Cardiff University)*

*To view the complete list of the published volumes in the series, please visit:
http://www.worldscientific.com/series/css

CATALYTIC SCIENCE SERIES — VOL. 14

Series Editor: Graham J. Hutchings

Enantioselective Titanium-Catalysed Transformations

Hélène Pellissier

CNRS, France

Imperial College Press

ICP

Published by

Imperial College Press
57 Shelton Street
Covent Garden
London WC2H 9HE

Distributed by

World Scientific Publishing Co. Pte. Ltd.

5 Toh Tuck Link, Singapore 596224

USA office: 27 Warren Street, Suite 401-402, Hackensack, NJ 07601

UK office: 57 Shelton Street, Covent Garden, London WC2H 9HE

Library of Congress Cataloging-in-Publication Data
Pellissier, Hélène.
    Enantioselective titanium-catalysed transformations / Hélène Pellissier (CNRS, France).
      pages cm. -- (Catalytic science series ; vol. 14)
    Includes bibliographical references.
    ISBN 978-1-78326-894-8 (alk. paper)
    1. Enantioselective catalysis. 2. Catalysts. 3. Titanium. I. Title.
    QD505.P463 2015
    541'.395--dc23
                   2015030954

British Library Cataloguing-in-Publication Data
A catalogue record for this book is available from the British Library.

In-house Editors: Tasha D'Cruz/Dipasri Sardar

Typeset by Stallion Press
Email: enquiries@stallionpress.com

Printed in Singapore

Contents

General Introduction

The catalysis of organic reactions by metals still constitutes one of the most useful and powerful tools in organic synthesis.[1] Although asymmetric synthesis is sometimes viewed as a sub-discipline of organic chemistry, this topical field actually transcends any narrow classification, and pervades all of chemistry. Of the methods available for preparing chiral compounds, catalytic asymmetric synthesis has attracted most attention. Asymmetric transition-metal catalysis has emerged as a powerful tool to perform reactions in a highly enantioselective fashion over the past few decades.

Previously, efforts to develop new asymmetric transformations focused primarily on the use of few metals, such as titanium, copper, ruthenium, rhodium, palladium, iridium, and more recently gold. However, the lower costs of titanium catalysts in comparison with other transition metals, and their nontoxicity (permitting their use for medical purposes such as prostheses), enantioselective titanium-mediated transformations have received ever-growing attention during the last decades, leading to exciting and fruitful research.[1j,2]

This interest is also related to the fact that titanium complexes are of high abundance, exhibit a remarkably diverse chemical reactivity, and constitute one of the most useful Lewis acids in asymmetric catalysis. The goal of this book is to provide a comprehensive overview of the major developments in enantioselective titanium-catalysed transformations reported since the beginning of 2008, when this field was reviewed by Yu and co-workers in a book chapter dealing with titanium Lewis acids,[2a] as it had been earlier by several groups.[1j,2b,d,e]

This book demonstrates the impressive amount of enantioselective synthetic uses that have been found for new and previously known titanium chiral catalysts in the last seven years — from basic organic transformations, such as nucleophilic additions, cycloadditions, oxidations and reductions, to completely novel methodologies including domino reactions. The excellent capability for the control of the stereochemistry demonstrated by titanium complexes in the catalysis can be attributed to their rich coordination chemistry and facile modification of titanium Lewis acid centre by structurally modular ligands. The advantages of titanium Lewis acids, such as high abundance, low cost and low toxicity, as well as impressively diverse chemical reactivity will definitely continue to stimulate future works on their applications in synthetic organic chemistry.

The book is divided into 10 chapters, according to the different types of reactions catalysed (or promoted) by chiral titanium catalysts, covering alkylation, arylation, alkynylation, allylation, and vinylation reactions of carbonyl compounds in the first chapter; cyanation reactions of carbonyl compounds and derivatives in the second; thioether oxidations in the third; epoxidation reactions in the fourth; cycloaddition reactions in the fifth; aldol-type reactions in the sixth; reduction reactions in the seventh; ring-opening reactions of epoxides and aziridines in the eighth; domino and tandem reactions in the ninth; and miscellaneous reactions for the tenth chapter.

References

(1) (a) R. Noyori, in *Asymmetric Catalysts in Organic Synthesis,* Wiley, New York, **1994**. (b) *Transition Metals for Organic Synthesis,* M. Beller, C. Bolm, eds. Wiley–VCH, Weinheim, **1998**, Vols I and II. (c) *Comprehensive Asymmetric Catalysis,* E. N. Jacobsen, A. Pfaltz, H. Yamamoto, eds.; Springer, Berlin, **1999**. (d) *Catalytic Asymmetric Synthesis,* 2nd ed., I. Ojima, ed., Wiley–VCH, New-York, **2000**. (e) G. Poli, G. Giambastiani, A. Heumann, *Tetrahedron* **2000**, *56,* 5959. (f) E. Negishi in *Handbook of Organopalladium Chemistry for Organic Synthesis,* John Wiley & Sons, Inc., Hoboken NJ, **2002**, *2,* 1689. (g) A. de Meijere, P. von Zezschwitz, H. Nüske, B. Stulgies, *J. Organomet. Chem.* **2002**, *653,* 129. (h) *Transition Metals for Organic Synthesis,* 2nd

ed., M. Beller, C. Bolm, eds., Wiley–VCH, Weinheim, **2004**. (i) L. F. Tietze, I. Hiriyakkanavar, H. P. Bell, *Chem. Rev.* **2004**, *104*, 3453. (j) D. J. Ramon, M. Yus, *Chem. Rev.* **2006**, *106*, 2126.

(2) (a) Y. Yu, K. Ding, G. Chen in *Acid Catalysis in Modern Organic Synthesis*, H. Yamamoto, K. Ishihara, eds., Wiley–VCH: Weinheim, **2008**, Chapter 14, p. 721. (b) F. Chen, X. Feng, Y. Jiang, *Arkivoc* **2003**, *ii*, 21. (c) J. Cossy, S. Bouz, F. Pradaux, C.Willis, V. Bellosta, *Synlett* **2002**, 1595. (d) K. Mikami, M. Terada in *Lewis Acids in Organic Synthesis*, H. Yamamoto, ed., Wiley–VCH: Weinheim, **2000**, Vol. 2, Chapter 16, p. 799. (e) K. Mikami, M. Terada in *Lewis Acid Reagents*, H. Yamamoto, ed., Wiley–VCH: Weinheim, **1999**, Vol. 1, Chapter 6, p. 93. (f) D. Ramon, M. Yus, *Rec. Res. Dev. Org. Chem.* **1998**, *2*, 489. (g) A. H. Hoveyda, J. P. Morken, *Angew. Chem., Int. Ed. Engl.* **1996**, *35*, 1262. (h) B. M. Trost in *Stereocontrolled Organic Synthesis*, B. M. Trost, ed., Blackwell, Oxford, **1994**, p. 17.

Chapter 1

Enantioselective Titanium-Promoted Alkylation, Arylation, Alkynylation, Allylation, and Vinylation Reactions of Carbonyl Compounds

1.1. Introduction

The catalysis of organic reactions by metals still constitutes one of the most useful and powerful tools in organic synthesis.[1] Although asymmetric synthesis is sometimes viewed as a sub-discipline of organic chemistry, actually this topical field transcends any narrow classification and pervades essentially all chemistry. Of the methods available for preparing chiral compounds, catalytic asymmetric synthesis has attracted most attention. In particular, asymmetric transition-metal catalysis has emerged over the past few decades as a powerful tool for performing reactions in a highly enantioselective fashion.

Previously, efforts to develop new asymmetric transformations focused preponderantly on the use of few metals, such as titanium, nickel, copper, ruthenium, rhodium, palladium, iridium, and more recently gold. However, the lower costs of titanium catalysts compared with other transition metals, and their nontoxicity — which has permitted their use for medical purposes (prostheses) — enantioselective titanium-mediated

1

transformations have received a growing attention during the last decades, leading to exciting and fruitful researches.[1j,2] This interest is also related to the fact that titanium complexes are highly abundant, exhibit a remarkably diverse chemical reactivity, and constitute one of the most useful Lewis acids in asymmetric catalysis.

This usefulness is particularly significant in the area of enantioselective 1,2-alkylation, 1,2-arylation, 1,2-alkynylation, 1,2-allylation, and 1,2-vinylation reactions of carbonyl compounds. These methodologies have a strategically synthetic advantage in forming a new carbon to carbon (C–C) bond, a new functionality (alcohol) with concomitant creation of a stereogenic centre in a single transformation.

Since the first enantioselective titanium-promoted addition of diethylzinc to benzaldehyde was reported in 1989 by Ohno and Yoshioka,[3] enantioselective titanium-promoted additions of organometallic reagents to prochiral aldehydes and ketones have been studied extensively. For example, impressive progress has been made in the design and development of chiral titanium Lewis acids for asymmetric catalysis of additions of dialkylzinc reagents to carbonyl compounds to reach various chiral functionalised alcohols under relatively mild conditions. This is based on the extraordinary ability of chiral titanium catalysts to control stereochemistry which can be attributed to their rich coordination chemistry and facile modification of titanium Lewis acid centre by structurally modular ligands. The enantioselective addition of organometallic alkynyl derivatives to carbonyl compounds is today the most expedient route toward chiral propargylic alcohols, which constitute strategic building blocks for the enantioselective synthesis of a range of complex and important molecules. It must be noted that it is only recently that the first highly efficient enantioselective alkylations of aldehydes with trialkylboranes and organolithium reagents have been developed. Moreover, highly enantioselective titanium-promoted alkynylations of aldehydes with special alkynes, such as diynes and enynes, have been successfully developed recently among other reactions.

This chapter provides a comprehensive overview of the major developments in enantioselective titanium-promoted alkylation, arylation, alkynylation, allylation, and vinylation reactions of carbonyl compounds reported since the beginning of 2008, when this general field was reviewed by Yu and co-workers in a book chapter dealing with titanium Lewis acids.[2a] (In addition, areas of this field were included in several reviews not especially based on titanium catalysis.)[4–5]

The chapter has been divided into three parts. The first part deals with enantioselective titanium-promoted alkylation and arylation reactions of carbonyl compounds; the second part includes enantioselective titanium-promoted alkynylation reactions of carbonyl compounds; while the third part collects enantioselective titanium-promoted allylation and vinylation reactions of carbonyl compounds.

1.2. Titanium-Promoted Alkylation and Arylation Reactions of Carbonyl Compounds

The formation of a C–C bond via nucleophilic addition of an organometallic reagent to a carbonyl substrate constitutes one of the most elementary transformations in organic synthesis, and has been studied extensively during the last several decades.[6] The dawn of organometallic chemistry dates back to 1849 with Frankland's early work on organozinc compounds.[6] By the turn of the 20th century, the routine use of organozinc reagents in organic synthesis was largely supplanted by main-group organometallics, thanks to the rapid growth of Grignard chemistry,[7] and the development of practical routes to organolithium compounds.[8] The genesis of enantioselective addition to carbonyl compounds dates from 1940, with a report by Betti and Lucchi on the reaction of methylmagnesium iodide with benzaldehyde in the presence of *N,N*-dimethylbornylamine as solvent to give enantioenriched 1-phenylethanol.[9] It was later demonstrated that the slight optical rotation observed apparently originated from an optically active by-product generated from the *N,N*-dimethylbornylamine solvent. In the 1950s, Wright *et al.* reported what appears to be the first successful enantioselective addition of Grignard reagents to carbonyl compounds, using chiral ethers as cosolvents, providing enantioselectivities of 17% ee.[10] The nucleophilic addition reactions of organometallic reagents (such as organozinc, aluminum, magnesium, titanium, and lithium species in addition to boron reagents) to carbonyl compounds can be mediated by titanium complexes, through transmetalation of organometallic reagents, or by enhancing the electrophilicity of the carbonyl compounds via titanium coordination. Alkyltitanium complexes can be obtained from metal carbanions via titanation. Introduction of chirality at the titanium centre or on the ligand (or a combination of both) enables the possibility of asymmetric induction in the carbonyl addition reaction. After the pioneering work by Seebach and Weber on the asymmetric catalytic phenyl additions to aldehydes employing a highly reactive $PhTi(Oi-Pr)_3$ reagent in 1994,[11] chemists have shown a continual interest in developing highly enantioselective catalysts for the asymmetric alkyl and aryl transfer to aldehydes.

1.2.1. Additions of Dialkylzinc Reagents to Carbonyl Compounds

1.2.1.1. Aldehydes as electrophiles

In 1983, Oguni and Omi reported the first reaction of diethylzinc with benzaldehyde performed in the presence of a catalytic amount of (S)-leucinol achieved with a moderate enantioselectivity (49% ee).[12] Later, in 1986, Noyori and co-workers discovered (–)-DAIB, the first highly enantioselective ligand for the dialkylzinc addition to aldehydes, providing enantioselectivities of up to 95% ee.[13] In 1989, Yoshioka and Ohno[3] reported the first titanium-promoted enantioselective addition of dialkylzinc reagents[14] to aldehydes,[4b–d,6g] using chiral *trans*-1,2-*bis*(trifluoromethanesulfonyl) aminocyclohexane as ligand of titanium tetraisopropoxide. The titanium catalyst was prepared *in situ* in the presence of the diorganozinc. In 1991, Seebach and Schmidt demonstrated that TADDOL-derived titanium complexes also functioned as efficient asymmetric catalysts.[15] Ever since, a number of chiral titanium catalysts have been developed, and high enantioselectivities have been reached.[4,16]

Studies reported in 2004 by Gau and Wu on the mechanism of the addition of alkylzinc reagents in the presence of $Ti(Oi-Pr)_4$ suggested that the role of $Ti(Oi-Pr)_4$ was to receive an alkyl group from the zinc reagent, and to bind the chiral titanium catalyst in a preset form.[17] Indeed, $Ti(Oi-Pr)_4$ first aids in the transfer of the nucleophilic alkyl group to the aldehyde. Shortly after mixing, R_2Zn is consumed and $RZn(Oi-Pr)$ is produced, along with various ethyltitanium species that may serve as alkylating agents.[6e,18] Second, $Ti(Oi-Pr)_4$ often reacts with the chiral ligand to form a Lewis acidic complex which activates the carbonyl group for the nucleophilic attack.[3a,19]

It must be noted that the exact mechanism of this enantioselective reaction performed in the presence of an excess of Ti(Oi-Pr)$_4$ and substoichiometric amounts of chiral ligands is still not well-known. A number of titanium chiral ligands have been successfully applied to induce chirality when dialkylzinc reagents are added to aldehydes including BINOL derivatives.[20] For example, Abdi *et al.* have reported good to high enantioselectivities of up to 94% ee for the reaction of various aldehydes with diethylzinc induced by silica-supported chiral BINOL ligand **1**, used at 5 mol% of catalyst loading in the presence of Ti(Oi-Pr)$_4$.[21] This heterogenised ligand was covalently anchored on two different, relatively large pore sized, mesoporous silicas (SBA-15) (7.5 nm) and mesocellular foams (MCF) (14 nm), by a covalent grafting method using *N*-methyl-3-aminopropyltriethoxysilane as the reactive surface modifier. As shown in Scheme 1.1, the reaction of small as well as bulkier aldehydes afforded the corresponding chiral secondary alcohols with excellent conversions of 94 to 99%, and good to very high enantioselectivities of up to 94% ee in the case of aromatic aldehydes, while cinnamaldehyde provided an 88% conversion combined with an enantioselectivity of 86% ee.

Note that the substituents on benzaldehyde derivatives had some influence on the reactivity and enantioselectivity of the reaction, since *p*-substituted aldehydes showed better reactivity with respect to conversion and enantioselectivity than *o*-substituted benzaldehyde (Scheme 1.1). This might probably due to the strong steric effect of the *o*-substituent, which could deteriorate the coordination of the substrate to the chiral catalyst, thus lowering the reactivity. The MCF-supported BINOL catalyst could be reused in several catalytic runs without significant drop of enantioselectivity. The pore size of silica supports, and capping of free silanol groups with trimethylsilyl (TMS) groups on the silica surface, were found to be important toward achieving high enantioselectivities.

In 2008, Li and co-workers reported the synthesis of a range of novel 3-substituted chiral BINOL ligands to be applied to enantioselective diethylzinc addition to aldehydes.[22] For example, three 3-aminomethyl-substituted BINOL ligands — such as (*S*)-3-(1H-imidazol-1-yl)methyl-1,1'-binaphthol, (*S*)-3-(1H-1,2,3-benzotriazol-1-yl)methyl-1,1'-binaphthol, and (*S*)-3-(2H-1,2,3-benzotriazol-2-yl)methyl-1,1'-binaphthol — were easily synthesised from (*S*)-2,2'-dimethoxymethyl-1,1'-binaphthol in four steps, and further investigated as chiral ligands of Ti(Oi-Pr)$_4$ in addition of

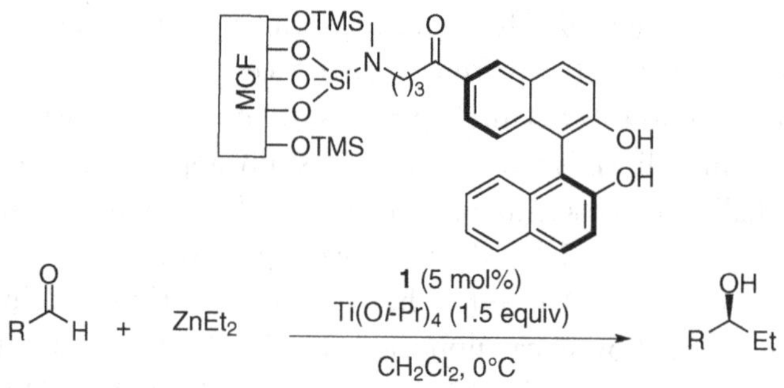

R = Ph: 99% conversion, ee = 94%
R = *p*-Tol: 98% conversion, ee = 87%
R = *p*-FC$_6$H$_4$: 98% conversion, ee = 90%
R = *m*-MeOC$_6$H$_4$: 99% conversion, ee = 85%
R = *o*-MeOC$_6$H$_4$: 94% conversion, ee = 70%
R = 1-Naph: 98% conversion, ee = 91%
R = (*E*)-PhCH=CH: 88% conversion, ee = 86%

Scheme 1.1. Silica-supported chiral BINOL ligand in addition of diethylzinc to aldehydes

diethylzinc to benzaldehyde. The best enantioselectivity of 77% ee combined with 91% yield was achieved by using (*S*)-3-(2H-1,2,3-benzotriazol-2-yl)methyl-1,1'-binaphthol **2** at 5 mol% of catalyst loading, as shown in Scheme 1.2. It is important to highlight that this work represented a rare example of using only catalytic amounts of Ti(O*i*-Pr)$_4$ (70 mol%) in enantioselective titanium-promoted alkylation of aldehydes.

Better enantioselectivities were reached by the same authors employing other chiral 3-substituted aminomethyl BINOL derivatives in the reaction of diethylzinc with a range of aliphatic as well as aromatic aldehydes.[23] As shown in Scheme 1.2, enantioselectivities of up to 92% ee were obtained by inducing a reaction with chiral 3-arylaminomethyl BINOLs, such as (*R*)-3-(naphthalene-1-ylamino)methyl-1,1'-binaphthol **3**. This ligand was selected among three novel chiral 3-substituted BINOL Schiff bases, which provided moderate enantioselectivities (≤77% ee) and their reductive 3-arylaminomethyl-BINOL derivatives. Using the most efficient ligand **3**, the best result was achieved for sterically hindered 1-naphthaldehyde, with an almost quantitative yield in combination with an enantioselectivity of 92% ee. Moreover, these authors investigated novel BINOL-based ligands bounded with both sulfur-contained heterocycle (such as thiazole or thiadiazole),

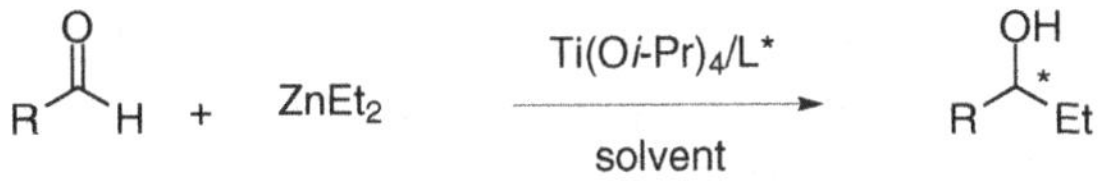

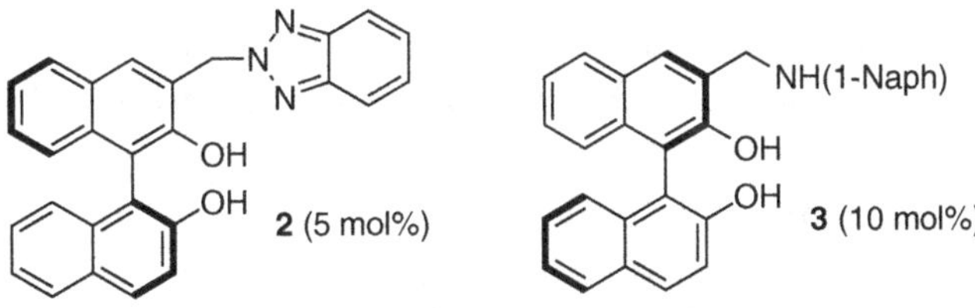

R–CHO + ZnEt₂ →[Ti(O*i*-Pr)₄/L*] [solvent] R–CH(OH)–Et (*)

$$\text{R-CHO} + \text{ZnEt}_2 \xrightarrow[\text{solvent}]{\text{Ti(O}i\text{-Pr)}_4/\text{L*}} \text{R-*CH(OH)-Et}$$

2 (5 mol%)

3 (10 mol%)

Ti(O*i*-Pr)₄ (70 mol%)
toluene, r.t.

R = Ph: 91% yield, ee = 77% (*S*)

Ti(O*i*-Pr)₄ (1.2 equiv)
CH₂Cl₂, 0 °C

R = Ph: 93% yield, ee = 87% (*R*)
R = *p*-ClC₆H₄: 96% yield, ee = 88% (*R*)
R = *p*-BrC₆H₄: 99% yield, ee = 89% (*R*)
R = *p*-MeOC₆H₄: 95% yield, ee = 80% (*R*)
R = *o*-MeOC₆H₄: 95% yield, ee = 82% (*R*)
R = 1-Naph: >99% yield, ee = 92% (*R*)
R = (*E*)-PhCH=CH: 99% yield, ee = 87% (*R*)

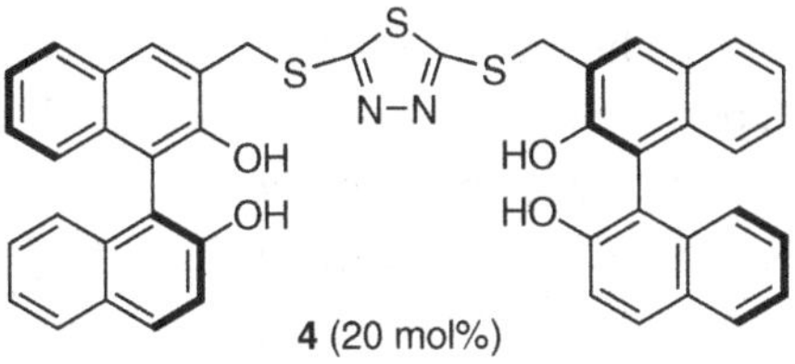

4 (20 mol%)
Ti(O*i*-Pr)₄ (1 equiv)
toluene, r.t.

R = Ph: 95% yield, ee = 81% (*S*)
R = *p*-ClC₆H₄: 94% yield, ee = 89% (*S*)
R = *p*-BrC₆H₄: 95% yield, ee = 91% (*S*)
R = *p*-MeOC₆H₄: 91% yield, ee = 87% (*S*)
R = *o*-MeOC₆H₄: 97% yield, ee = 93% (*S*)
R = (*E*)-PhCH=CH: 94% yield, ee = 87% (*S*)
R = 1-Naph: 92% yield, ee = 86% (*S*)

Scheme 1.2. 3-Substituted chiral BINOL ligands in addition of diethylzinc to aldehydes and thioether block, in which the sulfur could serve as a talent anchor.[24] The active chiral titanium catalyst in the enantioselective addition of diethylzinc to benzaldehyde was generated *in situ* from Ti(O*i*-Pr)$_4$ and the chiral ligand. Among four ligands tested, ligand (*S,S*)-2,5-bis(2,2′-dihydroxy-1,1′-binaphthalene-3-yl)-1,3,4-thiadiazole **4** was found to be the most efficient, providing (*S*)-secondary alcohols in enantioselectivities of up to 93% ee as well as generally excellent yields, ranging from 91–97% for a range of aldehydes, as shown in Scheme 1.2.

In 2010, the same authors described a novel 3-substituted chiral ligand derived from (*S*)-BINOL, such as (*S*)-3-dihydroxyborane-2,2′-bis(methoxymethoxy)-1,1′-binaphthyl **5**.[25] This ligand was easily synthesised in 82% overall yield, starting from commercially available (*S*)-BINOL through a six-step sequence. The authors further studied its catalytic activity in order to induce diethylzinc addition to aldehydes among a range of a series of (*S*)-BINOL-derived ligands substituted at the 3 position, with some five-membered nitrogen-containing aromatic heterocycles. They found that ligand **5** exhibited the best catalytic efficiency, providing enantioselectivities of up to 88% ee in combination with very high to quantitative yields, as shown in Scheme 1.3. The best results were achieved by using 5 mol% of catalyst loading. The scope of the procedure was extended to a range of aromatic, heteroaromatic, and α,β-unsaturated aldehydes with diverse electronic and steric properties (Scheme 1.3).

In addition to chiral 3-substituted BINOL ligands, Judeh and Gou have described the synthesis and applications of novel chiral 2,2′-disubstituted BINOL ligands on the same reactions.[26] As shown in Scheme 1.4, the use of ligand **6** bearing two free hydroxyl (OH) groups and synthesised in one step from (*S*)-BINOL allowed a range of chiral secondary aromatic alcohols to be achieved in good to high yields, with enantioselectivities of up to 89% ee. The authors have demonstrated that the presence of the free OHs were indispensable for catalytic activity.

Chiral periodic mesoporous organosilicas with chiral ligands in the framework are novel chiral porous solids which have demonstrated application potential in asymmetric catalysis, constituting a challenge in the field of heterogeneous asymmetric catalysis. In 2010, Yang *et al.* reported the synthesis of (*R*)-BINOL-functionalised mesoporous organosilicas PPB-30, based on co-condensation of 1,2-*bis*(trimethoxysilyl)ethane (BTME) and

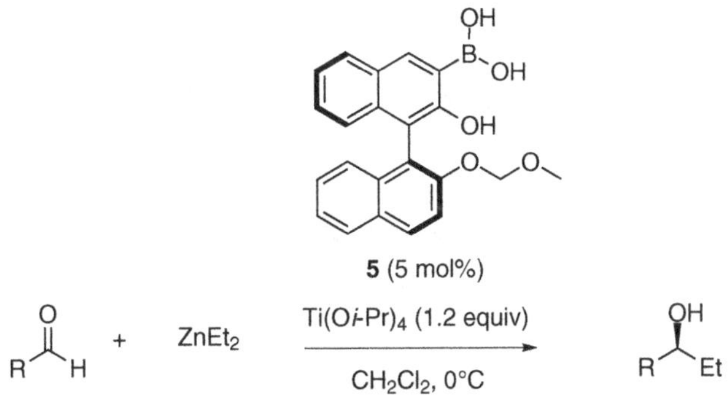

5 (5 mol%)

$$\underset{R}{\overset{O}{\underset{}{\|}}}\text{H} \quad + \quad \text{ZnEt}_2 \quad \xrightarrow[\text{CH}_2\text{Cl}_2,\ 0°C]{\text{Ti(O}i\text{-Pr)}_4\ (1.2\ \text{equiv})} \quad \underset{R}{\overset{OH}{\underset{}{\|}}}\text{Et}$$

R = Ph: >99% yield, ee = 83%
R = p-FC$_6$H$_4$: >99% yield, ee = 82%
R = p-ClC$_6$H$_4$: 93% yield, ee = 82%
R = p-BrOC$_6$H$_4$: 98% yield, ee = 84%
R = p-MeOC$_6$H$_4$: 97% yield, ee = 80%
R = p-Tol: 92% yield, ee = 83%
R = p-CF$_3$C$_6$H$_4$: 78% yield, ee = 85%
R = p-(NMe$_2$)C$_6$H$_4$: 90% yield, ee = 82%
R = p-((EtO)$_2$CH)C$_6$H$_4$: 98% yield, ee = 84%
R = m-MeOC$_6$H$_4$: 93% yield, ee = 88%
R = m-Tol: 83% yield, ee = 78%
R = o-MeOC$_6$H$_4$: >99% yield, ee = 76%
R = o-ClC$_6$H$_4$: 93% yield, ee = 70%
R = 2,4-Cl$_2$C$_6$H$_3$: >99% yield, ee = 74%
R = 2,4-(MeO)$_2$C$_6$H$_3$: 94% yield, ee = 64%
R = 3,4-Cl$_2$C$_6$H$_3$: 86% yield, ee = 76%
R = 1-Naph: >99% yield, ee = 91%
R = (E)-PhCH=CH: >99% yield, ee = 80%
R = 2-furyl: 98% yield, ee = 48%
R = 5-piperonyl: >99% yield, ee = 82%

Scheme 1.3. Another 3-substituted chiral BINOL ligand in addition of diethylzinc to aldehydes

(R)-2,2′-di(methoxymethyl)oxy-6,6′-di(1-propyltrimethoxysilyl)-1,1′-binaphthyl (BSBINOL) as a chiral silane precursor in an acidic medium, using the P123 surfactant as the template (Scheme 1.5).[27] When applied to the same reactions as above, this chiral heterogeneous ligand exhibited higher enantioselectivity, but lower catalytic activity, than its homogeneous counterpart in CH$_2$Cl$_2$ as solvent. A shown in Scheme 1.5, a range of chiral aromatic secondary alcohols was obtained, in good to high enantioselectivities of up to 92% ee. Moreover, this catalyst was shown to be

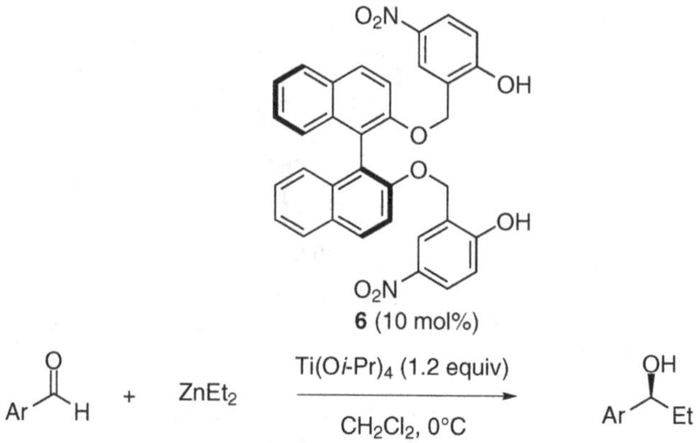

Ar = Ph: 90% yield, ee = 77%
Ar = *o*-Tol: 84% yield, ee = 73%
Ar = *m*-Tol: 88% yield, ee = 67%
Ar = *p*-Tol: 92% yield, ee = 68%
Ar = *p*-EtC_6H_4: 87% yield, ee = 83%
Ar = *p*-PhC_6H_4: 89% yield, ee = 67%
Ar = *o*-$MeOC_6H_4$: 77% yield, ee = 78%
Ar = *m*-$MeOC_6H_4$: 73% yield, ee = 61%
Ar = *p*-$MeOC_6H_4$: 80% yield, ee = 79%
Ar = *m*-$BnOC_6H_4$: 71% yield, ee = 83%
Ar = *p*-FC_6H_4: 70% yield, ee = 79%
Ar = *p*-ClC_6H_4: 93% yield, ee = 84%
Ar = *p*-BrC_6H_4: 88% yield, ee = 69%
Ar = *p*-IC_6H_4: 81% yield, ee = 89%
Ar = *p*-$CF_3C_6H_4$: 76% yield, ee = 64%
Ar = *o*-IC_6H_4: 76% yield, ee = 81%
Ar = *m*-IC_6H_4: 84% yield, ee = 82%
Ar = 1-Naph: 92% yield, ee = 72%
Ar = 2-Naph: 95% yield, ee = 75%

Scheme 1.4. 2,2′-Disubstituted chiral BINOL ligand in addition of diethylzinc to aromatic aldehydes

recyclable, since an 88% conversion combined with 89% ee was obtained in the second run.

In 2010, Pereira *et al.* reported the preparation of novel chiral BINOL-derived ligands consisting of two BINOL or H_8-BINOL fragments, joined by diverse linkages through the oxygen at the 2′-position of the arylic fragments.[28] These ligands were further investigated as promotors in the addition of $ZnEt_2$ to benzaldehyde, in the presence of $Ti(O\textit{i}\text{-}Pr)_4$. It was shown preparation of mesoporous ligand PBB-30:

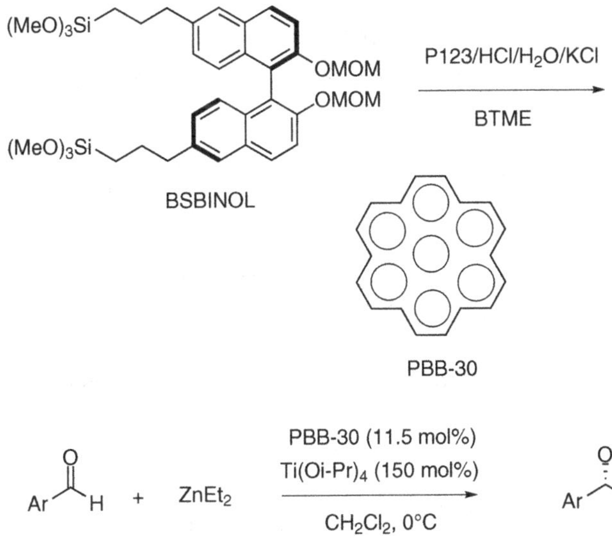

BSBINOL

PBB-30

$$Ar \overset{O}{\underset{H}{\bigvee}} + ZnEt_2 \xrightarrow[\substack{\text{PBB-30 (11.5 mol\%)} \\ \text{Ti(Oi-Pr)}_4 \text{ (150 mol\%)}}]{\text{CH}_2\text{Cl}_2, 0°C} Ar \overset{OH}{\underset{Et}{\bigvee}}$$

Ar = Ph: 99% conversion, ee = 92%
Ar = *o*-Tol: 60% conversion, ee = 93%
Ar = *m*-Tol: 61% conversion, ee = 88%
Ar = *p*-Tol: 40% conversion, ee = 91%
Ar = *m*-MeOC$_6$H$_4$: 98% conversion, ee = 89%
Ar = *p*-MeOC$_6$H$_4$: 88% conversion, ee = 88%
Ar = *m*-ClC$_6$H$_4$: 72% conversion, ee = 84%
Ar = *m*-BrC$_6$H$_4$: 57% conversion, ee = 83%
Ar = *p*-CF$_3$C$_6$H$_4$: 50% conversion, ee = 68%

Scheme 1.5. (R)-BINOL-Functionalised mesoporous organosilica ligand in addition of diethylzinc to aromatic aldehydes

that the performance of these catalysts was very sensitive to the nature of the ether linkage. A ligand with a propylene link provided a better enantioselectivity (70% ee) than those with two or four carbon atoms joining the BINOL fragments. Furthermore, using the propylene link, but replacing (R)-BINOL with (R)-H$_8$-BINOL in ligand 7, a significant improvement in the enantioselectivity of the reaction was achieved (81% ee), as shown in Scheme 1.6. The scope of this methodology was extended to several other aromatic aldehydes, which provided the corresponding chiral alcohols in comparable enantioselectivities of up to 79% ee (Scheme 1.6).[29]

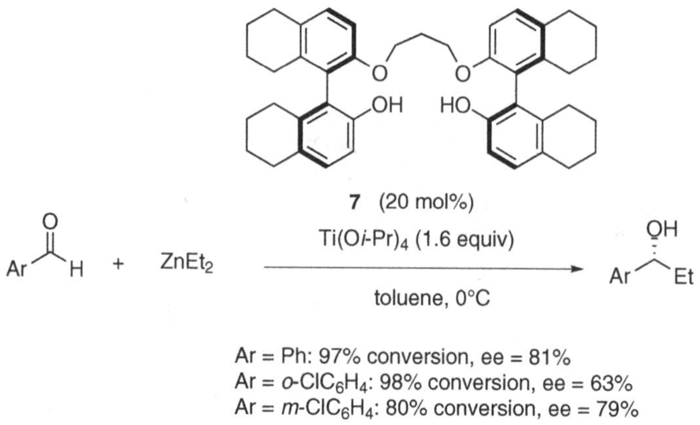

Scheme 1.6. Chiral *bis*-H_8-BINOL-2,2′-propylether ligand in addition of diethylzinc to aromatic aldehydes

In 2011, a series of chiral cross-linked titanium polymers based on the 1,1′-binaphthyl building blocks were synthesised by Lin *et al.* via cobalt-catalysed trimerisation reaction of terminal alkyne groups.[30] These highly porous cross-linked polymers containing chiral dihydroxy functionalities were treated with Ti(O*i*-Pr)₄, to generate chiral Lewis acid catalysts for asymmetric addition of $ZnEt_2$ to aromatic aldehydes. Along with excellent conversions, the observed enantioselectivities were moderate to good, since the most efficient ligand **8** provided enantioselectivities of 68 to 81% ee, as shown in Scheme 1.7. Notably, this polymer could be readily recycled and reused up to ten times without loss of conversion or enantioselectivity.

In 2012, a new class of easily tunable chiral 1,2,3-triazole-BINOL ligands were developed by Mancheno and Beckendorf, and their activity in asymmetric Lewis acid catalysis explored for the first time in the diethylzinc addition to aldehydes.[31] Among them, ligand **9** (Scheme 1.8) showed an interesting catalytic behaviour, which suggested the non-innocent participation of the triazole units in both the formation and reactivity of the active titanium catalyst. Good enantioselectivities of up to 86% ee were obtained by both the right selection of the substitution pattern at the triazole ring (phenyl at 4-position), and the fine tuning of the reaction conditions (10 mol% of catalyst loading in toluene at room temperature). Remarkably, only 10 mol% of Ti(O*i*-Pr)₄ was employed in this process.

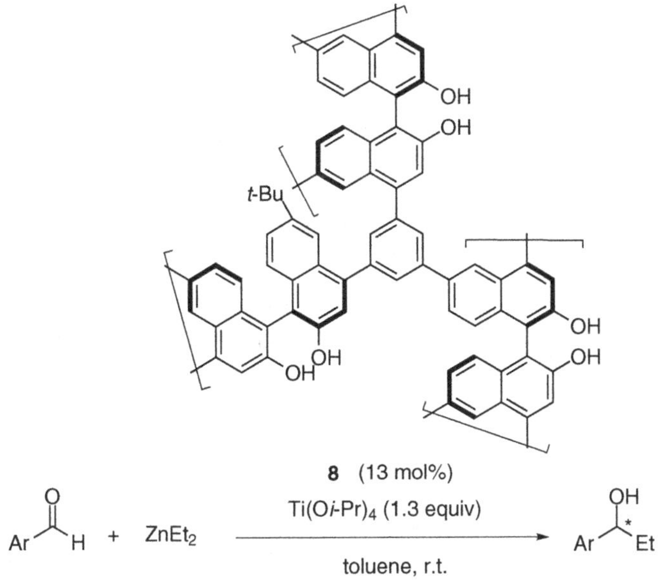

8 (13 mol%)

Ti(O*i*-Pr)$_4$ (1.3 equiv)

$$Ar\overset{O}{\underset{}{\diagdown}}H\ +\ ZnEt_2 \xrightarrow{\quad\text{toluene, r.t.}\quad} Ar\overset{OH}{\underset{}{\diagdown}}\overset{*}{}Et$$

Ar = Ph: 94% conversion, ee = 68%
Ar = *p*-ClC$_6$H$_4$: >99% conversion, ee = 71%
Ar = *p*-BrC$_6$H$_4$: >99% conversion, ee = 71%
Ar = Ph: >99% conversion, ee = 81%

Scheme 1.7. Chiral cross-linked polymer ligand in addition of diethylzinc to aromatic aldehydes

Always in the area of BINOL-derived ligands, Harada *et al.* have studied the enantioselective titanium-promoted alkylation of aldehydes by using functionalised alkylzinc bromides for the first time.[32] It was shown that the reactivity of these organozinc halide reagents was enhanced by mixing them with Ti(O*i*-Pr)$_4$ and MgBr$_2$. In the presence of (*R*)-DPP-H$_8$-BINOL, a variety of functionalised alkylzinc reagents, prepared from readily available bromide precursors, underwent enantioselective addition to aromatic, heteroaromatic, and α,β-unsaturated aldehydes, giving the corresponding functionalised chiral alcohols in good to high enantioselectivities. For example, enantioselectivities of up to 93% ee were achieved, in combination with general high yields. by using a range of variously functionalised zinc reagents **10** prepared from the corresponding bromide precursors by treatment with zinc dust in the presence of LiCl. The same

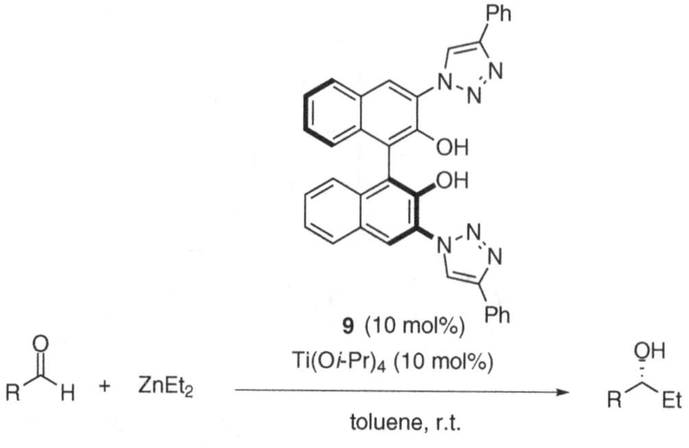

R = Ph: 79% yield, ee = 86%
R = 2-Naph: 99% yield, ee = 86%
R = 1-Naph: 86% yield, ee = 86%
R = p-MeOC$_6$H$_4$: 96% yield, ee = 77%
R = p-FOC$_6$H$_4$: 75% yield, ee = 78%
R = p-ClC$_6$H$_4$: 94% yield, ee = 81%
R = m-ClC$_6$H$_4$: 66% yield, ee = 46%
R = o-ClC$_6$H$_4$: 86% yield, ee = 54%
R = Cy: 39% yield, ee = 50%

Scheme 1.8. Chiral 1,2,3-triazole-BINOL ligand in addition of diethylzinc to aldehydes

reaction conditions were applied to the enantioselective addition of BuZnBr to aldehydes, which provided the corresponding chiral alcohols **11** in generally lower enantioselectivities — except for naphthaldehyde (93% ee) — as shown in Scheme 1.9.

In addition to BINOL-derived ligands, a range of other types of chiral ligands have been investigated in enantioselective titanium-promoted addition of zinc reagents to aldehydes. For example, Ando *et al.* have developed novel recyclable fluorous chiral ligands designed as the first fluorinated analogues of TADDOL.[33] Unlike TADDOL,[34] which has four aromatic substituents, these ligands have only three perfluoroalkyl substituents. Applied to induce chirality in addition of ZnMe$_2$ to aromatic as well as aliphatic aldehydes, these novel ligands provided excellent results. Among them, diol **12** seems to be the most effective, since very high enantioselectivities of up to 98% ee were achieved, along with general almost quantitative yields for various aldehydes, as shown in Scheme 1.10. Moreover, the

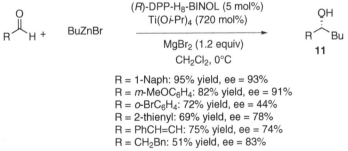

(*R*)-DPP-H$_8$-BINOL (5 mol%)
Ti(O*i*-Pr)$_4$ (720 mol%)

$$R^1\!-\!CHO + R^2ZnBr\cdot LiCl \quad\xrightarrow[\text{toluene, r.t. or CH}_2\text{Cl}_2,\ 0\ °C]{\text{MgBr}_2\ (1.2\ \text{equiv})}\quad R^1\!-\!CH(OH)\!-\!R^2$$

10

R^1 = Ph, R^2 = *n*-Bu: 87% yield, ee = 86%
R^1 = *p*-Tol, R^2 = *n*-Bu: 86% yield, ee = 89%
R^1 = *p*-ClC$_6$H$_4$, R^2 = *n*-Bu: 92% yield, ee = 92%
R^1 = 1-Naph, R^2 = C$_{12}$H$_{25}$: 76% yield, ee = 91%
R^1 = 1-Naph, R^2 = (CH$_2$)$_2$CH=CH$_2$: 96% yield, ee = 93%
R^1 = *p*-ClC$_6$H$_4$, R^2 = (CH$_2$)$_3$CH=CH$_2$: 81% yield, ee = 91%
R^1 = 1-Naph, R^2 = (CH$_2$)$_4$Cl: 91% yield, ee = 86%
R^1 = *m*-MeOC$_6$H$_4$, R^2 = (CH$_2$)$_3$OTIPS: 96% yield, ee = 91%
R^1 = *p*-ClC$_6$H$_4$, R^2 = (CH$_2$)$_6$OTIPS: 92% yield, ee = 90%
R^1 = 1-Naph, R^2 = (CH$_2$)$_6$OTr: 77% yield, ee = 92%
R^1 = 1-Naph, R^2 = C$_8$H$_{16}$CN: 80% yield, ee = 86%
R^1 = PhCH=CH, R^2 = Bu: 84% yield, ee = 83%

(*R*)-DPP-H$_8$-BINOL (5 mol%)
Ti(O*i*-Pr)$_4$ (720 mol%)

$$R\!-\!CHO + BuZnBr \quad\xrightarrow[\text{CH}_2\text{Cl}_2,\ 0°C]{\text{MgBr}_2\ (1.2\ \text{equiv})}\quad R\!-\!CH(OH)\!-\!Bu$$

11

R = 1-Naph: 95% yield, ee = 93%
R = *m*-MeOC$_6$H$_4$: 82% yield, ee = 91%
R = *o*-BrC$_6$H$_4$: 72% yield, ee = 44%
R = 2-thienyl: 69% yield, ee = 78%
R = PhCH=CH: 75% yield, ee = 74%
R = CH$_2$Bn: 51% yield, ee = 83%

Scheme 1.9. Chiral 3,5-diphenylphenyl-H$_8$-BINOL ligand in additions of organozinc halide reagents to aldehydes

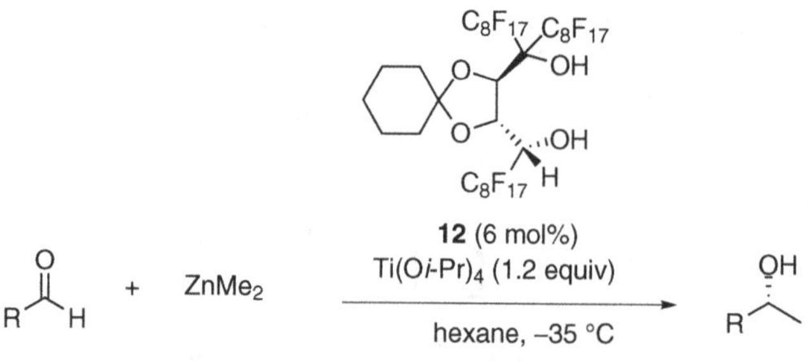

$$C_8F_{17} \quad C_8F_{17}$$

(TADDOL-based fluorinated ligand structure)

$$C_8F_{17} \quad H$$

12 (6 mol%)

$$\underset{R}{\overset{O}{\underset{\quad}{\parallel}}}\!\!\!\!\!\!{H} \quad + \quad ZnMe_2 \quad \xrightarrow[\text{hexane, } -35\,^{\circ}C]{Ti(Oi\text{-}Pr)_4 \ (1.2 \text{ equiv})} \quad \underset{R}{\overset{OH}{\diagup\!\!\!\diagdown}}$$

R = Ph: 99% yield, ee = 95%
R = 2-Naph: 95% yield, ee = 95%
R = o-ClC$_6$H$_4$: 96% yield, ee = 93%
R = o-FC$_6$H$_4$: 99% yield, ee = 91%
R = 2-furyl: 98% yield, ee = 98%
R = (E)-PhCH=CH: 92% yield, ee = 92%
R = n-Hept: 97% yield, ee = 91%
R = Cy: 95% yield, ee = 93%

Scheme 1.10. Chiral TADDOL-based fluorinated ligand in addition of dimethylzinc to aldehydes

authors have studied the recyclability of ligand **12**, taking advantage of its low solubility in cold toluene. Upon cooling a solution of the crude reaction product in toluene, the ligand precipitated, and was separated from the products by filtration. By this simple method, the authors could recycle ligand **12** up to four times without decreasing the enantioselectivity.

In 2008, Hitchcock and Dean investigated (R,R)-hydrobenzoin **13** as chiral ligand in asymmetric addition of diethylzinc to aromatic and α,β-unsaturated aldehydes, in the presence or absence of Ti(Oi-Pr)$_4$.[35] The enantioselectivity of the process involving no titanium catalyst was as high as 85% ee in the case of 2-naphthaldehyde, favouring the formation of the (S)-enantiomer of the corresponding alcohol. Surprisingly, the enantioselectivities of the reaction when performed in the presence of Ti(Oi-Pr)$_4$ were of up to 68% ee, favouring the formation of the (R)-enantiomers. The formation of the opposite enantiomers was attributed to the different transition states **14** and **15**, mediated by either zinc or titanium (Scheme 1.11). As shown in Scheme 1.11, the use of 100 mol% of Ti(Oi-Pr)$_4$ allowed moderate to good enantioselectivities to be achieved (44–68% ee), whereas a low enantioselectivity of 23% ee was obtained in the case of cinnamaldehyde as substrate.

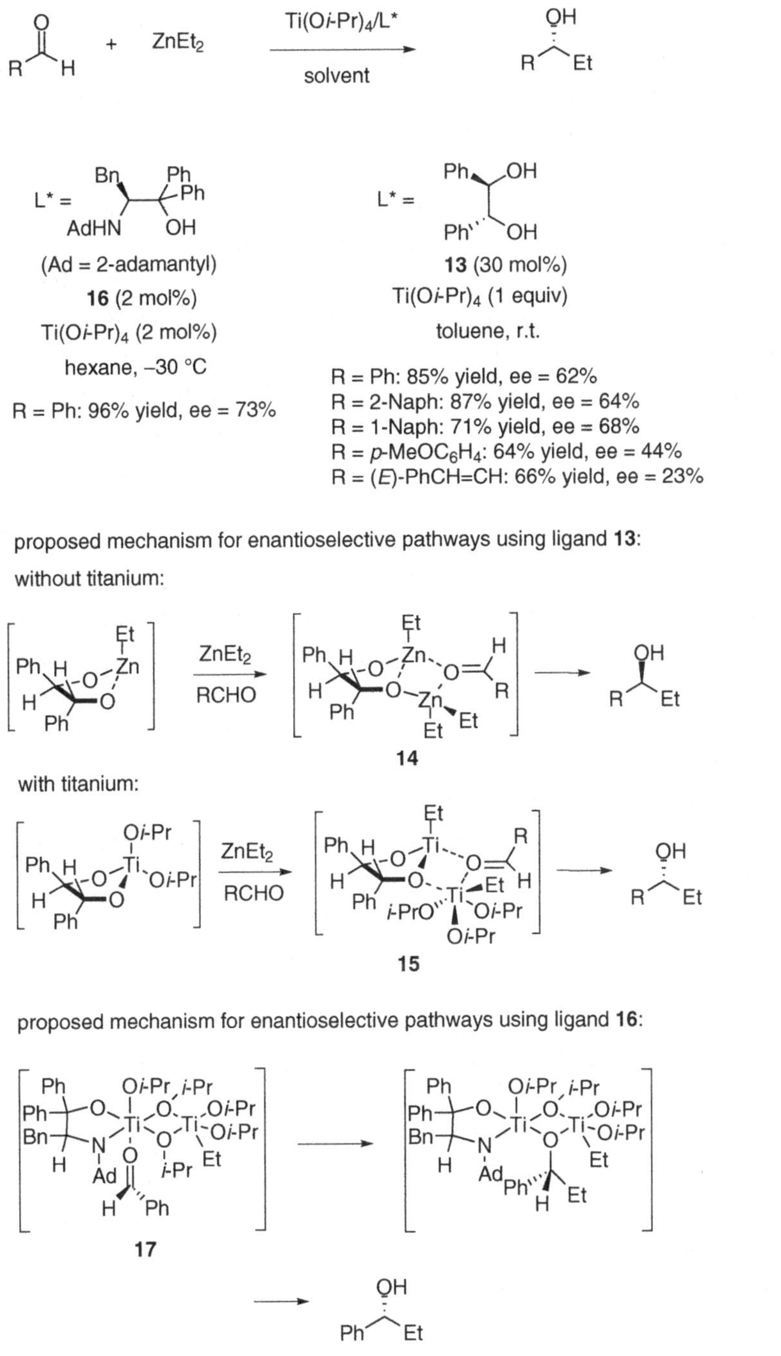

$$\underset{R}{\overset{O}{\|}}\!\!\!\underset{H}{} + ZnEt_2 \xrightarrow[\text{solvent}]{\text{Ti(O}i\text{-Pr)}_4/L^*} \underset{R}{\overset{OH}{\|}}\!\!\!\underset{Et}{}$$

L* =

Bn, Ph
AdHN OH
(Ad = 2-adamantyl)
16 (2 mol%)
Ti(O*i*-Pr)$_4$ (2 mol%)
hexane, –30 °C

R = Ph: 96% yield, ee = 73%

L* =

Ph, OH
Ph'' OH
13 (30 mol%)
Ti(O*i*-Pr)$_4$ (1 equiv)
toluene, r.t.

R = Ph: 85% yield, ee = 62%
R = 2-Naph: 87% yield, ee = 64%
R = 1-Naph: 71% yield, ee = 68%
R = *p*-MeOC$_6$H$_4$: 64% yield, ee = 44%
R = (*E*)-PhCH=CH: 66% yield, ee = 23%

proposed mechanism for enantioselective pathways using ligand **13**:

without titanium:

14

with titanium:

15

proposed mechanism for enantioselective pathways using ligand **16**:

17

Scheme 1.11. Chiral 1,2-diol and 1,2-amino alcohol ligands in addition of diethylzinc to aldehydes

Later, in 2012, Johnson *et al.* studied sterically encumbered chiral L-amino alcohols with secondary amines and tertiary alcohols as ligands in addition of diethylzinc to benzaldehyde.[36] The ligands were substituted at the amino nitrogen with isopropyl, cyclohexyl, or adamantyl groups, and at the position α to the alcohol with hydrogen, methyl, butyl, or phenyl groups. The catalyst which gave the highest enantioselectivities was ligand **16**, exhibiting the most steric hindrance, containing adamantyl substituent on the nitrogen and phenyl groups α to the oxygen. Performing the reaction in the presence of 2 mol% of ligand **16** without Ti(O*i*-Pr)$_4$ provided the corresponding (*R*)-1-phenylpropanol with an enantioselectivity of 58% ee. On the other hand, when the reaction was induced by a catalytic amount (2 mol%) of Ti(O*i*-Pr)$_4$, it afforded the same product with the similar configuration in 96% yield, and a better enantioselectivity of 73% ee, as shown in Scheme 1.11. Actually, in almost all cases, the addition catalysed by the titanium complexes exhibited higher enantioselectivity than that of the amino alcohol ligand alone. A steric argument could be employed to rationalise these results. The dimeric titanium complex **17** had significant steric bulk near the ligand nitrogen due to the *N*-adamantyl substituent. Therefore, when benzaldehyde bound to the titanium centre, it did so in avoiding the steric environment due to the chiral backbone substituent (Bn), and orientating the phenyl group away from the *N*-alkyl substituent. The addition of the ethyl group from the other titanium centres or incoming zinc reagent was then directed to the *Re* face, as shown in Scheme 1.11. It must be highlighted that the reaction catalysed by ligand **16** constituted one rare example of alkylation of aldehydes by zinc reagents involving only a catalytic amount (2 mol%) of titanium.

In 2010, the enantioselective titanium-promoted addition of diethylzinc to benzaldehyde was also performed in the presence of chiral tetrakis(sulfonamides) as ligands by de Parrodi and Somanathan.[37] Indeed, these authors reported the synthesis of a series of novel C_2-symmetric tetrakis(sulfonamides), with the aim of surrounding the Lewis acid titanium metal centre with four chiral nitrogen atoms in a *cisoid* conformation, hoping that the additional chirality could enhance the enantioselectivity of the addition reaction in comparison with the corresponding more simple bis(sulfonamide) ligands. Among various tetrakis(sulfonamide) ligands investigated, ligands **18** and **19** provided the best enantioselectivities of 71 and 81% ee, respectively, along with excellent yields of up to 98%. In the context of sulfonamide ligands, Viso *et al.* have developed a family of chiral sulfinamido-sulfonamide ligands synthesised from sulfinimines, which were evaluated in the same reaction.[38] Interestingly, experimental evidences showed a crucial cooperation between the sulfinyl and sulfonyl functionalities to reach enantiocontrol in the alkylation process, since suppressing the sulfur chiral atom by oxidation into the sulfonamide led to the corresponding bis(sulfonamide) ligands, which provided racemic 1-phenylpropanol when tested under the same conditions. Among a range of various sulfinamido-sulfonamides investigated, the best enantioselectivity of 74% ee, associated with a complete conversion, was obtained with ligand **20** (Scheme 1.12).

In 2010, Watanabe *et al.* reported a total synthesis of paleic acid, an antimicrobial agent effective against *Mannheimia* and *Pasteurella*, which was based on an enantioselective titanium-promoted alkylation of 7-hydroxyheptylaldehyde protected as a *tert*-butyldiphenylsilyl ether **21**, to give the corresponding almost enantiopure alcohol **22** in 62% yield (Scheme 1.13).[39] The process employed a catalytic amount of (1*S*,2*S*)-*bis*(trifluoromethanesulfonylamino)cyclohexane **23** as chiral ligand. Product **22** was subsequently converted into expected paleic acid through five supplementary steps.

Excellent results were described by Bauer and Smolinski in 2010, in the enantioselective addition of diethylzinc to aliphatic and aromatic aldehydes using D-glucosamine-derived ligands.[40] The obvious advantage of these types of ligands was their modular synthesis. Indeed, three sites of this type of ligands were easily altered during the synthesis, thus leading to various ligands having the same chiral precursor. Among them, D-glucosamine-derived β-hydroxy *N*-trifluoromethylsulfonamide **24** was found to be the most active ligand when employing at a catalyst loading as low as 1 mol%, providing remarkable enantioselectivities of up to >99% ee especially with aromatic aldehydes while aliphatic aldehydes gave enantioselectivities of up to 88% ee (Scheme 1.14). It must be noted that generally high yields were achieved, except in the cases of 2-, 3-, and 4-nitrobenzaldehydes and 4-*N*,*N*-dimethylaminobenzaldehyde, which gave yields ranging from 15 to 38%. Finally, a series of novel camphorsulfonylated ligands derived from L-camphor and chiral NOBIN were synthesised by Song *et al.*, to be tested

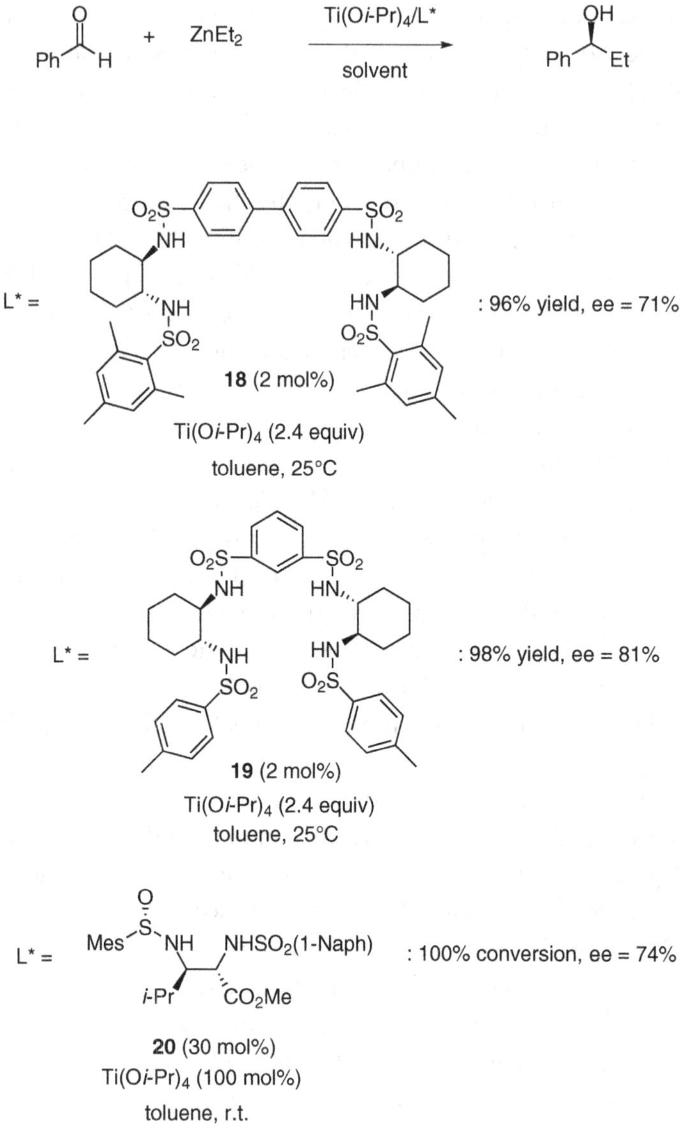

Scheme 1.12. Chiral *bis*(sulfonamide) and sulfinamido-sulfonamide ligands in addition of diethylzinc to benzaldehyde

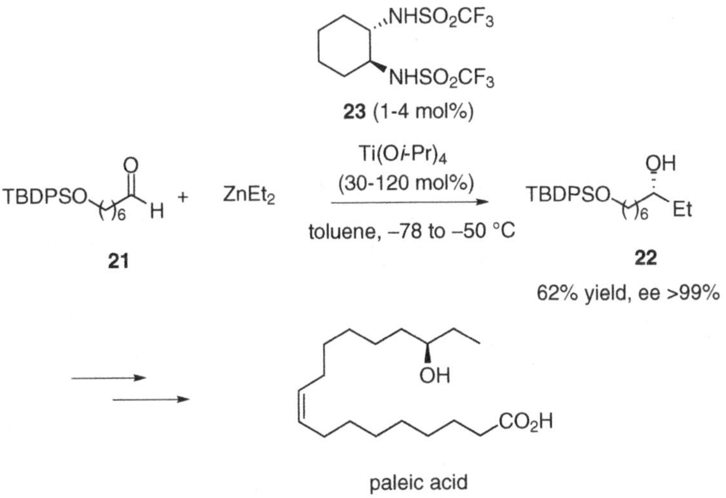

Scheme 1.13. Chiral *bis*(sulfonamide) ligand in addition of diethylzinc to an aliphatic functionalised aldehyde and synthesis of paleic acid

in the titanium-promoted addition of dialkylzinc reagents to aromatic and aliphatic aldehydes.[41] The highest catalytic efficiency was obtained with mono-N-hydroxycamphorsulfonylated (S)-NOBIN **25** in toluene, which gave (S)-addition products with high yields of up to 98%, and enantiose-lectivities of up to 87% ee, as shown in Scheme 1.14. It must be noted that better enantioselectivities were observed for the addition of diethylzinc in comparison with that of dimethylzinc, which provided enantioselectivities ranging from 11 to 44% ee.

1.2.1.2. *Ketones as electrophiles*

In comparison with aldehydes, the catalytic asymmetric addition of alkyl group to ketones is a more challenging task for synthetic chemists, owing to their low electrophilicity, the reduced propensity of ketone carbonyl to coordinate with Lewis acids, and the difficult discrimination of the both faces of the double bond. This goal was achieved by Yus and Ramon in 2000s, using a stoichiometric amount of Ti(Oi-Pr)$_4$ combined with a cata-lytic amount of camphorsulfonamide derivatives as chiral ligands.[42] In this work, the corresponding tertiary alcohols were achieved in

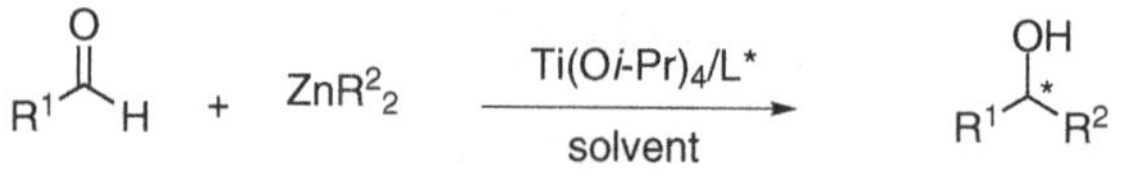

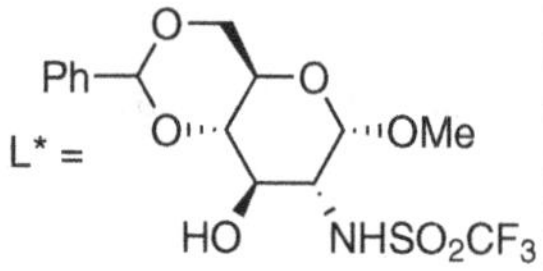

$$R^1\text{-CHO} + ZnR^2_2 \xrightarrow[\text{solvent}]{\text{Ti}(Oi\text{-Pr})_4/L^*} R^1\text{-CH(OH)-}R^2$$

L* = structure **24** (1 mol%)
Ti(O*i*-Pr)$_4$ (1.4 equiv)
CH$_2$Cl$_2$, r.t.

R^1 = Ph, R^2 = Et: 83% yield, ee = 93% (*R*)
R^1 = *o*-MeOC$_6$H$_4$, R^2 = Et: 83% yield, ee = 75% (*R*)
R^1 = *m*-MeOC$_6$H$_4$, R^2 = Et: 74% yield, ee >99% (*R*)
R^1 = *p*-MeOC$_6$H$_4$, R^2 = Et: 98% yield, ee = 77% (*R*)
R^1 = *o*-FC$_6$H$_4$, R^2 = Et: 84% yield, ee = 94% (*R*)
R^1 = *p*-FC$_6$H$_4$, R^2 = Et: 73% yield, ee = 78% (*R*)
R^1 = *o*-ClC$_6$H$_4$, R^2 = Et: 94% yield, ee = 98% (*R*)
R^1 = *p*-ClC$_6$H$_4$, R^2 = Et: 78% yield, ee = 88% (*R*)
R^1 = *p*-BrC$_6$H$_4$, R^2 = Et: 74% yield, ee = 81% (*R*)
R^1 = *p*-(NMe$_2$)C$_6$H$_4$, R^2 = Et: 22% yield, ee = 10% (*R*)
R^1 = (*E*)-PhCH=CH, R^2 = Et: 82% yield, ee = 87% (*R*)
R^1 = Cy, R^2 = Et: 97% yield, ee = 85% (*R*)
R^1 = *n*-Hex, R^2 = Et: 84% yield, ee = 88% (*R*)

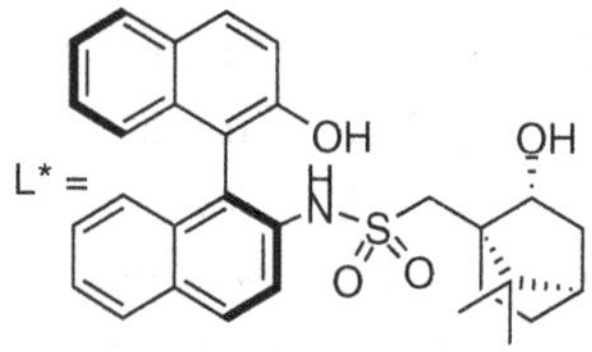

L* = structure **25** (10 mol%)
Ti(O*i*-Pr)$_4$ (1.5 equiv)
toluene, 0 °C

R^1 = Ph, R^2 = Et: 95% yield, ee = 87% (*S*)
R^1 = *p*-ClC$_6$H$_4$, R^2 = Et: 98% yield, ee = 85% (*S*)
R^1 = *p*-MeOC$_6$H$_4$, R^2 = Et: 96% yield, ee = 83% (*S*)
R^1 = *p*-CF$_3$C$_6$H$_4$, R^2 = Et: 95% yield, ee = 84% (*S*)
R^1 = 2-Naph, R^2 = Et: 93% yield, ee = 78% (*S*)
R^1 = 1-Naph, R^2 = Et: 90% yield, ee = 77% (*S*)
R^1 = *o*-Tol, R^2 = Et: 96% yield, ee = 22% (*S*)
R^1 = (*E*)-PhCH=CH, R^2 = Et: 75% yield, ee = 55% (*S*)
R^1 = *o*-MeOC$_6$H$_4$, R^2 = Et: 97% yield, ee = 64% (*S*)
R^1 = Cy, R^2 = Et: 90% yield, ee = 84% (*S*)
R^1 = Ph, R^2 = Me: 95% yield, ee = 44% (*S*)
R^1 = *o*-Tol, R^2 = Me: 98% yield, ee = 11% (*S*)
R^1 = 1-Naph, R^2 = Me: 97% yield, ee = 36% (*S*)
R^1 = *p*-CF$_3$C$_6$H$_4$, R^2 = Me: 97% yield, ee = 29% (*S*)

Scheme 1.14. Chiral ligands derived from D-glucosamine and L-camphor in addition of dialkylzincs to aldehydes

remarkable enantioselectivities of up to 99% ee, with a variety of ketones. Efficient chiral tertiary alcohols synthesis currently constitutes one of the most rapidly advancing fields in organic chemistry, since these compounds are versatile building blocks for the synthesis of natural products and pharmaceuticals. More recently, Wang *et al.* designed novel chiral ligand **26** derived from L-tartaric acid, to be employed as promoter in the titanium-catalysed addition of diethylzinc to ketones.[43] As shown in Scheme 1.15, acetophenone provided the best enantioselectivity of 99% ee in combination with a good yield (73%). Variously substituted acetophenones were investigated under similar conditions, and the authors found that the presence of substituents at the *ortho-* and *meta*-positions of acetophenone were incompatible with the reaction, since no desired products were formed. These results were ascribed to steric repulsion of the substituent and the ethyl group. Moreover, ketones containing heteroaromatic groups such as 2-acetyl furan and 2-acetyl thiophene, generated the corresponding products with low enantioselectivities of 12 and 15% ee, respectively, along with moderate yields (68 and 70%, respectively). These results could stem from the binding of the heteroatom in the substrate with the titanium centre. On the other hand, 2-acetyl naphthalene, having more steric hindrance than other ketones, gave a high enantioselectivity of 82% ee. In another context, de Parrodi *et al.* later developed novel chiral ligand **27**, derived from L-camphor and based on a C_2-symmetric 11,12-diamino-9,10-dihydro-9,10-ethanoanthracene backbone.[44] This ligand exhibited a large N−C−C−N dihedral angle, and a larger bite angle than *trans*-1,2-diaminocyclohexane. When applied at 5 mol% of catalyst loading to the titanium-promoted addition of diethylzinc of a variety of aryl alkyl ketones, it afforded enantioselectivities of up to 99% ee, as summarised in Scheme 1.15.

In 2012, Ramon reported an enantioselective total synthesis of biologically active (+)-gossonorol, the key step of which was the titanium-promoted addition of dimethylzinc to 5-methyl-1-(2-methylphenyl) hex-4-en-1-one **28**, to give the corresponding chiral tertiary alcohol (+)-gossonorol.[45] This process was performed with 5 mol% of chiral isoborneolsulfonamide ligand **29** in the presence of Ti(O*i*-Pr)$_4$ at 110 mol%, providing the key alcohol in 81% yield and enantioselectivity of 82% ee, as shown in Scheme 1.16. Notably, the synthesis of (+)-gossonorol was accomplished in 60% yield, through a three-step process from commercially available reagents. When applying the same conditions to the addition of diethylzinc, the corresponding tertiary alcohol was achieved in both lower yield (25%) and enantioselectivity (64% ee).

Finally, Ramon and Yus investigated the titanium-promoted additions of diethyl-, dimethyl-, and diphenyl-zincs to various methyl ketones in the

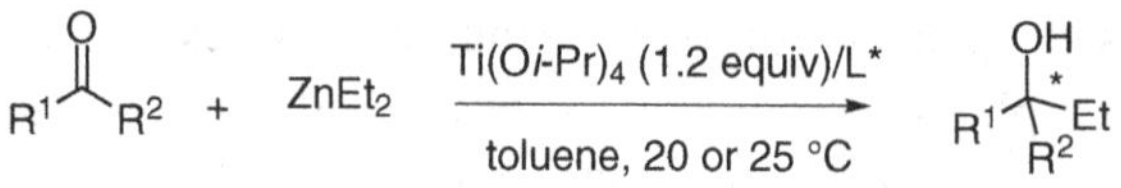

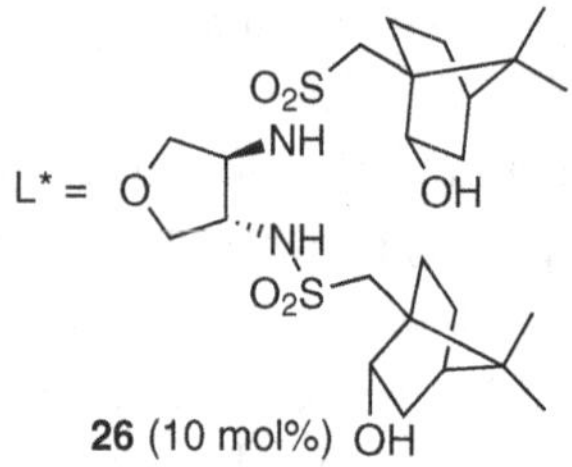

$$R^1 \underset{O}{\overset{O}{\|}} R^2 \;+\; ZnEt_2 \quad \xrightarrow[\text{toluene, 20 or 25 °C}]{\text{Ti(O}i\text{-Pr)}_4\ (1.2\ \text{equiv})/L^*} \quad R^1 \underset{R^2}{\overset{OH}{\underset{*}{\|}}} Et$$

R^1 = Ph, R^2 = Me: 73% yield, ee = 99% (S)
R^1 = p-FC$_6$H$_4$, R^2 = Me: 75% yield, ee = 80% (S)
R^1 = p-ClC$_6$H$_4$, R^2 = Me: 77% yield, ee = 80% (S)
R^1 = p-BrC$_6$H$_4$, R^2 = Me: 74% yield, ee = 77% (S)
R^1 = p-PhC$_6$H$_4$, R^2 = Me: 78% yield, ee = 70% (S)
R^1 = p-Tol, R^2 = Me: 58% yield, ee = 65% (S)
R^1 = 2-furyl, R^2 = Me: 68% yield, ee = 12% (S)
R^1 = 2-thienyl, R^2 = Me: 70% yield, ee = 15% (S)
R^1 = 2-Naph, R^2 = Me: 79% yield, ee = 82% (S)

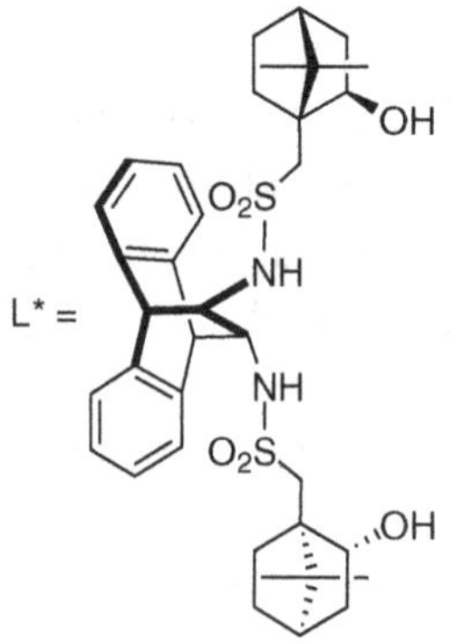

26 (10 mol%)

R^1 = Ph, R^2 = Me: 50% yield, ee = 99% (S)
R^1 = m-Tol, R^2 = Me: 91% yield, ee = 96% (S)
R^1 = m-CF$_3$C$_6$H$_4$, R^2 = Me: 88% yield, ee = 99% (S)
R^1 = Ph, R^2 = n-Bu: 85% yield, ee = 23% (S)
R^1 = Ph, R^2 = n-Pr: 98% yield, ee = 24% (S)
R^1 = Ph, R^2 = (CH$_2$)$_2$Cl: 20% yield, ee = 76% (S)
R^1 = (E)-PhCH=CH, R^2 = Me: 51% yield, ee = 53% (S)
R^1 = Ph, R^2 = (CH$_2$)$_3$: 98% yield, ee = 99% (S)

27 (5 mol%)

Scheme 1.15. Chiral ligands derived from L-Tartaric acid and L-Camphor in addition of diethylzinc to ketones

presence of novel grafted isoborneolsulfonamide polymer **30** employed at 5 mol% of catalyst loading along with 110 mol% of Ti(Oi-Pr)$_4$.[46] Whereas the highest and most remarkable enantioselectivities of up to >99% ee were achieved in the ethylation process (Scheme 1.17), the highest chemical yields of up to 98% were obtained in the phenylation process, albeit associated to moderate enantioselectivities (28–66% ee). It should be noted that this heterogeneous ligand could be reused at least three times without any significant loss of activity.

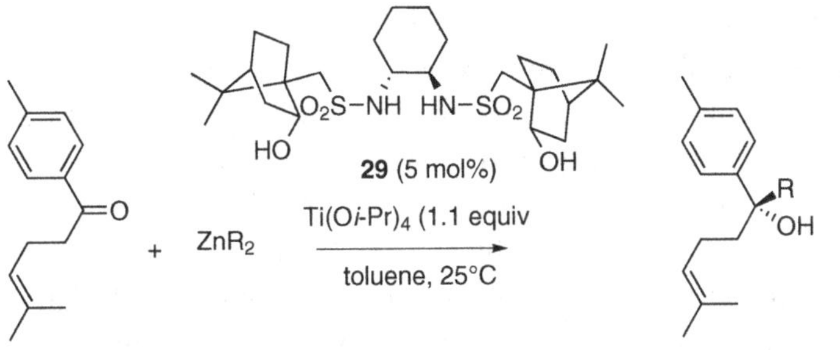

R = Me: (+)-gossonorol: 81% yield, ee = 82%
R = Et: 25% yield, ee = 64%

Scheme 1.16. Chiral isoborneolsulfonamide ligand in addition of dialkylzinc to 5-methyl-1-(2-methylphenyl)hex-4-en-1-one and total synthesis of (+)-gossonorol

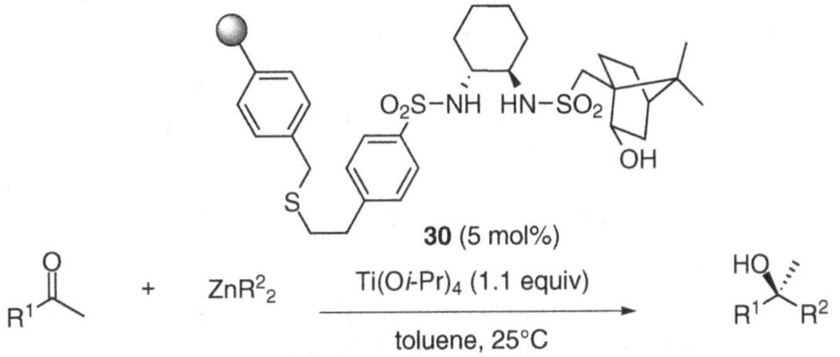

R^1 = Ph, R^2 = Et: 40% yield, ee >99%
R^1 = p-Tol, R^2 = Et: 29% yield, ee = 96%
R^1 = p-FC$_6$H$_4$, R^2 = Et: 42% yield, ee = 94%
R^1 = PhC≡C, R^2 = Et: 94% yield, ee >99%
R^1 = Ph, R^2 = Me: 25% yield, ee >99%

Scheme 1.17. Chiral grafted isoborneolsulfonamide polymer ligand in addition of dialkylzinc to ketones

1.2.2. Additions of Organoaluminum Reagents to Carbonyl Compounds

1.2.2.1. Aldehydes as electrophiles

Unfortunately, few organozinc reagents are commercially available, and their preparation is not always straightforward. To circumvent such limitations, attention has been focused on the use of other organometallic reagents. For example, trialkylaluminum reagents are readily available, and constitute valuable alkylating reagents for the enantioselective addition to aldehydes and ketones. Additional advantages of organoaluminium compounds include low toxicities and considerable stabilities. In most cases, chiral aluminates are first generated by reaction of chiral ligands with trialkylaluminum reagents, enabling the *in situ* formation of chiral titanium catalysts through transmetalation. The first example of asymmetric addition of AlEt$_3$ to aldehydes catalysed by a chiral titanium complex of BINOL was developed by Chan *et al.* in 1997.[47] This work was followed by several other reports, by the Gau and Carreira groups among others, dealing with titanium-catalysed additions of other organoaluminum reagents, including arylaluminum reagents. Although the enantioselective stoichiometric addition of chiral aryltitanium reagents was reported in 1987,[48] the catalytic enantioselective aryl addition to carbonyl compounds with titanium catalysts was only achieved by Walsh in 2003, using a catalyst prepared from bis(sulfonamide) ligand and Ti(Oi-Pr)$_4$ with diarylzinc as the nucleophiles, to afford the corresponding chiral tertiary alcohols in good to excellent enantioselectivities.[49] More recently, Gau and co-workers developed remarkable enantioselective additions of AlPh$_3$(THF) to both aliphatic and aromatic aldehydes in the presence of Ti(Oi-Pr)$_4$, and a catalytic amount of chiral disulfonamide ligand **31** in THF.[50] As shown in Scheme 1.18, the corresponding chiral secondary alcohols were produced with enantioselectivities of ≥94% ee except for two substrates, such as *n*-butanal and *trans*-cinnamaldehyde, which provided enantioselectivities of 87% and 85% ee, respectively. Furthermore, the products were obtained in general excellent yields of up to 98%.

In order to further explore arylaluminum reagents, the same authors later reported the synthesis of arylaluminum reagents containing various adducts of Et$_2$O, OPPh$_3$, DMAP, in addition to THF, as well as their asymmetric titanium-promoted aryl additions to aldehydes, employing a catalyst loading of 5 mol% of titanium complex **32** bearing chiral N-sulfonylated amino alcohols as a catalyst precursor.[51] It was demonstrated that the adduct ligand had a strong influence on the reactivity and enantioselectivity of the arylation reactions. Indeed, the phenylaluminum reagents with OPPh$_3$ or DMAP were unreactive toward aldehydes, and AlPh$_3$(THF) was

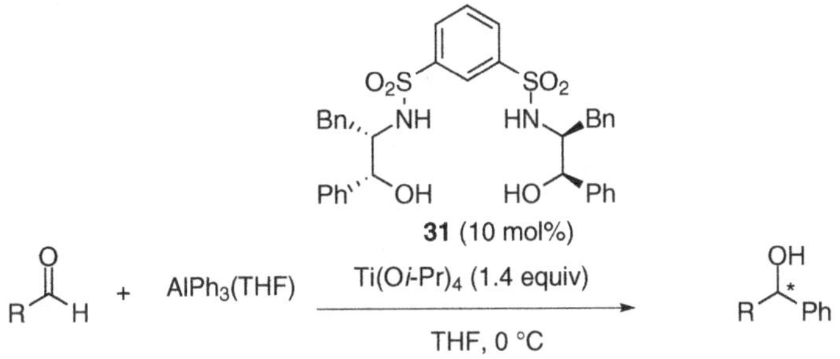

$$
R-\overset{O}{\underset{H}{\parallel}}C-H \;+\; AlPh_3(THF) \;\xrightarrow[\text{THF, 0 °C}]{\text{31 (10 mol\%)} \atop \text{Ti(O}i\text{-Pr)}_4 \text{ (1.4 equiv)}}\; R-\overset{OH}{\underset{*}{C}}-Ph
$$

R = o-MeOC$_6$H$_4$: 95% yield, ee = 96% (R)
R = p-MeOC$_6$H$_4$: 98% yield, ee = 95% (R)
R = o-Tol: 97% yield, ee = 98% (R)
R = p-Tol: 97% yield, ee = 98% (R)
R = o-ClC$_6$H$_4$: 95% yield, ee = 95% (R)
R = m-ClC$_6$H$_4$: 97% yield, ee = 97% (R)
R = p-ClC$_6$H$_4$: 96% yield, ee = 96% (R)
R = p-CF$_3$C$_6$H$_4$: 97% yield, ee = 96% (R)
R = o-BrC$_6$H$_4$: 97% yield, ee = 95% (R)
R = 1-Naph: 96% yield, ee = 95% (R)
R = 2-Naph: 97% yield, ee = 96% (R)
R = (E)-PhCH=CH: 93% yield, ee = 85% (R)
R = 1-thiophene: 97% yield, ee = 95% (R)
R = n-Bu: 90% yield, ee = 87% (R)
R = i-Pr: 90% yield, ee = 97% (R)
R = t-Bu: 95% yield, ee = 99% (R)
R = Cy: 85% yield, ee = 95% (R)

Scheme 1.18. Chiral disulfonamide ligand in addition of AlPh$_3$(THF) to aldehydes

found to be superior to AlPh$_3$(OEt$_2$) or AlPhEt$_2$(THF). In the presence of 150 mol% of Ti(Oi-Pr)$_4$ and 5 mol% of complex **32**, the asymmetric additions of AlAr$_3$(THF) to aldehydes afforded the corresponding chiral secondary alcohols in high yields and enantioselectivities of up to 94% ee, as shown in Scheme 1.19. The best results were achieved in the cases of aromatic aldehydes.

In addition, these authors have developed an easy preparation of AlArEt$_2$(THF) reagents which were added to aldehydes in the presence of 10 mol% of a titanium complex of (R)-H$_8$-BINOL and 150 mol% of Ti(Oi-Pr)$_4$ in toluene.[52] As shown in Scheme 1.20, the process afforded the corresponding chiral secondary aryl alcohols **33** as exclusive products in high yields, and excellent enantioselectivities of up to 98% ee, except for 2-naphthaldehyde, 4-methoxybenzaldehyde, 4-methylbenzaldehyde, and 4-bromobenzaldehyde for which minor ethylation products **34** were obtained with yields of 8 to 13%.

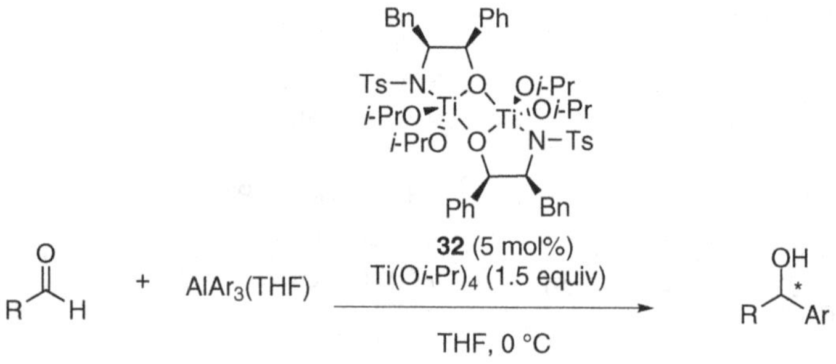

Bn   Ph

Ts—N   O   O*i*-Pr
i-PrO⁀Ti   Ti⁘O*i*-Pr
i-PrO   O   N—Ts

Ph   Bn

32 (5 mol%)
Ti(O*i*-Pr)$_4$ (1.5 equiv)

$$\underset{R}{\overset{O}{\Vert}}{\diagdown}H \;+\; AlAr_3(THF) \quad \xrightarrow[\text{THF, 0 °C}]{} \quad \underset{R}{\overset{OH}{\diagup}}\overset{*}{\diagdown}Ar$$

R = *o*-ClC$_6$H$_4$, Ar = Ph: 92% yield, ee = 93% (*R*)
R = *p*-ClC$_6$H$_4$, Ar = Ph: 91% yield, ee = 90% (*R*)
R = *o*-BrC$_6$H$_4$, Ar = Ph: 94% yield, ee = 93% (*R*)
R = *p*-BrC$_6$H$_4$, Ar = Ph: 96% yield, ee = 90% (*R*)
R = *o*-MeOC$_6$H$_4$, Ar = Ph: 94% yield, ee = 94% (*R*)
R = *p*-MeOC$_6$H$_4$, Ar = Ph: 98% yield, ee = 90% (*R*)
R = *o*-Tol, Ar = Ph: 95% yield, ee = 91% (*R*)
R = *p*-Tol, Ar = Ph: 93% yield, ee = 92% (*R*)
R = *p*-CF$_3$C$_6$H$_4$, Ar = Ph: 80% yield, ee = 90% (*R*)
R = 1-Naph, Ar = Ph: 96% yield, ee = 91% (*R*)
R = 2-Naph, Ar = Ph: 94% yield, ee = 92% (*R*)
R = (*E*)-PhCH=CH, Ar = Ph: 92% yield, ee = 87% (*S*)
R = 2-furyl, Ar = Ph: 93% yield, ee = 90% (*R*)
R = *t*-Bu, Ar = Ph: 93% yield, ee = 74% (*S*)
R = *i*-Pr, Ar = Ph: 92% yield, ee = 77% (*S*)
R = Ph, Ar = *p*-Tol: 91% yield, ee = 91% (*S*)

Scheme 1.19. Chiral *N*-sulfonylated amino alcohol titanium complex in addition of AlAr$_3$(THF) to aldehydes

On the other hand, Yus *et al.* have developed asymmetric addition of alkylaluminum reagents to a range of aromatic as well as aliphatic aldehydes by using an excess of Ti(O*i*-Pr)$_4$ and a catalytic amount of chiral readily available BINOL-derived ligand **35** in Et$_2$O.[53] As shown in Scheme 1.21, the asymmetric methylation, ethylation, and propargylation of a wide variety of aldehydes proceeded with good yields and high enantioselectivities of up to 94% ee. In 2014, Chen *et al.* reinvestigated the asymmetric ethylation of aromatic aldehydes by using 5 mol% of (*S*)- or (*R*)-BIQOL as chiral ligand in the presence of 160 mol% of Ti(O*i*-Pr)$_4$ in THF.[54] The reactions provided the corresponding aromatic alcohols in remarkably complete conversions with good enantioselectivities of up to 87% ee (Scheme 1.21).

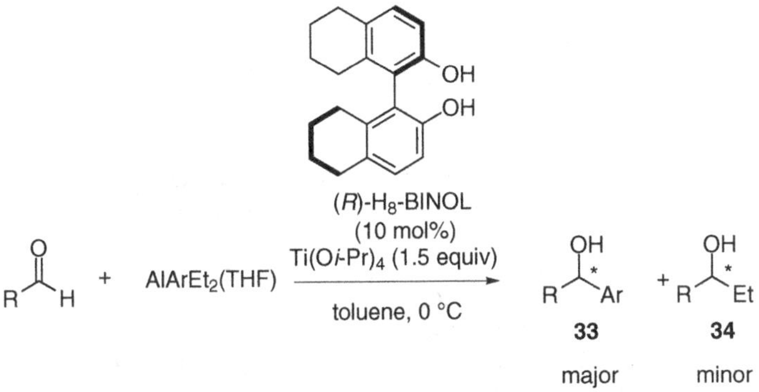

(*R*)-H$_8$-BINOL
(10 mol%)
Ti(O*i*-Pr)$_4$ (1.5 equiv)

$$R\overset{O}{\underset{H}{\diagdown}} + \text{AlArEt}_2(\text{THF}) \xrightarrow[\text{toluene, 0 °C}]{} R\overset{OH}{\underset{\text{Ar}}{\diagup}} + R\overset{OH}{\underset{\text{Et}}{\diagup}}$$

33   **34**

major   minor

R = 1-Naph, Ar = Ph: 93% yield, **33/34** >99:1, ee = 98% (*R*)
R = 2-Naph, Ar = Ph: 85% yield, **33/34** = 87:13, ee = 83% (*R*)
R = *o*-ClC$_6$H$_4$, Ar = Ph: 91% yield, **33/34** >99:1, ee = 90% (*R*)
R = *p*-ClC$_6$H$_4$, Ar = Ph: 93% yield, **33/34** >99:1, ee = 92% (*R*)
R = *o*-MeOC$_6$H$_4$, Ar = Ph: 90% yield, **33/34** >99:1, ee = 74% (*R*)
R = *p*-MeOC$_6$H$_4$, Ar = Ph: 86% yield, **33/34** = 90:10, ee = 87% (*R*)
R = *p*-CF$_3$C$_6$H$_4$, Ar = Ph: 93% yield, **33/34** >99:1, ee = 90% (*R*)
R = *o*-Tol, Ar = Ph: 94% yield, **33/34** >99:1, ee = 96% (*R*)
R = *p*-Tol, Ar = Ph: 89% yield, **33/34** = 92:8, ee = 91% (*R*)
R = *o*-BrC$_6$H$_4$, Ar = Ph: 92% yield, **33/34** >99:1, ee = 86% (*R*)
R = *p*-BrC$_6$H$_4$, Ar = Ph: 86% yield, **33/34** = 91:9, ee = 80% (*R*)
R = (*E*)-PhCH=CH, Ar = Ph: 93% yield, **33/34** >99:1, ee = 91% (*S*)
R = *t*-Bu, Ar = Ph: 90% yield, **33/34** >99:1, ee = 94% (*S*)
R = *i*-Pr, Ar = Ph: 91% yield, **33/34** >99:1, ee = 95% (*S*)
R = Ph, Ar = *p*-MeOC$_6$H$_4$: 85% yield, **33/34** = 90:10, ee = 62% (*S*)

Scheme 1.20. (*R*)-H$_8$-BINOL ligand in addition of AlArEt$_2$(THF) to aldehydes

1.2.2.2. *Ketones as electrophiles*

In 2008, Gau *et al.* reported asymmetric additions of AlAr$_3$(THF) to ketones promoted by a titanium catalyst generated *in situ* from a large excess of Ti(O*i*-Pr)$_4$ and 20 mol% of *trans*-1,2-bis(hydroxycamphorsulfonylamino) cyclohexane **29**, in the presence of MgBr$_2$ as an additive.[55] The study demonstrated several important features. First, a novel aspect of the inorganic salt MgBr$_2$ as a key additive for promoting the aryl addition of AlAr$_3$(THF) to ketones was demonstrated. Second, the catalytic system worked very well for aromatic ketones bearing either an electron-withdrawing or an electron-donating substituent on the aromatic group, to afford the corresponding chiral tertiary alcohols in enantioselectivities of ≥90% ee except

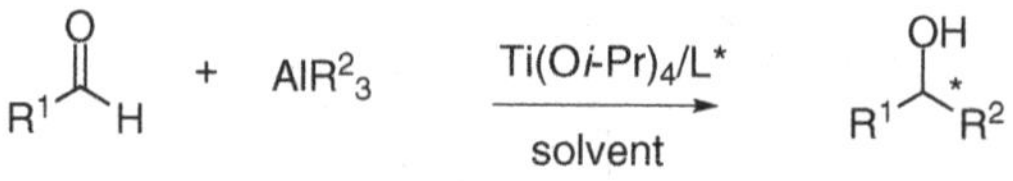

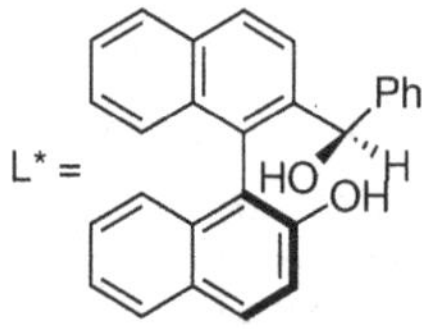

$$R^1 \overset{O}{\underset{H}{\bigwedge}} \quad + \quad AlR^2_3 \quad \xrightarrow[\text{solvent}]{Ti(Oi\text{-}Pr)_4/L^*} \quad R^1 \overset{OH}{\underset{*}{\bigwedge}} R^2$$

R^1 = Ph, R^2 = Me: >99% yield, ee = 94% (*S*)
R^1 = *o*-Tol, R^2 = Me: 92% yield, ee = 80% (*S*)
R^1 = *m*-Tol, R^2 = Me: 98% yield, ee = 94% (*S*)
R^1 = *p*-Tol, R^2 = Me: 99% yield, ee = 94% (*S*)
R^1 = *p*-CF$_3$C$_6$H$_4$, R^2 = Me: >99% yield, ee = 94% (*S*)
R^1 = *p*-CNC$_6$H$_4$, R^2 = Me: 99% yield, ee = 94% (*S*)
R^1 = *p*-AcC$_6$H$_4$, R^2 = Me: 99% yield, ee = 94% (*S*)
R^1 = 2-Naph, R^2 = Me: 99% yield, ee = 94% (*S*)
R^1 = 2-furyl, R^2 = Me: 75% yield, ee = 88% (*S*)
R^1 = (*E*)-PhCH=CH, R^2 = Me: 98% yield, ee = 90% (*S*)
R^1 = Bn, R^2 = Me: 92% yield, ee = 94% (*S*)
R^1 = Ph, R^2 = Me: >99% yield, ee = 94% (*S*)
R^1 = Ph, R^2 = Et: 77% yield, ee = 90% (*S*)
R^1 = *p*-Tol, R^2 = Et: 65% yield, ee = 87% (*S*)
R^1 = *p*-ClC$_6$H$_4$, R^2 = Et: 70% yield, ee = 92% (*S*)
R^1 = *p*-ClC$_6$H$_4$, R^2 = *n*-Pr: 35% yield, ee = 94% (*S*)
R^1 = *p*-MeOC$_6$H$_4$, R^2 = *n*-Pr: 26% yield, ee = 92% (*S*)
R^1 = 2-Naph, R^2 = Ph: 90% yield, ee = 20% (*S*)
R^1 = *t*-Bu, R^2 = Ph: 61% yield, ee = 72% (*S*)

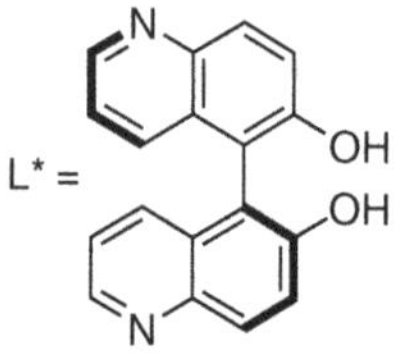

L* =

35 (10 mol%)
Ti(O*i*-Pr)$_4$ (4 equiv)
Et$_2$O, 0°C

L* =

(*S*)-BIQOL (5 mol%)
Ti(O*i*-Pr)$_4$ (1.6 equiv)
THF, 0°C

with L* = (*S*)-BIQOL:
R^1 = Ph, R^2 = Et: >99% conversion, ee = 86% (*S*)
R^1 = *p*-MeOC$_6$H$_4$, R^2 = Et: >99% conversion, ee = 85% (*S*)
R^1 = *p*-ClC$_6$H$_4$, R^2 = Et: >99% conversion, ee = 85% (*S*)
R^1 = *p*-NO$_2$C$_6$H$_4$, R^2 = Et: >99% conversion, ee = 87% (*S*)
with L* = (*R*)-BIQOL:
R^1 = *o*-MeOC$_6$H$_4$, R^2 = Et: >99% conversion, ee = 74% (*R*)
R^1 = *o*-FC$_6$H$_4$, R^2 = Et: >99% conversion, ee = 74% (*R*)

Scheme 1.21. Chiral BINOL-derived ligand and chiral BIQOL ligand in addition of AlR$_3$ to aldehydes

for 2′-methoxyacetophenone, as shown in Scheme 1.22. Third, longer reaction times were required for *ortho*-substituted aromatic ketones to furnish products in good yields. Fourth, the reactions of PhTi(O*i*-Pr)$_3$ in additions to 2′-acetonaphthone promoted by the same catalyst provided the product in low yield and low enantioselectivity, suggesting that AlAr$_3$(THF)

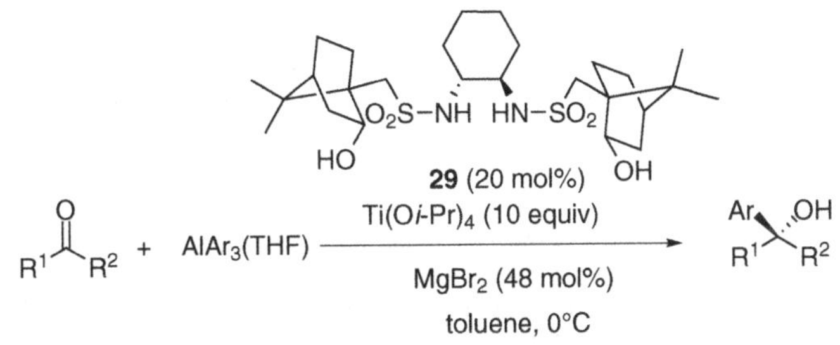

R^1 = 2-Naph, R^2 = Me, Ar = Ph: 99% yield, ee = 92%
R^1 = 1-Naph, R^2 = Me, Ar = Ph: 80% yield, ee = 93%
R^1 = o-Tol, R^2 = Me, Ar = Ph: 85% yield, ee = 96%
R^1 = o-ClC$_6$H$_4$, R^2 = Me, Ar = Ph: 95% yield, ee = 97%
R^1 = p-ClC$_6$H$_4$, R^2 = Me, Ar = Ph: 98% yield, ee = 92%
R^1 = o-BrC$_6$H$_4$, R^2 = Me, Ar = Ph: 87% yield, ee = 98%
R^1 = p-BrC$_6$H$_4$, R^2 = Me, Ar = Ph: 97% yield, ee = 92%
R^1 = o-MeOC$_6$H$_4$, R^2 = Me, Ar = Ph: 99% yield, ee = 18%
R^1 = m-CF$_3$C$_6$H$_4$, R^2 = Me, Ar = Ph: 98% yield, ee = 97%
R^1 = p-CF$_3$C$_6$H$_4$, R^2 = Me, Ar = Ph: 97% yield, ee = 93%
R^1 = p-NO$_2$C$_6$H$_4$, R^2 = Me, Ar = Ph: 99% yield, ee = 92%
R^1 = n-Bu, R^2 = Me, Ar = Ph: 97% yield, ee = 52%
R^1 = i-Pr, R^2 = Me, Ar = Ph: 82% yield, ee = 83%
R^1 = Cy, R^2 = Me, Ar = Ph: 87% yield, ee = 81%
R^1 = 2-Naph, R^2 = Me, Ar = p-Tol: 97% yield, ee = 90%
R^1 = Ph, R^2 = Me, Ar = p-TMSC$_6$H$_4$: 83% yield, ee = 81%
R^1 = Ph, R^2 = Me, Ar = 2-Naph: 95% yield, ee = 91%

Scheme 1.22. Chiral *trans*-1,2-bis(hydroxycamphorsulfonylamino)cyclohexane ligand in addition of AlR$_3$(THF) to ketones

addition reactions could not proceed via aryltitanium species. Phenyl additions to aliphatic ketones and 1-acetyl-1-cyclohexene were also examined. The resulting tertiary chiral alcohols were obtained in good to excellent yields and good enantioselectivities of 75 to 83% ee, except for the alcohol derived from linear 2-hexanone which gave 52% ee.

Later, the same authors studied the first asymmetric addition of a (2-furyl)aluminum reagent to aromatic ketones and one α,β-unsaturated ketone promoted by a titanium catalyst of (S)-BINOL to afford the corresponding chiral tertiary 2-furyl alcohols in good to excellent enantioselectivities of 87 to 93% ee (Scheme 1.23).[56] Although the furylaluminum reagent employed was prepared as a mixture of three species of formulas (2-furyl)$_x$AlEt$_{3-x}$(THF) (x = 0, 1, or 2), the addition reactions gave only chiral furyl alcohols with no observations of ethylation products.

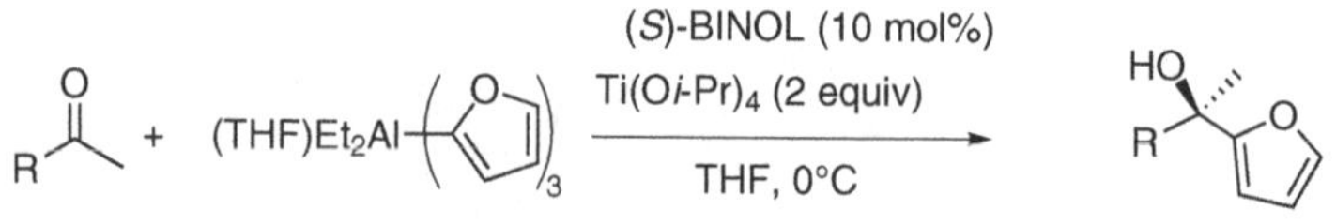

(S)-BINOL (10 mol%)

Ti(Oi-Pr)$_4$ (2 equiv)

THF, 0°C

R = Ph: 90% yield, ee = 93%
R = o-Tol: 74% yield, ee = 90%
R = p-Tol: 78% yield, ee = 93%
R = p-MeOC$_6$H$_4$: 74% yield, ee = 92%
R = o-ClC$_6$H$_4$: 70% yield, ee = 92%
R = p-ClC$_6$H$_4$: 86% yield, ee = 91%
R = o-BrC$_6$H$_4$: 40% yield, ee = 93%
R = p-BrC$_6$H$_4$: 92% yield, ee = 90%
R = p-CF$_3$C$_6$H$_4$: 90% yield, ee = 91%
R = p-NO$_2$C$_6$H$_4$: 88% yield, ee = 92%
R = 1-Naph: 28% yield, ee = 87%
R = 2-Naph: 94% yield, ee = 90%
R = (E)-BrC=CHPh: 88% yield, ee = 88%

Scheme 1.23. (S)-BINOL as ligand in addition of (2-furyl)AlEt$_2$(THF) to ketones

Similarly, the authors applied the same catalyst system to the first asymmetric thienylaluminum addition to a variety of ketones.[57] As shown in Scheme 1.24, the additions of Al(2-thienyl)$_3$(THF) to aromatic ketones, having either an electron-donating or an electron-withdrawing substituent on the aromatic ring and for 1-acetylcyclohexene, provided the corresponding chiral tertiary 2-thienyl alcohols in excellent enantioselectivities of up to 97% ee. In contrast, the additions of 2-thienyl to aliphatic ketones produced the corresponding alcohols in low enantioselectivities of 8 to 17% ee. Importantly, this methodology was applied to a concise synthesis of (S)-tiemonium iodide in three steps.

Finally, these authors developed an easy preparation of AlArEt$_2$(THF) reagents which were added to a variety of ketones in the presence of 10 mol% of a titanium complex of (R)-H$_8$-BINOL and 150 mol% of Ti(Oi-Pr)$_4$ in toluene.[55] The results collected in Scheme 1.25 show that the catalytic system worked very well in terms of stereocontrol for a wide range of aromatic ketones, regardless of the electronic nature or the steric effect of the substituents on the aryl groups, affording the corresponding chiral aryl tertiary alcohols as the sole products with enantioselectivities of ≥90% ee. However, for aliphatic ketones, such as 3-methyl-2-butanone and 2-hexanone, the phenyl additions afforded the corresponding alcohols

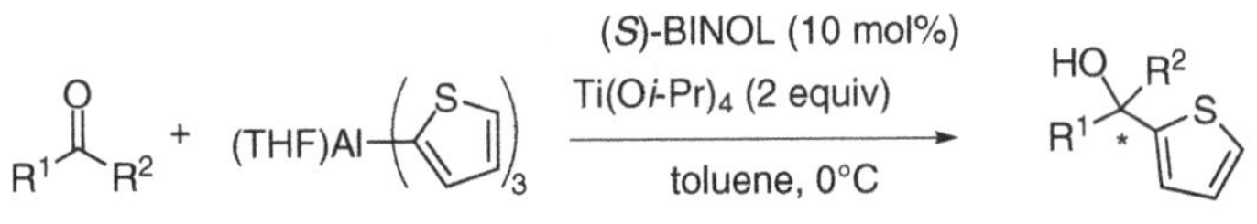

(S)-BINOL (10 mol%)

Ti(Oi-Pr)$_4$ (2 equiv)

toluene, 0°C

R^1 = Ph, R^2 = Me: 94% yield, ee = 93%
R^1 = o-Tol, R^2 = Me: 75% yield, ee = 91%
R^1 = m-Tol, R^2 = Me: 96% yield, ee = 90%
R^1 = p-Tol, R^2 = Me: 95% yield, ee = 93%
R^1 = o-MeOC$_6$H$_4$, R^2 = Me: 96% yield, ee = 45%
R^1 = p-MeOC$_6$H$_4$, R^2 = Me: 92% yield, ee = 92%
R^1 = o-ClC$_6$H$_4$, R^2 = Me: 92% yield, ee = 97%
R^1 = m-ClC$_6$H$_4$, R^2 = Me: 95% yield, ee = 90%
R^1 = p-ClC$_6$H$_4$, R^2 = Me: 96% yield, ee = 95%
R^1 = o-BrC$_6$H$_4$, R^2 = Me: 93% yield, ee = 96%
R^1 = p-BrC$_6$H$_4$, R^2 = Me: 95% yield, ee = 93%
R^1 = p-CF$_3$C$_6$H$_4$, R^2 = Me: 96% yield, ee = 93%
R^1 = p-NO$_2$C$_6$H$_4$, R^2 = Me: 95% yield, ee = 94%
R^1 = 2-Naph, R^2 = Me: 94% yield, ee = 92%
R^1 = 1-Naph, R^2 = Me: 70% yield, ee = 93%
R^1 = 2-Naph, R^2 = CH$_2$Br: 95% yield, ee = 80%
R^1 = Ph, R^2 = (CH$_2$)$_2$Br: 90% yield, ee = 94%
R^1 = i-Pr, R^2 = Me: 96% yield, ee = 17%

Scheme 1.24. (S)-BINOL as ligand in addition of AL(2-thienyl)$_3$(THF) to ketones

in low yields (38 and 60%, respectively) and poor enantioselectivities of 48 and 15% ee, respectively.

1.2.3. *Additions of Grignard Reagents to Aldehydes*

Grignard reagents are among the least expensive and most commonly used organometallic reagents in both laboratory and industry. Because of the high reactivity of these compounds, direct highly enantioselective Grignard addition to aldehydes has rarely been reported. The recent procedures using Grignard reagents as starting materials in addition to aldehydes often focused on transmetalation to form less reactive intermediates, such as RTi(Oi-Pr)$_3$, generated *in situ* from RMgX and Ti(Oi-Pr)$_4$. For example, in the 1990s Weber and Seebach were the first authors to report the successful asymmetric addition of Grignard reagents to ketones performed in the presence of chiral titanium complexes of TADDOLs, providing chiral tertiary alcohols in enantioselectivity greater than 95% ee.[58]

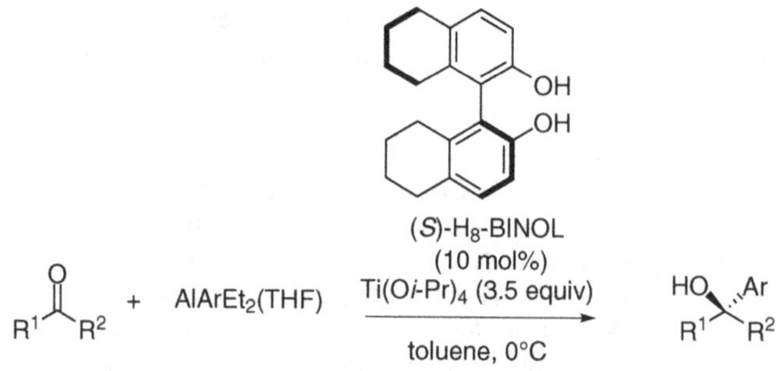

(S)-H$_8$-BINOL
(10 mol%)
Ti(O*i*-Pr)$_4$ (3.5 equiv)

toluene, 0°C

R^1 = 2-Naph, R^2 = Me, Ar = Ph: 87% yield, ee = 90%
R^1 = 1-Naph, R^2 = Me, Ar = Ph: 39% yield, ee = 88%
R^1 = *o*-ClC$_6$H$_4$, R^2 = Me, Ar = Ph: 82% yield, ee = 94%
R^1 = *p*-ClC$_6$H$_4$, R^2 = Me, Ar = Ph: 90% yield, ee = 91%
R^1 = *o*-BrC$_6$H$_4$, R^2 = Me, Ar = Ph: 49% yield, ee = 93%
R^1 = *p*-BrC$_6$H$_4$, R^2 = Me, Ar = Ph: 88% yield, ee = 92%
R^1 = *o*-Tol, R^2 = Me, Ar = Ph: 35% yield, ee = 90%
R^1 = *m*-CF$_3$C$_6$H$_4$, R^2 = Me, Ar = Ph: 90% yield, ee = 80%
R^1 = *p*-CF$_3$C$_6$H$_4$, R^2 = Me, Ar = Ph: 96% yield, ee = 90%
R^1 = *p*-NO$_2$C$_6$H$_4$, R^2 = Me, Ar = Ph: 88% yield, ee = 92%
R^1 = 2-Naph, R^2 = CH$_2$Br, Ar = Ph: 93% yield, ee = 77%
R^1 = 2-furyl, R^2 = Me, Ar = Ph: 90% yield, ee = 83%
R^1 = *n*-Bu, R^2 = Me, Ar = Ph: 60% yield, ee = 15%
R^1 = *i*-Pr, R^2 = Me, Ar = Ph: 38% yield, ee = 48%
R^1 = Ph, R^2 = Me, Ar = *p*-MeOC$_6$H$_4$: 89% yield, ee = 93%
R^1 = Ph, R^2 = Me, Ar = *p*-TMSC$_6$H$_4$: 77% yield, ee = 91%

Scheme 1.25. (S)-H$_8$-BINOL as ligand in addition of AlArEt$_2$(THF) to ketones

In the last few years, various enantioselective titanium-catalysed additions of Grignard reagents to aldehydes have been developed by several groups, all based on the use of BINOL derivatives as chiral ligands of Ti(O*i*-Pr)$_4$. As an example, Harada *et al.* have reported general excellent enantioselectivities of up to 97% ee for ethylation, butylation, propylation, and even phenylation of aromatic, α,β-unsaturated, and aliphatic aldehydes, starting from the corresponding Grignard reagents, which were previously treated by titanium tetraisopropoxide at −78°C and then introduced to the reaction mixture (Scheme 1.26).[59] The reaction proceeded in the presence of an excess of Ti(O*i*-Pr)$_4$ and a catalytic amount (2 mol%) of (*R*)-3-(3,5-diphenylphenyl)-2,2′-dihydroxy-1,1′-binaphthyl ((*R*)-DPP-BINOL), affording the corresponding chiral secondary alcohols in good to very high yields of up to 94%. It must be noted that chloromagnesium

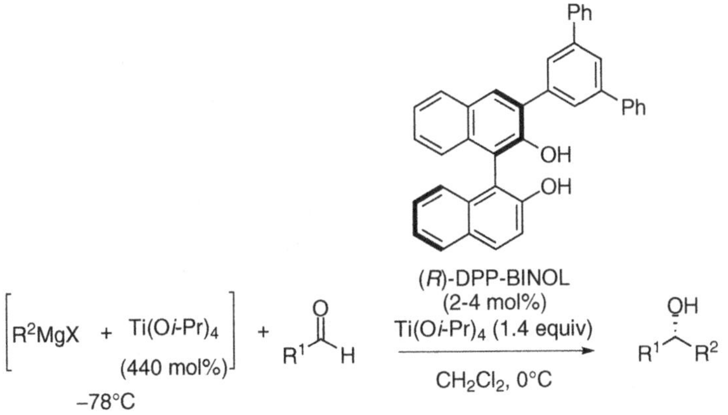

$$\left[R^2MgX \ + \ \underset{(440\ mol\%)}{Ti(O\text{-}Pr)_4} \right] \ + \ \underset{R^1}{\overset{O}{\underset{\ }{\|}}}_{\ \ \ H} \quad \xrightarrow[\text{CH}_2\text{Cl}_2,\ 0°C]{\begin{array}{c}(R)\text{-DPP-BINOL}\\(2\text{-}4\ mol\%)\\ \text{Ti(O-Pr)}_4\ (1.4\ equiv)\end{array}} \quad \underset{R^1}{\overset{OH}{\underset{\ }{\ }}}_{\ \ \ R^2}$$

−78°C

R^1 = Ph, R^2 = *n*-Bu, X = Cl: 83% yield, ee = 97%
R^1 = Ph, R^2 = *n*-Pr, X = Cl: 82% yield, ee = 94%
R^1 = Ph, R^2 = Et, X = Cl: 38% yield, ee = 97%
R^1 = *p*-Tol, R^2 = *n*-Bu, X = Cl: 70% yield, ee = 96%
R^1 = *p*-Tol, R^2 = Et, X = Cl: 41% yield, ee = 88%
R^1 = *p*-CF$_3$C$_6$H$_4$, R^2 = *n*-Bu, X = Cl: 62% yield, ee = 95%
R^1 = *m*-MeOC$_6$H$_4$, R^2 = *n*-Bu, X = Cl: 86% yield, ee = 95%
R^1 = *o*-MeOC$_6$H$_4$, R^2 = *n*-Bu, X = Cl: 89% yield, ee = 98%
R^1 = *o*-ClC$_6$H$_4$, R^2 = *n*-Bu, X = Cl: 79% yield, ee = 94%
R^1 = 1-Naph, R^2 = *n*-Bu, X = Cl: 90% yield, ee = 96%
R^1 = 1-Naph, R^2 = Et, X = Br: 63% yield, ee = 93%
R^1 = 1-Naph, R^2 = Ph, X = Br: 94% yield, ee = 86%
R^1 = 1-Naph, R^2 = Et, X = Cl: 56% yield, ee = 94%
R^1 = 1-Naph, R^2 = Me, X = Cl: 89% yield, ee = 28%
R^1 = 2-Naph, R^2 = *n*-Bu, X = Cl: 92% yield, ee = 91%
R^1 = (*E*)-PhCH=CH, R^2 = *n*-Bu, X = Cl: 76% yield, ee = 96%
R^1 = CH$_2$=C(Me), R^2 = *n*-Bu, X = Cl: 39% yield, ee = 95%
R^1 = BnCH$_2$, R^2 = Et, X = Cl: 36% yield, ee = 92%

Scheme 1.26. (R)-DPP-BINOL as ligand in addition of alkylmagnesium halides to aldehydes

reagents and bromomagnesium reagents could be employed with comparable efficiency and selectivity. On the other hand, the reaction of 1-naphthaldehyde with MeMgCl resulted in a low enantioselectivity (28% ee). In contrast, a relatively high enantioselectivity (86% ee) was obtained for the phenylation of the same aldehyde. Furthermore, α,β-unsaturated aldehydes provided high enantioselectivities of up to 96% ee, while, although sluggish, the reaction of aliphatic aldehydes also provided high enantioselectivities of up to 92% ee.

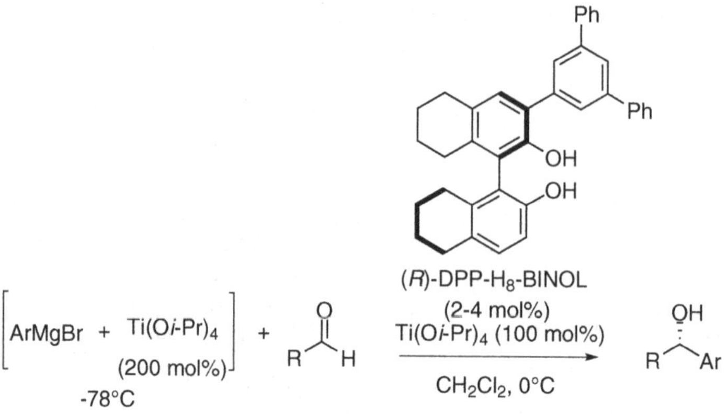

$$\left[\text{ArMgBr} + \underset{(200\ \text{mol\%})}{\text{Ti(O}i\text{-Pr})_4} \right] + \underset{R}{\overset{O}{\underset{}{\|}}}\!\!\!\!\!\!\!\!\!\!_{H} \xrightarrow[\substack{\text{CH}_2\text{Cl}_2,\ 0\,°\text{C}}]{\substack{(R)\text{-DPP-H}_8\text{-BINOL} \\ (2\text{-}4\ \text{mol\%}) \\ \text{Ti(O}i\text{-Pr})_4\ (100\ \text{mol\%})}} \underset{R}{\overset{\text{OH}}{\underset{}{\wedge}}}\!\!_{\text{Ar}}$$

-78°C

R = Ph, Ar = *p*-Tol: 99% yield, ee = 91%
R = Ph, Ar = *p*-ClC$_6$H$_4$: 95% yield, ee = 91%
R = Ph, Ar = *p*-FC$_6$H$_4$: 94% yield, ee = 97%
R = Ph, Ar = *o*-MeOC$_6$H$_4$: 66% yield, ee = 9%
R = Ph, Ar = 2,4,6-Me$_3$C$_6$H$_2$: 89% yield, ee = 96%
R = *p*-Tol, Ar = Ph: 93% yield, ee = 90%
R = *p*-FC$_6$H$_4$, Ar = Ph: 97% yield, ee = 95%
R = *p*-ClC$_6$H$_4$, Ar = Ph: 96% yield, ee = 94%
R = *p*-CNC$_6$H$_4$, Ar = Ph: 96% yield, ee = 92%
R = *p*-PhC$_6$H$_4$, Ar = Ph: 99% yield, ee = 91%
R = *m*-MeOC$_6$H$_4$, Ar = Ph: 90% yield, ee = 95%
R = 1-Naph, Ar = Ph: 97% yield, ee = 95%
R = 2-Naph, Ar = Ph: 96% yield, ee = 91%
R = 2-furyl, Ar = Ph: 83% yield, ee = 89%
R = 2-thienyl, Ar = Ph: 82% yield, ee = 90%
R = 4-pyridyl, Ar = Ph: 85% yield, ee = 82%
R = CH$_2$=C(Me), Ar = Ph: 87% yield, ee = 97%
R = *n*-Bu, Ar = Ph: 88% yield, ee = 88%
R = Cy, Ar = Ph: 86% yield, ee = 80%

Scheme 1.27. (R)-DPP-BINOL as ligand in addition of arylmagnesium bromides to aldehydes

As an extension of the preceding methodology, these authors applied a closely related catalyst system based on (R)-DPP-H$_8$-BINOL to the asymmetric arylation of a range of aldehydes, starting from the corresponding aryl Grignard reagents in combination with Ti(Oi-Pr)$_4$.[62b,60] As shown in Scheme 1.27, the results produced high enantioselectivities, and yields of up to 97% ee and 99%, respectively, for various combinations of aromatic, aliphatic, and α,β-unsaturated aldehydes and aryl bromomagnesium reagents, including those with functional groups.

In 2011, the same authors reported an efficient novel method using the same catalyst system as above for the enantioselective arylation of aldehydes albeit starting from aryl bromides.[61] Indeed, in this case, functionalised aryl Grignard reagents were prepared *in situ* by bromine–magnesium exchange of the corresponding aryl bromides with *i*-PrMgCl (Scheme 1.28). This novel method was based on the fact that aryl bromides constitute preferable precursors for the preparation of functionalised Grignard reagents, in light of their stability, good availability, and low price in comparison to the corresponding iodides. The method was applicable to aryl bromides bearing CF_3, Br, and CN groups, affording a range

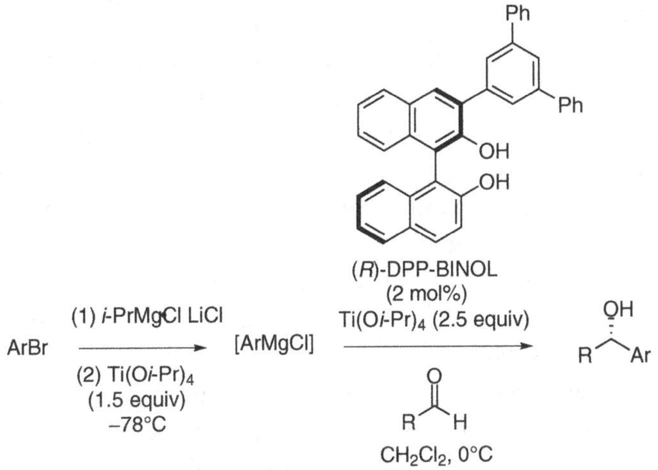

R = 1-Naph, Ar = *p*-CF$_3$C$_6$H$_4$: 95% yield, ee = 96%
R = *p*-NCC$_6$H$_4$, Ar = *p*-CF$_3$C$_6$H$_4$: 94% yield, ee = 99%
R = 1-Naph, Ar = *m*-BrC$_6$H$_4$: 71% yield, ee = 94%
R = 1-Naph, Ar = *m*-NCC$_6$H$_4$: 68% yield, ee = 93%
R = Ph, Ar = *m*-NCC$_6$H$_4$: 68% yield, ee = 93%
R = 1-furyl, Ar = *m*-NCC$_6$H$_4$: 87% yield, ee = 87%
R = Cy, Ar = *m*-NCC$_6$H$_4$: 54% yield, ee = 63%
R = Ph, Ar = *p*-NCC$_6$H$_4$: 61% yield, ee = 86%

Scheme 1.28. (*R*)-DPP-BINOL as ligand in addition of *in situ*-generated arylmagnesium chlorides from aryl bromides to aldehydes of chiral functionalised aryl secondary alcohols of synthetic importance in good to high yields and enantioselectivities of up to 99% ee.

A drawback of the method reported by Harada *et al.* (Schemes 1.26–1.27) was the need to add the Grignard reagent to titanium tetraisopropoxide at −78°C, and then to introduce the resulting mixture into the reaction for two hours. Although very useful on a laboratory scale, large-scale reactions at very low temperatures are impractical. A significant operational and economic improvement was reported by Da and co-workers, consisting in converting the Grignard reagents into less reactive triarylaluminum intermediates *in situ* by treatment with $AlCl_3$.[62] In this process, $MgBr_2$ and $MgBr(Oi\text{-}Pr)$ were formed, and could promote as Lewis acids the background reaction to form the racemic product, and lower the enantioselectivity of the reaction. In this context, 2,2′-oxy-bis(N,N-dimethylethanamine) (BDMAEE) was used as an additive to chelate the *in situ* generated Lewis acids $MgBr_2$ and $MgBr(Oi\text{-}Pr)$ and to suppress their activity so that the asymmetric additions promoted by 10 mol% of (S)-H_8-BINOL as chiral ligand of $Ti(Oi\text{-}Pr)_4$ were remarkably highly enantioselective for a variety of aromatic as well as aliphatic aldehydes with enantioselectivities of up to 99% ee (Scheme 1.29). Moreover, the chiral alcohols were obtained in general excellent yields of up to 97%. The authors have proposed the mechanism depicted in Scheme 1.29 to explain the role of $AlCl_3$ in the reaction. It consisted of accepting the aryl group from the Grignard reagent to generate $AlAr_3$, which ultimately transferred the aryl to the aldehyde. BDMAEE was believed to sequester the magnesium salts to prevent them from promoting the racemic background process.

Another methodology developed by the same authors, to deactivate alkyl Grignard reagents in the presence of $Ti(Oi\text{-}Pr)_4$, was to perform their enantioselective additions to aldehydes in the presence of 2,2′-oxy-bis(N,N-dimethylethanamine) (BDMAEE).[63] As in the preceding method, BDMAEE was supposed to chelate the *in situ* generated salts $MgBr_2$ of an equilibrium of RMgBr and $MgBr(Oi\text{-}Pr)$ from transmetalation of RMgBr with $Ti(Oi\text{-}Pr)_4$. When this process was promoted by 15 mol% of (S)-BINOL as chiral ligand of $Ti(Oi\text{-}Pr)_4$ employed in a large excess (890 mol%), it afforded a range of chiral secondary alcohols with remarkable enantioselectivities of up to >99% ee, combined with good yields, as shown in Scheme 1.30. The wide scope of the reaction must be highlighted, since

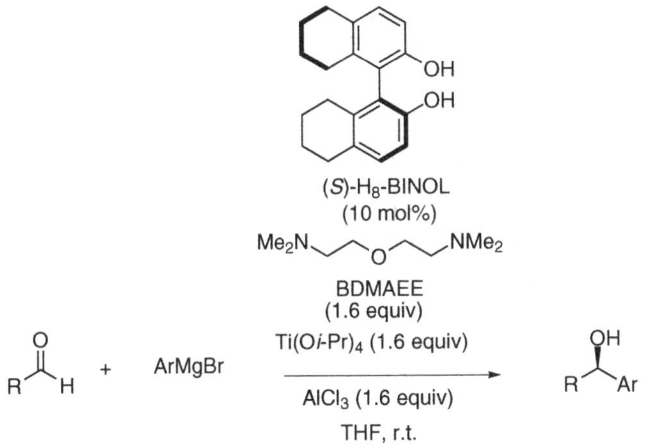

(S)-H$_8$-BINOL
(10 mol%)

Me$_2$N⌒⌒O⌒⌒NMe$_2$

BDMAEE
(1.6 equiv)

Ti(O*i*-Pr)$_4$ (1.6 equiv)
─────────────────
AlCl$_3$ (1.6 equiv)

THF, r.t.

$$\underset{R}{\overset{O}{\|}}\!\!-\!\!H \; + \; ArMgBr \longrightarrow \underset{R}{\overset{OH}{\|}}\!\!-\!\!Ar$$

R = *o*-ClC$_6$H$_4$, Ar = Ph: 94% yield, ee = 94%
R = *p*-FC$_6$H$_4$, Ar = Ph: 94% yield, ee = 97%
R = *p*-ClC$_6$H$_4$, Ar = Ph: 92% yield, ee = 96%
R = *p*-BrC$_6$H$_4$, Ar = Ph: 95% yield, ee = 95%
R = *m*-MeOC$_6$H$_4$, Ar = Ph: 93% yield, ee = 97%
R = *p*-Tol, Ar = Ph: 94% yield, ee = 97%
R = *p*-CF$_3$C$_6$H$_4$, Ar = Ph: 95% yield, ee = 98%
R = 1-Naph, Ar = Ph: 95% yield, ee = 99%
R = 2-Naph, Ar = Ph: 96% yield, ee = 95%
R = Cy, Ar = Ph: 91% yield, ee = 95%
R = *n*-Non, Ar = Ph: 91% yield, ee = 94%
R = 1-Naph, Ar = 2-Naph: 92% yield, ee = 95%
R = Cy, Ar = 2-Naph: 90% yield, ee = 98%
R = 1-Naph, Ar = *m*-Tol: 93% yield, ee = 95%
R = *p*-Tol, Ar = 1-Naph: 96% yield, ee = 96%
R = *p*-FC$_6$H$_4$, Ar = 1-Naph: 94% yield, ee = 99%
R = *p*-FC$_6$H$_4$, Ar = Ph: 93% yield, ee = 95%
R = *p*-ClC$_6$H$_4$, Ar = 2-thienyl: 95% yield, ee = 97%
R = 1-Naph, Ar = benzofuran-5-yl: 83% yield, ee = 87%

proposed mechanism:

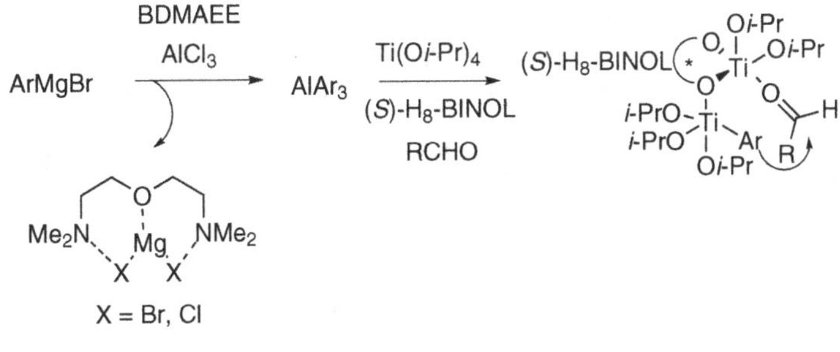

Scheme 1.29. (S)-H$_8$-BINOL as ligand in addition of arylmagnesium bromides to aldehydes with BDMAE and AlCl$_3$

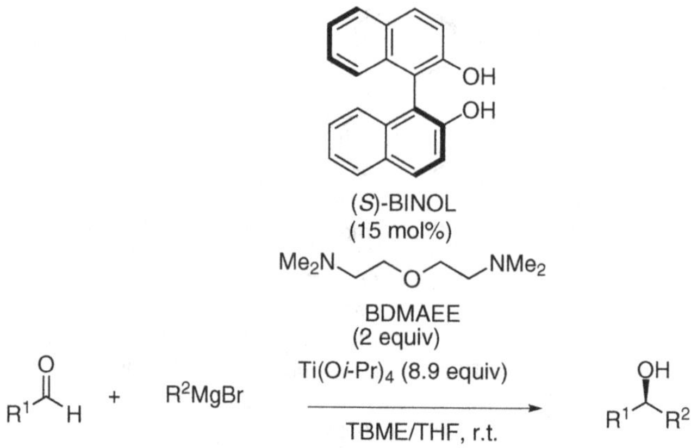
OH
OH
(S)-BINOL
(15 mol%)
Me2N
O
NMe2
BDMAEE
(2 equiv)
Ti(Oi-Pr)4 (8.9 equiv)
TBME/THF, r.t.
O
R1
H
+
R2MgBr
OH
R1
R2

R^1 = *o*-MeOC_6H_4, R^2 = *i*-Bu: 86% yield, ee = 97%
R^1 = *m*-MeOC_6H_4, R^2 = *i*-Bu: 63% yield, ee = 97%
R^1 = *m*-ClC_6H_4, R^2 = *i*-Bu: 91% yield, ee = >99%
R^1 = *p*-FC_6H_4, R^2 = *i*-Bu: 69% yield, ee = >99%
R^1 = 1-Naph, R^2 = *i*-Bu: 93% yield, ee = 98%
R^1 = 2-Naph, R^2 = *i*-Bu: 86% yield, ee = 97%
R^1 = *p*-Tol, R^2 = *i*-Bu: 81% yield, ee = 97%
R^1 = PhCH=CH, R^2 = *i*-Bu: 70% yield, ee = 90%
R^1 = Cy, R^2 = *i*-Bu: 27% yield, ee = 98%
R^1 = Ph, R^2 = *n*-Bu: 68% yield, ee = 92%
R^1 = 1-Naph, R^2 = *n*-Bu: 60% yield, ee = 91%
R^1 = Ph, R^2 = *n*-Pent: 53% yield, ee = 90%
R^1 = 2-Naph, R^2 = *n*-Pent: 77% yield, ee = 90%
R^1 = 1-Naph, R^2 = *n*-Pent: 79% yield, ee = 90%
R^1 = 1-Naph, R^2 = *n*-Hept: 52% yield, ee = 87%
R^1 = *p*-MeOC_6H_4, R^2 = $Me_2C=CH(CH_2)_2$: 50% yield, ee = 92%
R^1 = Ph, R^2 = CH_2Bn: 62% yield, ee = 96%
R^1 = *p*-MeOC_6H_4, R^2 = CH_2Bn: 63% yield, ee = 88%
R^1 = *p*-Tol, R^2 = CH_2Bn: 63% yield, ee = 90%
R^1 = PhCH=CH, R^2 = CH_2Bn: 70% yield, ee = 79%
R^1 = Cy, R^2 = CH_2Bn: 43% yield, ee = 96%
R^1 = *n*-Pr, R^2 = CH_2Bn: 46% yield, ee = 90%
R^1 = 1-Naph, R^2 = $(CH_2)_2$(2-thienyl): 78% yield, ee = 97%
R^1 = 2-thienyl, R^2 = $(CH_2)_2$(2-thienyl): 53% yield, ee = 90%
R^1 = 1-Naph, R^2 = $(CH_2)_2$(2-thienyl): 78% yield, ee = 97%

Scheme 1.30. (*S*)-BINOL as ligand in addition of alkylmagnesium bromides to aldehydes with BDMAEE butylation, pentylation, heptylation, arylethylation, as well as alkenylethylation could be successfully achieved with homogeneous excellent results with aromatic, aliphatic, as well as α,β-unsaturated aldehydes.

In 2011, Yus *et al.* reported the use of another efficient catalyst for the direct addition of alkylmagnesium bromides to aliphatic and aromatic aldehydes in the presence of 15 equivalents of Ti(O*i*-Pr)$_4$.[64] This chiral ligand **35** was derived from (*S*)-BINOL and employed at 20 mol% of catalyst loading in toluene, providing various chiral secondary alcohols in good yields and moderate to high enantioselectivities of up to 96% ee, as shown in Scheme 1.31.

In the same context, these authors have more recently developed a related readily available BINOL-derived chiral ligand **36**, bearing a pyridine which was applied to the enantioselective direct additions of alkylmagnesium

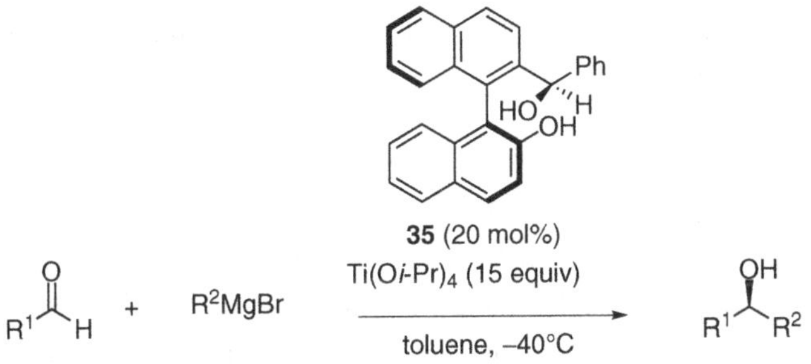

35 (20 mol%)
Ti(O*i*-Pr)$_4$ (15 equiv)

$$R^1\text{-CHO} + R^2\text{MgBr} \xrightarrow{\text{toluene, } -40^\circ\text{C}} R^1\text{-CH(OH)-}R^2$$

R^1 = Ph, R^2 = Me: 92% yield, ee = 88%
R^1 = *m*-Tol, R^2 = Me: 99% yield, ee = 88%
R^1 = *p*-CF$_3$C$_6$H$_4$, R^2 = Me: 88% yield, ee = 88%
R^1 = *p*-ClC$_6$H$_4$, R^2 = Me: 98% yield, ee = 84%
R^1 = 2-Naph, R^2 = Me: 92% yield, ee = 90%
R^1 = Bn, R^2 = Me: 70% yield, ee = 70%
R^1 = PhCH=CH, R^2 = Me: 90% yield, ee = 68%
R^1 = 2-thienyl, R^2 = Me: 53% yield, ee = 58%
R^1 = Ph, R^2 = Et: 95% yield, ee = 86%
R^1 = Ph, R^2 = *n*-Bu: 90% yield, ee = 96%
R^1 = Cy, R^2 = *n*-Bu: 98% yield, ee = 50%
R^1 = Ph, R^2 = *i*-Bu: 91% yield, ee = 86%
R^1 = 2-Naph, R^2 = Ph: 98% yield, ee = 15%

Scheme 1.31. Chiral BINOL derivative as ligand in addition of alkylmagnesium bromides to aldehydes

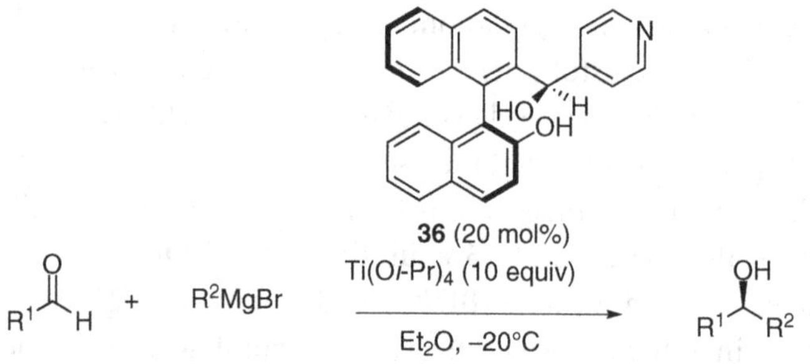

36 (20 mol%)

Ti(O*i*-Pr)$_4$ (10 equiv)

$$ R^1\text{CHO} + R^2MgBr \xrightarrow[\text{Et}_2O,\ -20°C]{} R^1\text{CH(OH)}R^2 $$

R^1 = Cy, R^2 = *n*-Bu: 97% yield, ee = 90%
R^1 = *i*-Pent, R^2 = *n*-Bu: 97% yield, ee = 80%
R^1 = CH$_2$=CH, R^2 = *n*-Bu: 53% yield, ee = 96%
R^1 = *c*-Pent, R^2 = Et: 80% yield, ee = 86%
R^1 = *n*-Oct, R^2 = Me: 98% yield, ee = 88%
R^1 = BnCH$_2$, R^2 = Me: 81% yield, ee = 86%
R^1 = *i*-Pent, R^2 = Me: 99% yield, ee = 83%
R^1 = Cy, R^2 = Me: 61% yield, ee = 99%
R^1 = (*E*)-PhCH=CH, R^2 = Me: >99% yield, ee = 82%

Scheme 1.32. Another chiral BINOL derivative as ligand in addition of alkylmagnesium bromides to aliphatic aldehydes

bromides to aliphatic aldehydes in the presence of one equivalent of Ti(O*i*-Pr)$_4$.[65] As shown in Scheme 1.32, this method allowed the synthesis of a range of chiral secondary aliphatic alcohols to be achieved in good to quantitative yields, and enantioselectivities of up to 99% ee. This novel method overcame the main problems associated with the use of aliphatic substrates, such as their multiple conformations, the absence of possible π-stacking interactions with the catalyst and/or their highly enolisable character.

1.2.4. Additions of Organotitanium Reagents to Aldehydes

Roles of excess of Ti(O*i*-Pr)$_4$ in titanium-promoted asymmetric additions of organometallic compounds to carbonyl compounds have been suggested, to generate a dititanium active species bearing a chiral ligand, and also to facilitate the removal of the product. It has been suggested that the reactions involved the additions of organotitanium species, which were generated *in situ* from reactions of organometallic compounds with

Ti(O*i*-Pr)$_4$. However, direct asymmetric additions of organotitanium reagents have been demonstrated only rarely since the publication by Seebach and Weber who, in 1994, described the first catalytic asymmetric addition of RTi(O*i*-Pr)$_3$ to aldehydes by using a TADDOL-derived chiral titanium catalyst.[14] In 2011, Gau and Li reported highly enantioselective additions of alkyltitanium reagents RTi(O*i*-Pr)$_3$ to aldehydes promoted by an *in situ* generated titanium catalyst of (*R*)-H$_8$-BINOL.[66] This ligand was employed at 10 mol% of catalyst loading, along with one equivalent of Ti(O*i*-Pr)$_4$ in hexane at room temperature, allowing the formation of a wide variety of chiral secondary alcohols in good to high yields and enantioselectivities of up to 94% ee, as shown in Scheme 1.33. Aromatic, heteroaromatic as well as aliphatic aldehydes were compatible with this protocol, and three different alkyltitanium reagents RTi(O*i*-Pr)$_3$ (R = Cy, *n*-Bu, and *i*-Bu) were successfully employed, with reactivity and enantioselectivity differences in

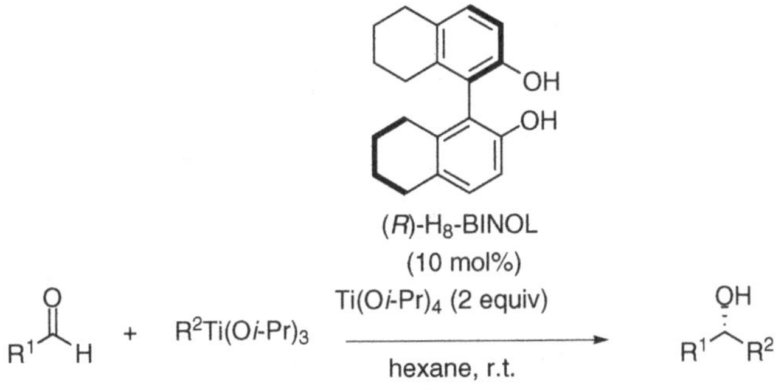

(*R*)-H$_8$-BINOL
(10 mol%)

$$R^1\text{CHO} + R^2Ti(Oi\text{-Pr})_3 \xrightarrow[\text{hexane, r.t.}]{\text{Ti}(Oi\text{-Pr})_4 \text{ (2 equiv)}} R^1\text{CH(OH)}R^2$$

R^1 = Ph, R^2 = Cy: 90% yield, ee = 92%
R^1 = *p*-Tol, R^2 = Cy: 83% yield, ee = 90%
R^1 = *o*-FC$_6$H$_4$, R^2 = Cy: 66% yield, ee = 93%
R^1 = *o*-CF$_3$C$_6$H$_4$, R^2 = Cy: 77% yield, ee = 94%
R^1 = 2-furyl, R^2 = Cy: 85% yield, ee = 90%
R^1 = (*E*)-PhCH=CH, R^2 = Cy: 77% yield, ee = 76%
R^1 = Ph, R^2 = *i*-Bu: 78% yield, ee = 94%
R^1 = *p*-Tol, R^2 = *i*-Bu: 78% yield, ee = 92%
R^1 = *m*-ClC$_6$H$_4$, R^2 = *i*-Bu: 80% yield, ee = 91%
R^1 = (*E*)-PhCH=CH, R^2 = *i*-Bu: 76% yield, ee = 83%
R^1 = Ph, R^2 = *n*-Bu: 78% yield, ee = 84%
R^1 = *m*-ClC$_6$H$_4$, R^2 = *n*-Bu: 80% yield, ee = 86%
R^1 = (*E*)-PhCH=CH, R^2 = *n*-Bu: 74% yield, ee = 75%

Scheme 1.33. (*R*)-H$_8$-BINOL as ligand in addition of alkyltitanium reagents to aldehydes terms of steric bulkiness of the R nucleophiles. Therefore, the additions of secondary *c*-hexyl to aldehydes were slower than those of primary *i*-butyl or *n*-butyl nucleophiles. For the primary alkyls, lower enantioselectivities were obtained for products from additions of the linear *n*-butyl as compared with the enantioselectivities of products from additions of the branched *i*-butyl group.

The same authors also described remarkable highly enantioselective additions of aryltitanium reagents $ArTi(Oi\text{-}Pr)_3$ to aromatic as well as aliphatic aldehydes, based on the use of a catalytic amount (3–10 mol%) of a preformed chiral titanium catalyst **37** derived from (R)-H_8-BINOL.[67] These reactions proceeded instantaneously at room temperature, affording chiral aromatic secondary alcohols in general excellent enantioselectivities, always ≥90% ee and up to 99% ee, as summarised in Scheme 1.34. In addition to the mild reaction conditions and the rapidity of the reaction, another great advantage of this process was that excess amounts of

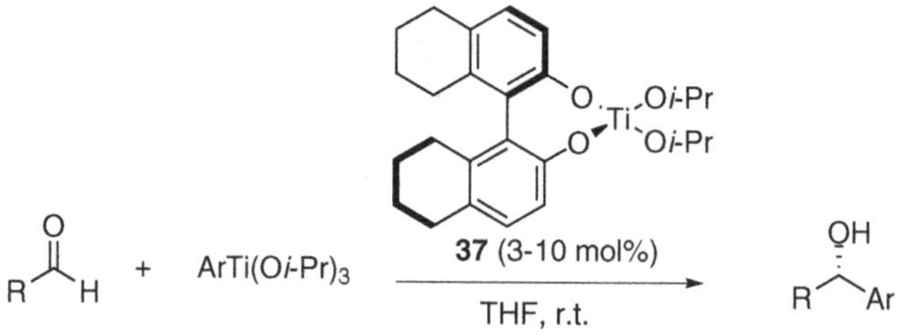

R = *o*-Tol, Ar = Ph: 94% yield, ee = 99%
R = *p*-Tol, Ar = Ph: 95% yield, ee = 94%
R = *m*-MeOC$_6$H$_4$, Ar = Ph: 95% yield, ee = 95%
R = *p*-MeOC$_6$H$_4$, Ar = Ph: 91% yield, ee = 90%
R = *p*-ClC$_6$H$_4$, Ar = Ph: 95% yield, ee = 98%
R = *p*-CF$_3$C$_6$H$_4$, Ar = Ph: 96% yield, ee = 97%
R = (*E*)-PhCH=CH, Ar = Ph: 93% yield, ee = 90%
R = 2-furyl, Ar = Ph: 94% yield, ee = 93%
R = *n*-Bu, Ar = Ph: 90% yield, ee = 92%
R = *i*-Pr, Ar = Ph: 88% yield, ee = 95%
R = Ph, Ar = *p*-Tol: 96% yield, ee = 95%
R = Ph, Ar = *p*-MeOC$_6$H$_4$: 94% yield, ee = 95%
R = Ph, Ar = 2-Naph: 86% yield, ee = 94%
R = Ph, Ar = *p*-TMSC$_6$H$_4$: 95% yield, ee = 94%
R = Ph, Ar = *p*-MeOC$_6$H$_4$: 94% yield, ee = 95%

Scheme 1.34. (R)-H_8-BINOL as ligand in addition of aryltitanium reagents to aldehydes

Ti(Oi-Pr)$_4$ were not necessary. Furthermore, in all cases of substrates studied, the yields were very high, up to 96%.

More recently, Harada *et al.* reported a highly efficient, practical, and original method for the enantioselective arylation and heteroarylation of aldehydes, with organotitanium reagents prepared *in situ* through the reaction of aryl- and heteroaryllithium reagents with ClTi(Oi-Pr)$_3$.[68] Chiral titanium complexes generated *in situ* from (R)-DPP-H$_8$-BINOL ligand and (R)-3-aryl-H$_8$-BINOL ligand **38** exhibited an excellent catalytic activity in terms of enantioselectivity and turnover efficiency for the reaction, providing chiral diaryl-, aryl heteroaryl-, and diheteroaryl-methanol derivatives in high enantioselectivities of up to 98% ee, as shown in Scheme 1.35. In most cases of substrates, a catalyst loading as low as 2 mol% was sufficient to reach these results.

1.2.5. Additions of Alkylboron Reagents to Aldehydes

Because a variety of alkylboranes are commercially available and can be readily prepared by hydroboration of alkenes, they are promising candidates for practical alkylating reagents. Indirect use of trialkylboranes in asymmetric alkylation reactions has been reported by a boron-zinc exchange reaction.[69] Meanwhile, Harada and Ukon reported, in 2008, the direct use of trialkylboranes without converting them into alkylmetallic species.[70] As shown in Scheme 1.36, triethylborane was able to be directly and enantioselectively added to aromatic, aliphatic, and α,β-unsaturated aldehydes in the presence of three equivalents of Ti(Oi-Pr)$_4$, and a catalytic amount of (R)-DPP-H$_8$-BINOL as chiral ligand in THF. In most cases of substrates studied, excellent enantioselectivities of up to 97% ee were obtained for the corresponding chiral secondary alcohols formed. Remarkably, a low catalyst loading of only 2 mol% was sufficient to reach both excellent yields and enantioselectivities.

Later, the same authors reported another type of method for the synthesis of chiral secondary alcohols based on the use of *in situ* generated 1-alkenylboron reagents.[71] As shown in Scheme 1.37, this one pot procedure started from terminal alkynes and aldehydes. Hydroboration of these terminal alkynes with dicyclohexylborane and subsequent reaction of the

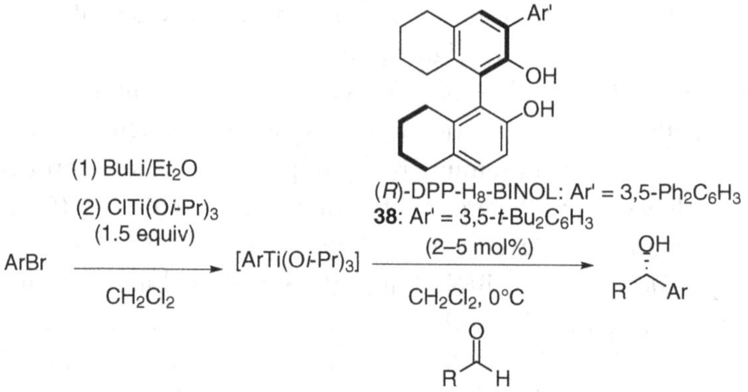

$$(1) \text{ BuLi/Et}_2\text{O}$$

$$\text{ArBr} \xrightarrow[\text{CH}_2\text{Cl}_2]{\begin{array}{c}(2)\ \text{ClTi(O}i\text{-Pr)}_3\\(1.5\ \text{equiv})\end{array}} [\text{ArTi(O}i\text{-Pr)}_3]$$

(R)-DPP-H$_8$-BINOL: Ar' = 3,5-Ph$_2$C$_6$H$_3$
38: Ar' = 3,5-t-Bu$_2$C$_6$H$_3$

$$\xrightarrow[\text{CH}_2\text{Cl}_2,\ 0°\text{C}]{(2\text{–}5\ \text{mol\%})}$$

with ligand (R)-DPP-H$_8$-BINOL (2 mol%):
R = 1-Naph, Ar = Ph: 96% yield, ee = 98%
R = p-ClC$_6$H$_4$, Ar = Ph: 92% yield, ee = 95%
R = CH$_2$=C(Me), Ar = p-ClC$_6$H$_4$: 88% yield, ee = 98%
R = n-Bu, Ar = p-ClC$_6$H$_4$: 82% yield, ee = 72%
R = Cy, Ar = p-ClC$_6$H$_4$: 97% yield, ee = 39%
R = Ph, Ar = p-MeOC$_6$H$_4$: 75% yield, ee = 94%
R = Ph, Ar = o-MeOC$_6$H$_4$: 92% yield, ee = 12%
R = Ph, Ar = 2-Naph: 90% yield, ee = 90%
R = p-Tol, Ar = 2-thienyl: 95% yield, ee = 89%
R = Cy, Ar = 2-thienyl: 91% yield, ee = 20%
R = 2-furyl, Ar = 2-thienyl: 96% yield, ee = 86%
R = Ph, Ar = 3-furyl: 84% yield, ee = 95%
R = 2-furyl, Ar = 3-furyl: 74% yield, ee = 91%
R = 2-thienyl, Ar = 3-thienyl: 94% yield, ee = 96%
with ligand **38** (2 mol%):
R = p-ClC$_6$H$_4$, Ar = 2-thienyl: 93% yield, ee = 86%
R = Ph, Ar = 2-thienyl: 95% yield, ee = 95%
with ligand **38** (0.5 mol%):
R = Ph, Ar = 3-thienyl: 90% yield, ee = 95%
with ligand **38** (5 mol%):
R = 1-Naph, Ar = 2-benzothienyl: 87% yield, ee = 90%
R = Ph, Ar = 2-benzothienyl: 94% yield, ee = 87%
R = Ph, Ar = 2-furyl: 68% yield, ee = 55%

Scheme 1.35. (R)-3-Aryl-H$_8$-BINOL derivatives as ligands in addition of *in situ*-generated (hetero)aryltitanium reagents to aldehydes

resulting generated alkenylboron reagents with aldehydes in the presence of a catalytic amount (5 mol%) of (R)-DPP-H$_8$-BINOL as chiral ligand and an excess of Ti(Oi-Pr)$_4$ afforded the corresponding chiral allylic alcohols in good to high enantioselectivities of up to 94% ee. The scope of the process was found to be wide, since a range of aromatic, aliphatic, and α,β-unsaturated aldehydes were compatible as well as various aliphatic

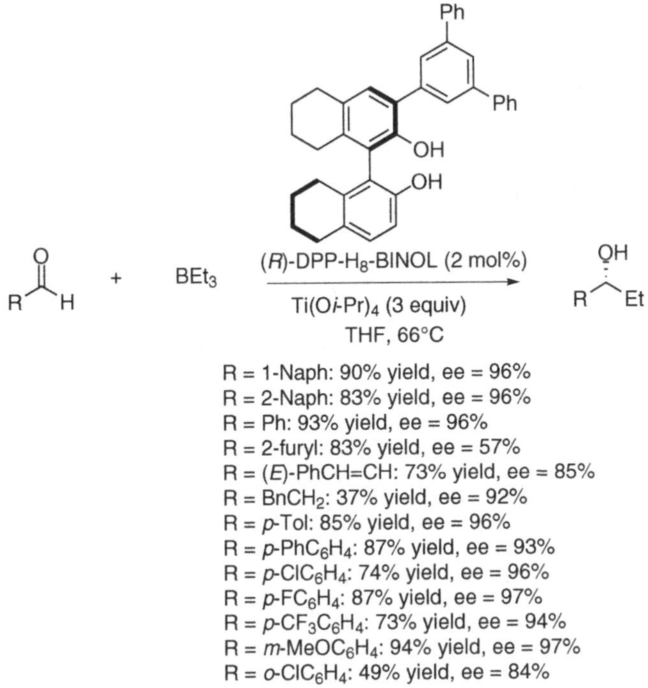

Scheme 1.36. (R)-DPP-H$_8$-BINOL as ligand in addition of BEt$_3$ to aldehydes

alkynes including those containing a chlorine atom, a protected alcohol, a nitrile, and an amide.

In 2013, a closely related method was applied by the same authors to the synthesis of chiral functionalised secondary alcohols through enantioselective additions of *in situ* generated functionalised alkylboron reagents to aromatic, heteroaromatic, and α,β-unsaturated aldehydes.[72] As shown in Scheme 1.38, the required functionalised alkylboron reagents were *in situ* generated through hydroboration of the corresponding functionalised terminal olefins with BH$_3$·SMe$_2$. The latter subsequently underwent addition to aldehydes in the presence of 5 mol% of (R)-DPP-H$_8$-BINOL and three equivalents of Ti(Oi-Pr)$_4$, to afford the corresponding alcohols in remarkable enantioselectivities of up to 99% ee. A range of starting functionalised olefins were tolerated, since terminal alkenes bearing aromatic, aliphatic, TIPS protected alcohol, phthalimide, bromide, isopropyl ester, and cyano groups could be successfully used in the reaction.

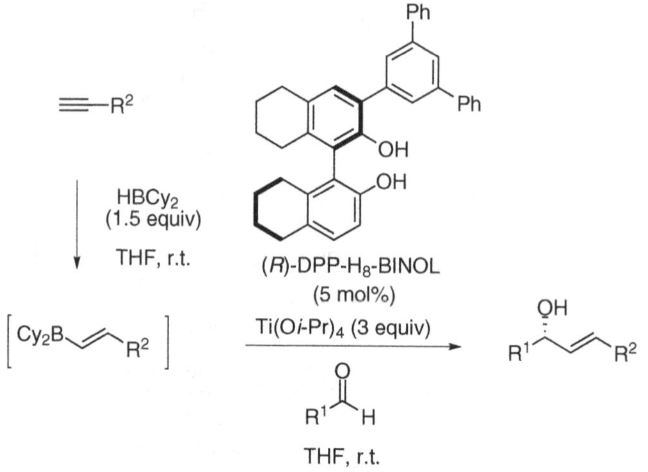

$R^1 = p\text{-}ClC_6H_4$, $R^2 = n\text{-Hex}$: 72% yield, ee = 91%
$R^1 = Ph$, $R^2 = n\text{-Hex}$: 60% yield, ee = 90%
$R^1 = 1\text{-Naph}$, $R^2 = n\text{-Hex}$: 62% yield, ee = 94%
$R^1 = CH_2=C(Me)$, $R^2 = n\text{-Hex}$: 46% yield, ee = 93%
$R^1 = CH_2Bn$, $R^2 = n\text{-Hex}$: 43% yield, ee = 82%
$R^1 = Ph$, $R^2 = i\text{-Pent}$: 66% yield, ee = 92%
$R^1 = Ph$, $R^2 = CH_2Bn$: 67% yield, ee = 94%
$R^1 = Ph$, $R^2 = c\text{-Pr}$: 61% yield, ee = 78%
$R^1 = Ph$, $R^2 = (CH_2)_3Cl$: 66% yield, ee = 90%
$R^1 = Ph$, $R^2 = (CH_2)_2OTr$: 70% yield, ee = 91%
$R^1 = Ph$, $R^2 = (CH_2)_3CN$: 74% yield, ee = 93%

Scheme 1.37. (R)-DPP-H$_8$-BINOL as ligand in addition of *in situ*-generated 1-alkenylboron reagents to aldehydes

1.2.6. *Additions of Organolithium Reagents to Aldehydes*

Organolithium compounds are common bench reagents found in any organic synthetic laboratory, and are widely used in industry to produce numerous materials from pharmaceuticals to polymers.[73] In 1969, Seebach *et al.* reported the first comprehensive investigation of the addition of organolithium reagents in the presence of various chiral ligands derived from diethyl tartrate.[74] It was only in 2011, that a substoichiometric enantioselective addition of methyllithium to *o*-tolylbenzaldehyde was reported by Maddaluno *et al.*[75] Yus *et al.* later reported the first efficient catalytic system for the asymmetric alkylation of aldehydes with organolithium

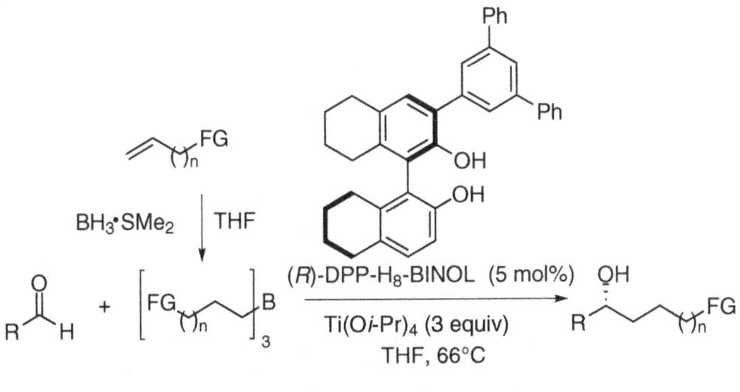

R = 1-Naph, FG = Me, n = 3: 64% yield, ee = 94%
R = *p*-ClC$_6$H$_4$, FG = Me, n = 3: 70% yield, ee = 96%
R = *m*-MeOC$_6$H$_4$, FG = Me, n = 3: 64% yield, ee = 96%
R = 2-furyl, FG = Me, n = 3: 53% yield, ee = 91%
R = (*E*)-PhCH=CH, FG = Me, n = 3: 61% yield, ee = 91%
R = *p*-ClC$_6$H$_4$, FG = Ph, n = 1: 72% yield, ee = 99%
R = *p*-ClC$_6$H$_4$, FG = *p*-MeOC$_6$H$_4$, n = 0: 48% yield, ee = 96%
R = *p*-ClC$_6$H$_4$, FG = Br, n = 2: 68% yield, ee = 91%
R = Ph, FG = Br, n = 2: 71% yield, ee = 93%
R = *p*-ClC$_6$H$_4$, FG = Br, n = 3: 55% yield, ee = 94%
R = 2-thienyl, FG = OTIPS, n = 3: 54% yield, ee = 94%
R = *p*-ClC$_6$H$_4$, FG = OTIPS, n = 2: 70% yield, ee = 93%
R = *o*-BrC$_6$H$_4$, FG = OTIPS, n = 2: 44% yield, ee = 64%
R = 1-Naph, FG = OTIPS, n = 9: 70% yield, ee = 94%
R = *p*-ClC$_6$H$_4$, FG = OTIPS, n = 9: 70% yield, ee = 93%
R = *p*-ClC$_6$H$_4$, FG = phthalimidyl, n = 3: 74% yield, ee = 96%
R = *p*-ClC$_6$H$_4$, FG = COO*i*-Pr, n = 8: 76% yield, ee = 98%
R = 2-thienyl, FG = CN, n = 9: 74% yield, ee = 96%

Scheme 1.38. (*R*)-DPP-H$_8$-BINOL as ligand in addition of functionalised alkylboron reagents to aldehydes

reagents performed in the presence of an excess of Ti(O*i*-Pr)$_4$.[76] The process involved readily available BINOL-derived chiral ligand **36** empolyed at a catalyst loading of 20 mol% in toluene. A variety of alkyllithium reagents were able to be added to aromatic aldehydes, providing the corresponding chiral aromatic secondary alcohols in good to high enantioselectivities of up to 96% ee, as shown in Scheme 1.39. Lower enantioselectivities (62–68% ee) were achieved in the cases of aliphatic aldehydes, and also by using an aryllithium reagent, such as phenyllithium, which provided an enantioselectivity of 17% ee by reaction with 2-naphthaldehyde. Notably, the potential problems associated with the high reactivity of organolithium compounds were overcome under the reaction conditions, demonstrating that this methodology was compatible with functionalised substrates.

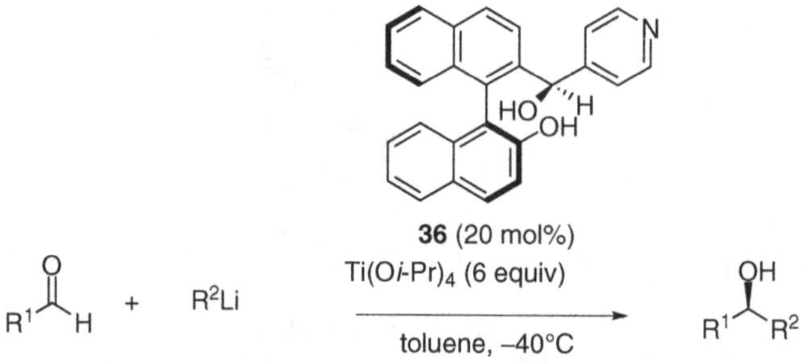

36 (20 mol%)
Ti(O*i*-Pr)$_4$ (6 equiv)

$$R^1\text{CHO} + R^2Li \xrightarrow{\text{toluene, }-40°C} R^1\text{CH(OH)}R^2$$

R^1 = Ph, R^2 = Me: 87% yield, ee = 90%
R^1 = *o*-Tol, R^2 = Me: 78% yield, ee = 62%
R^1 = *p*-Tol, R^2 = Me: 81% yield, ee = 88%
R^1 = *p*-MeOC$_6$H$_4$, R^2 = Me: 91% yield, ee = 89%
R^1 = *p*-CF$_3$C$_6$H$_4$, R^2 = Me: 82% yield, ee = 88%
R^1 = *p*-ClC$_6$H$_4$, R^2 = Me: 74% yield, ee = 84%
R^1 = *p*-CNC$_6$H$_4$, R^2 = Me: 84% yield, ee = 82%
R^1 = 2-Naph, R^2 = Me: 85% yield, ee = 90%
R^1 = 2-thienyl, R^2 = Me: 57% yield, ee = 88%
R^1 = 2-furyl, R^2 = Me: 56% yield, ee = 72%
R^1 = (*E*)-PhCH=CH, R^2 = Me: 84% yield, ee = 68%
R^1 = Bn, R^2 = Me: 23% yield, ee = 62%
R^1 = Ph, R^2 = Et: 78% yield, ee = 92%
R^1 = *p*-ClC$_6$H$_4$, R^2 = Et: 62% yield, ee = 92%
R^1 = Ph, R^2 = *n*-Bu: 90% yield, ee = 96%
R^1 = *p*-BrC$_6$H$_4$, R^2 = *n*-Bu: 89% yield, ee = 94%
R^1 = *p*-MeOC$_6$H$_4$, R^2 = *n*-Bu: 90% yield, ee = 96%
R^1 = 2-Naph, R^2 = Ph: 96% yield, ee = 17%

Scheme 1.39. (*S*)-BINOL derivative as ligand in addition of alkyllithium reagents to aldehydes

Finally, Harada and Muramatsu described the enantioselective addition of phenyllithium to 1-naphthaldehyde, providing the corresponding chiral alcohol in 85% yield and excellent enantioselectivity of 95% ee (Scheme 1.40).[63] In this process, the phenyllithium reagent could be employed after conversion into PhMgBr by treatment with MgBr$_2$. It must be noted that the reaction was carried out without removing concomitantly produced LiBr. It was simply performed by mixing PhLi with MgBr$_2$ (1.2 equiv) and Ti(O*i*-Pr)$_4$ (2 equiv). The chirality arose from using a catalytic amount (2 mol%) of (*R*)-DPP-BINOL as chiral ligand.

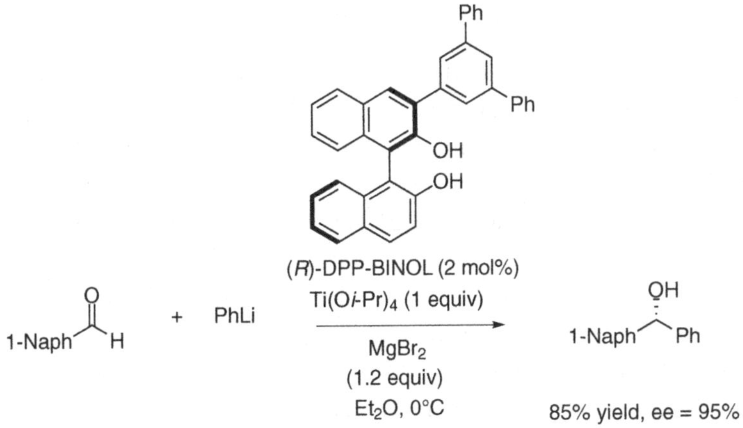

Ph

Ph

OH
OH

(*R*)-DPP-BINOL (2 mol%)

Ti(O*i*-Pr)$_4$ (1 equiv)

$$\text{1-Naph} \overset{O}{\underset{H}{\bigwedge}} + \text{PhLi} \xrightarrow[\substack{\text{Mgbr}_2 \\ (1.2\ \text{equiv}) \\ \text{Et}_2\text{O, }0°\text{C}}]{} \text{1-Naph} \overset{OH}{\underset{Ph}{\bigwedge}}$$

85% yield, ee = 95%

Scheme 1.40. (*R*)-DPP-BINOL as ligand in addition of PhLi to naphthylcarboxaldehyde

1.3. Titanium-Promoted Alkynylation Reactions of Carbonyl Compounds

1.3.1. Aldehydes as Electrophiles

Chiral propargylic alcohols are useful building blocks for the enantioselective synthesis of a number of important chiral complex molecules. As the alkylation reaction, the alkynyl addition has a strategically synthetic advantage in forming a new C–C bond, with concomitant creation of a stereogenic centre in a single operation. Alkynyl-metal reagents are ideal functional carbon nucleophiles, which can be prepared easily owing to the acidity of terminal alkynyl protons. Therefore, the enantioselective addition[7] of these intermediates to carbonyl compounds constitutes an attractive alternative to the synthesis of the corresponding propargylic alcohols.[77] The first examples of enantioselective catalytic alkynylation of aldehydes using chiral titanium catalysts were reported in 2002 by Chan and Pu.[78] In this study, the authors reported enantioselectivities of up to >99% ee for the produced propargylic alcohols, using BINOL derivatives as chiral ligands. Ever since, a number of other modified BINOL-derived ligands have been successfully used to induce these reactions, as well as various other types of ligands.

1.3.1.1. *Addition of phenylacetylene*

Among recently investigated new chiral ligands in enantioselective titanium-catalysed alkynylations of aldehydes, Mao and Zhang reported the use of a readily available and inexpensive novel chiral oxazolidine **39** as ligand in the addition of phenylacetylene to aldehydes.[79] This ligand was derived from readily available (1*R*,2*S*)-*cis*-1-amino-2-indanol. When used at 1 mol% of catalyst loading in combination with a catalytic amount (2 mol%) of Ti(O*i*-Pr)$_4$ and four equivalents of ZnEt$_2$ in THF, the reaction of a variety of aromatic aldehydes afforded the corresponding chiral propargylic alcohols in excellent yields of up to 98%, and enantioselectivities of up to 95% ee, as shown in Scheme 1.41. Notably, lower enantioselectivities

R = Ph: 98% yield, ee = 90%
R = *p*-Tol: 87% yield, ee = 90%
R = *o*-MeOC$_6$H$_4$: 91% yield, ee = 92%
R = *p*-MeOC$_6$H$_4$: 91% yield, ee = 95%
R = *o*-ClC$_6$H$_4$: 96% yield, ee = 93%
R = *m*-ClC$_6$H$_4$: 97% yield, ee = 91%
R = 1-Naph: 96% yield, ee = 90%
R = 2-Naph: 97% yield, ee = 93%
R = 2,3-(MeO)$_2$C$_6$H$_3$: 91% yield, ee = 90%
R = 2-furyl: 91% yield, ee = 90%
R = BnCH$_2$: 81% yield, ee = 77%

39 (1 mol%)
Ti(O*i*-Pr)$_4$ (2 mol%)
THF, 0°C to r.t.

R = Ph: 81% yield, ee = 90%
R = *p*-Tol: 82% yield, ee = 90%
R = *p*-MeOC$_6$H$_4$: 96% yield, ee = 87%
R = *p*-ClC$_6$H$_4$: 87% yield, ee = 93%
R = *m*-ClC$_6$H$_4$: 94% yield, ee = 91%
R = *o*-ClC$_6$H$_4$: 93% yield, ee = 77%
R = *m*-NO$_2$C$_6$H$_4$: 71% yield, ee = 92%
R = *p*-BrC$_6$H$_4$: 84% yield, ee = 92%
R = 1-Naph: 82% yield, ee = 86%
R = 2,3-(MeO)$_2$C$_6$H$_3$: 84% yield, ee = 83%
R = 2-furyl: 76% yield, ee = 83%
R = 2-thienyl: 98% yield, ee = 95%

40 (28 mol%)
Ti(O*i*-Pr)$_4$ (56 mol%)
ZnEt$_2$ (4 equiv)
THF, r.t.

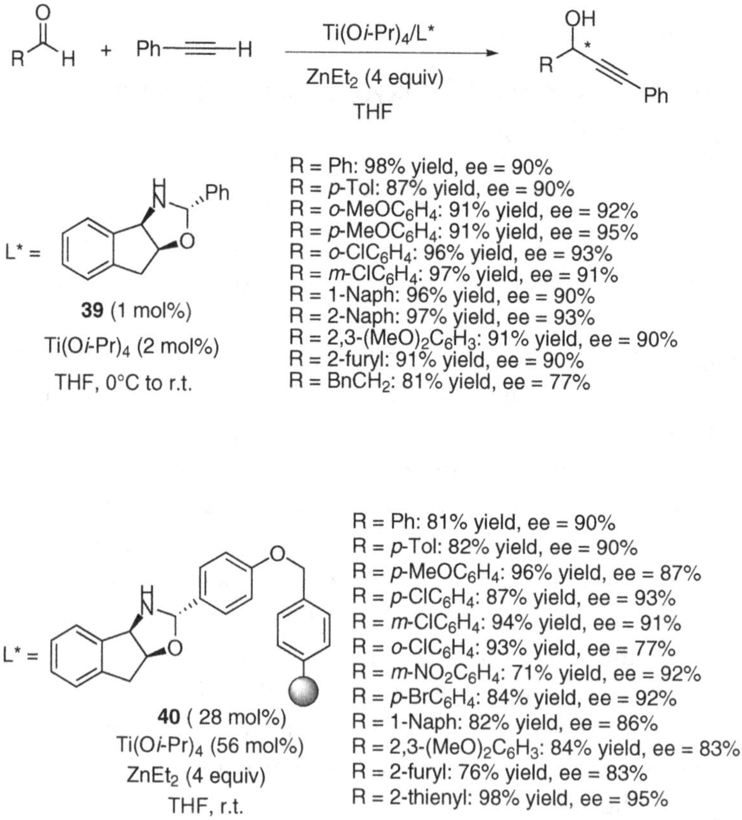

Scheme 1.41. Chiral oxazolidine and chiral resin-supported oxazolidine ligands in addition of phenylacetylene to aldehydes

were obtained in the cases of aliphatic aldehydes (77% ee for R = CH_2Bn for example). In order to render this type of economical ligand recyclable, Mao *et al.* developed resin-supported oxazolidine ligand **40** which was found to smoothly catalyse the same reactions with high yields of up to 98%, and enantioselectivities of up to 95% ee, as shown in Scheme 1.41.[80] In this case, the ligand was employed at 28 mol% of catalyst loading, in combination with 56 mol% of Ti(O*i*-Pr)$_4$ in the presence of four equivalents of ZnEt$_2$. Remarkably, this novel catalytic system could be reused for five times after a simple work-up. It was also suitable for alkynylation of heteroaromatic aldehydes, providing enantioselectivities of up to 95% ee.

Moreover, a soluble chiral polybinaphthol ligand **41** was synthesised through polymerisation of (*S*)-5,5'-dibromo-6,6'-dibutyl-2,2'-binaphthol with (*S*)-2,2'-bishexyloxy-1,1'-binaphthyl-6,6'-boronic acid, via palladium-catalysed Suzuki reaction, and further investigated in phenylacetylene addition, to both aromatic and aliphatic aldehydes.[81] Associated to four equivalents of ZnEt$_2$ and one equivalent of Ti(O*i*-Pr)$_4$ in toluene, THF, or CH$_2$Cl$_2$ as solvent, this novel polymer ligand provided a range of chiral propargylic alcohols in good to high yields, and enantioselectivities of up to 90% and 98% ee, respectively, even in the cases of aliphatic aldehydes (Scheme 1.42). This ligand was easily able to be recovered and reused without loss of activity or enantioselectivity.

In 2010, the same reactions were carried out by Gou *et al.* using other types of chiral ligands, such as camphor-derived sulfonylated amino alcohols.[82] As shown in Scheme 1.43, the best results were achieved employing 10 mol% of camphor sulfonylated amino alcohol **42**, associated to four equivalents of Ti(O*i*-Pr)$_4$ in the presence of three equivalents of ZnEt$_2$ in toluene. A variety of aromatic as well as α,β-unsaturated aldehydes were found to be suitable substrates, leading to the corresponding alcohols in good yields of up to 90%, and moderate enantioselectivities of up to 63% ee. Later, Bauer *et al.* promoted the same reactions for the first time with 20 mol% of β-hydroxy sulfonamide **43**, derived from D-glucosamine in combination with six equivalents of Ti(O*i*-Pr)$_4$.[83] Performed in CH$_2$Cl$_2$ with the presence of 120 mol% of ZnEt$_2$, the process provided various chiral aromatic, aliphatic, as well as α,β-unsaturated alcohols, in good to high yields of up to 95% and moderate to high enantioselectivities of up to 92% ee (Scheme 1.43). It was found that the enantioselectivity of the reaction was highly dependent on the nature of the aldehyde and substitution

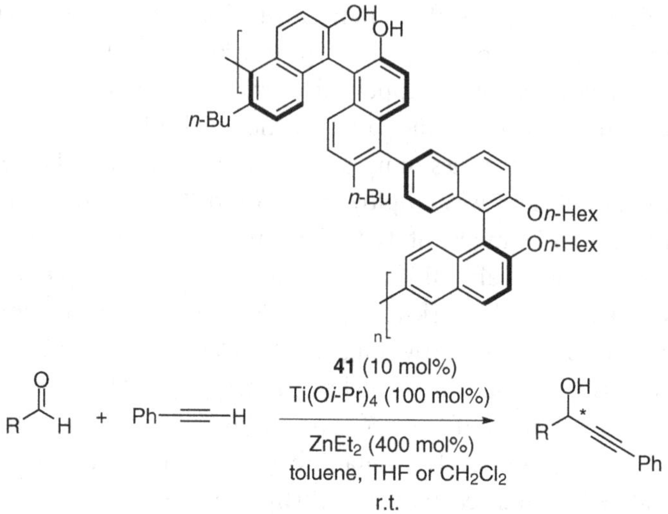

41 (10 mol%)
Ti(O*i*-Pr)$_4$ (100 mol%)
⟶
ZnEt$_2$ (400 mol%)
toluene, THF or CH$_2$Cl$_2$
r.t.

R = Ph: 84% yield, ee = 91%
R = *n*-Hept: 76% yield, ee = 96%
R = *o*-ClC$_6$H$_4$: 85% yield, ee = 67%
R = *p*-CF$_3$C$_6$H$_4$: 89% yield, ee = 92%
R = *p*-Tol: 84% yield, ee = 97%
R = Et: 80% yield, ee = 98%
R = *p*-FC$_6$H$_4$: 90% yield, ee = 98%

Scheme 1.42. Chiral polynaphthol ligand in addition of phenylacetylene to aldehydes

on its phenyl group. The best result was reached for 2-fluorobenzaldehyde (92% ee), while the reactions of 3- and 4-fluorobenzaldehydes gave very low enantioselectivities (12% ee and 22% ee, respectively). Moreover, a good enantioselectivity (85% ee) was observed for a cycloaliphatic aldehyde (R = Cy), whereas a simple linear aliphatic aldehyde (R = *n*-Pent) gave a lower enantiomeric excess of 49% ee.

Earlier, Wang *et al.* reported a novel catalytic system based on the combination of a catalytic amount (20 mol%) of a new readily available and inexpensive chiral β-sulfonamide alcohol **44** with a same catalytic amount of Ti(O*i*-Pr)$_4$.[84] The process also needed 3.5 equivalents of ZnEt$_2$ and 0.5 equivalent of a terminal base, such as diisopropylethylamine (DIPEA), as an additive. As summarised in Scheme 1.43, the reaction of phenylacetylene with benzaldehyde led to the corresponding propargylic alcohol in

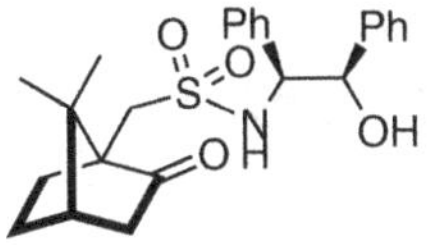

R = Ph: 90% yield, ee = 56%
R = p-MeOC$_6$H$_4$: 73% yield, ee = 61%
R = p-Tol: 87% yield, ee = 63%
R = p-FC$_6$H$_4$: 77% yield, ee = 56%
R = p-ClC$_6$H$_4$: 71% yield, ee = 54%
R = p-BrC$_6$H$_4$: 74% yield, ee = 52%
R = 1-Naph: 83% yield, ee = 51%
R = 2-furyl: 79% yield, ee = 45%
R = PhCH=CH: 57% yield, ee = 42%

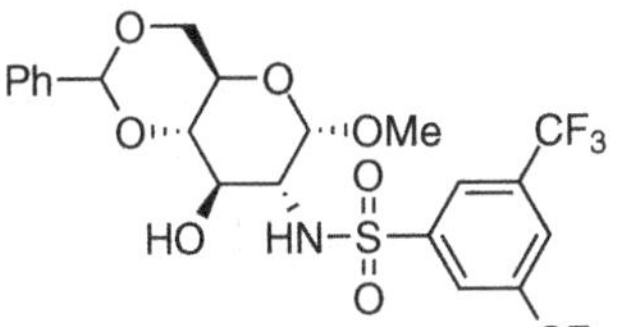

42 (10 mol%)
Ti(Oi-Pr)$_4$ (4 equiv)
ZnEt$_2$ (3 equiv)
toluene, 22 °C

R = Ph: 82% yield, ee = 90%
R = o-MeOC$_6$H$_4$: 63% yield, ee = 28%
R = m-MeOC$_6$H$_4$: 63% yield, ee = 23%
R = p-MeOC$_6$H$_4$: 85% yield, ee = 73%
R = o-FC$_6$H$_4$: 68% yield, ee = 92%
R = p-FC$_6$H$_4$: 59% yield, ee = 22%
R = m-FC$_6$H$_4$: 72% yield, ee = 12%
R = PhCH=CH: 92% yield, ee = 23%
R = Cy: 75% yield, ee = 85%
R = n-Pent: 95% yield, ee = 49%

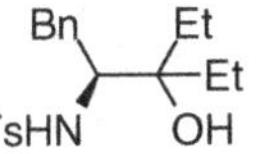

43 (20 mol%)
Ti(Oi-Pr)$_4$ (6 equiv)
ZnMe$_2$ (1.2 equiv)
CH$_2$Cl$_2$, r.t.

R = Ph: 78% yield, ee = 96%

44 (20 mol%)
Ti(Oi-Pr)$_4$ (20 mol%)
ZnEt$_2$ (350 mol%)
DIPEA (50 mol%)
toluene, 42 °C

Scheme 1.43. Chiral sulfonamide ligands in addition of phenylacetylene to aldehydes (RCHO)

good yield (78%), and very high enantioselectivity of 96% ee. The role of DIPEA was to facilitate the formation of the alkynylzinc reagents.

Alternatively, moderate enantioselectivities of up to 75% ee were reported by Kilic *et al.* for the same reactions of aromatic aldehydes, based on the use of novel chiral 4,4'-biquinazoline alcohols synthesised from readily accessible (*S*)-2-acetoxycarboxylic acids.[85] Even if the enantioselectivities remained moderate, the advantage of this process was the involvement of only 25 mol% of Ti(O*i*-Pr)$_4$ in combination with 10 mol% of the chiral ligand. These best results were reached with ligands **45** and **46** in the presence of two equivalents of ZnEt$_2$ in THF (Scheme 1.44). In addition, Dehaen *et al.* involved the same catalytic amount (25 mol%) of Ti(O*i*-Pr)$_4$ in combination with another type of novel chiral ligands, such as H$_8$-BIFOL, to induce the addition of phenylacetylene to benzaldehyde in the presence of two equivalents of ZnMe$_2$ in THF.[86] The process resulted, however, in a moderate enantioselectivity of 56% ee combined with 67% yield, as shown in Scheme 1.44.

Several β-hydroxy amide chiral ligands have also been investigated by the group of Hui and Xu. Among them, chiral C_2-symmetric *bis*(β-hydroxy amide) **47**, synthesised via the reaction of isophthaloyl dichloride

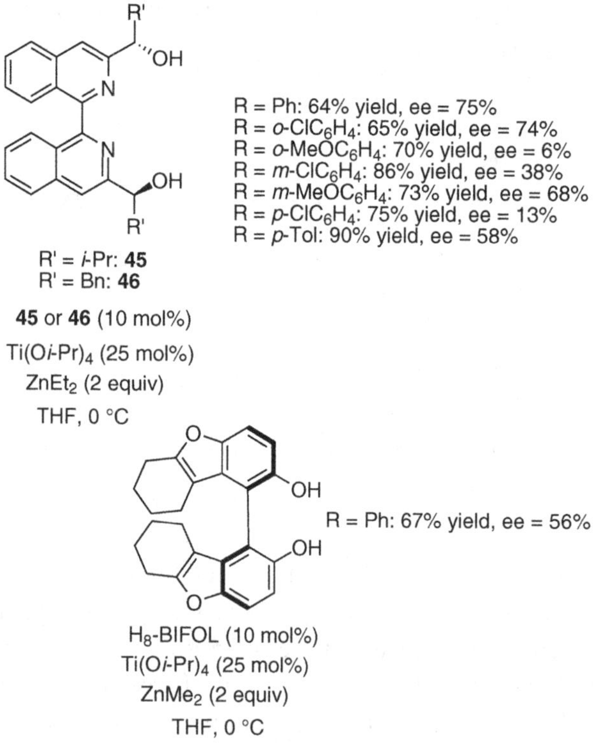

R = Ph: 64% yield, ee = 75%
R = *o*-ClC$_6$H$_4$: 65% yield, ee = 74%
R = *o*-MeOC$_6$H$_4$: 70% yield, ee = 6%
R = *m*-ClC$_6$H$_4$: 86% yield, ee = 38%
R = *m*-MeOC$_6$H$_4$: 73% yield, ee = 68%
R = *p*-ClC$_6$H$_4$: 75% yield, ee = 13%
R = *p*-Tol: 90% yield, ee = 58%

R' = *i*-Pr: **45**
R' = Bn: **46**

45 or **46** (10 mol%)
Ti(O*i*-Pr)$_4$ (25 mol%)
ZnEt$_2$ (2 equiv)
THF, 0 °C

R = Ph: 67% yield, ee = 56%

H$_8$-BIFOL (10 mol%)
Ti(O*i*-Pr)$_4$ (25 mol%)
ZnMe$_2$ (2 equiv)
THF, 0 °C

Scheme 1.44. Chiral 4,4′-biquinazoline alcohol ligand and H$_8$-BIFOL ligand in addition of phenylacetylene to aldehydes (RCHO)

and L-phenylalanine, provided excellent results for the addition of phenylacetylene to benzaldehyde when used at 10 mol% of catalyst loading, combined with a catalytic amount (30 mol%) of Ti(O*i*-Pr)$_4$ in the presence of three equivalents of ZnEt$_2$ in toluene.[87] Indeed, general excellent yields and enantioselectivities of up to 94% and 98% ee, respectively, were obtained with various aromatic aldehydes including benzaldehyde (Scheme 1.45) while cinnamaldehyde and *n*-pentaldehyde gave lower enantioselectivities of 87 and 52% ee, respectively. The authors have demonstrated that the two β-hydroxy amide moieties in this ligand behave as two independent ligands in the catalytic system. To further develop efficient catalysts for the phenylation of aliphatic and vinyl aldehydes, these authors later reported the synthesis of novel β-hydroxy amide ligands from L-tyrosine.[88] Among them, ligand **48** proved to be the most efficient, when used at 20 mol% of catalyst loading combined with 60 mol% of Ti(O*i*-Pr)$_4$. As shown in Scheme 1.45, the reaction of a range of aliphatic and α,β-unsaturated aldehydes performed in toluene in the presence of three equivalents of ZnEt$_2$ afforded the corresponding chiral propargylic alcohols in high yields of up to 86%, and very high enantioselectivities of up to 96% ee. Furthermore, an excellent enantioselectivity of 92% ee was also obtained in the case of benzaldehyde. With the aim of developing recyclable ligands of the same family, these authors synthesised a series of polymer-supported chiral β-hydroxy amides to be investigated in the same additions.[89] As shown in Scheme 1.45, the use of resin **49**, obtained through co-polymerisation of the corresponding monomer with styrene and divinyl benzene, at 20 mol% of catalyst loading in combination with 60 mol% of Ti(O*i*-Pr)$_4$, allowed the addition of phenylacetylene to aromatic and aliphatic aldehydes in the presence of one equivalent of ZnEt$_2$, to be achieved in high yields (up to 93%) and enantioselectivities of up to 93% ee. Moreover, this resin could be reused four times.

In 2010, Pu *et al.* used (*S*)-BINOL to induce chirality in the addition of phenylacetylene to enals **50**, which provided the corresponding chiral propargylic alcohol-based en-type precursors **51** for Pauson–Khand reactions.[90] As shown in Scheme 1.46, the combination of 20 mol% of (*S*)-BINOL with 100 mol% of Ti(O*i*-Pr)$_4$ in the presence of four equivalents of ZnEt$_2$ in CH$_2$Cl$_2$, provided both high yields (87–88%) and enantioselectivities of up to 94% ee. The formed chiral propargylic ene alcohols

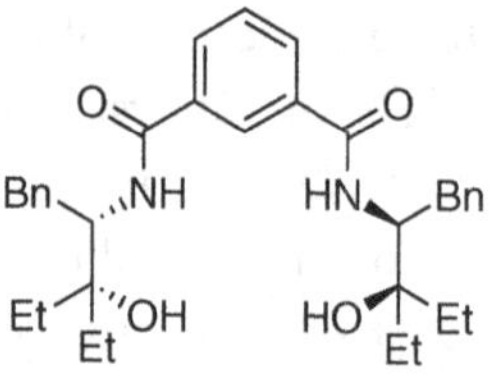

47 (10 mol%)

Ti(O*i*-Pr)$_4$ (30 mol%)

ZnEt$_2$ (3 equiv)

toluene, r.t.

R = Ph: 92% yield, ee = 95%
R = *o*-MeOC$_6$H$_4$: 93% yield, ee = 88%
R = *p*-MeOC$_6$H$_4$: 90% yield, ee = 93%
R = *o*-Tol: 91% yield, ee = 90%
R = *p*-Tol: 94% yield, ee = 98%
R = *o*-ClC$_6$H$_4$: 89% yield, ee = 93%
R = *o*-FC$_6$H$_4$: 93% yield, ee = 92%
R = *p*-ClC$_6$H$_4$: 91% yield, ee = 96%
R = 2-Naph: 82% yield, ee = 94%
R = (*E*)-PhCH=CH: 85% yield, ee = 87%
R = *n*-Bu: 83% yield, ee = 52%

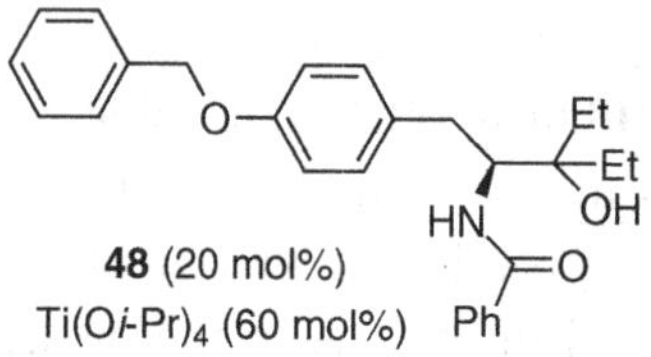

48 (20 mol%)

Ti(O*i*-Pr)$_4$ (60 mol%)

ZnEt$_2$ (3 equiv)

toluene, r.t.

R = Ph: 85% yield, ee = 92%
R = Et: 75% yield, ee = 90%
R = *n*-Pr: 85% yield, ee = 91%
R = *i*-Pr: 87% yield, ee = 92%
R = *n*-Hex: 80% yield, ee = 91%
R = *t*-Bu: 81% yield, ee = 88%
R = Cy: 81% yield, ee = 88%
R = (*E*)-PhCH=CH: 83% yield, ee = 96%
R = CH$_2$=CH: 71% yield, ee = 90%

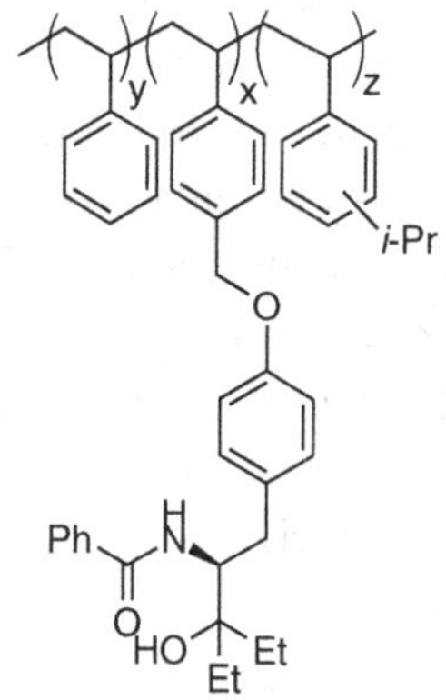

R = Ph: 85% yield, ee = 87%
R = *o*-ClC$_6$H$_4$: 84% yield, ee = 86%
R = *m*-ClC$_6$H$_4$: 87% yield, ee = 89%
R = *p*-ClC$_6$H$_4$: 89% yield, ee = 92%
R = *p*-Tol: 89% yield, ee = 90%
R = 2-Naph: 93% yield, ee = 88%
R = *n*-Pr: 85% yield, ee = 83%
R = Cy: 82% yield, ee = 80%

49 (20 mol%)

x = 0.1, y = 0.88, z = 0.02
Ti(O*i*-Pr)$_4$ (70 mol%)
ZnEt$_2$ (100 mol%)
toluene, r.t.

Scheme 1.45. Chiral β-hydroxy amide ligands in addition of phenylacetylene to aldehydes (RCHO)

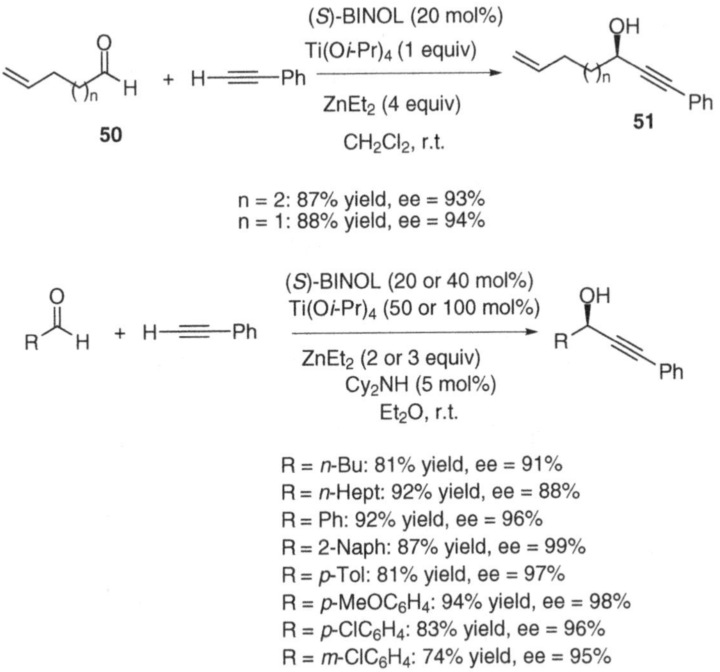

(S)-BINOL (20 mol%)

Ti(O*i*-Pr)$_4$ (1 equiv)

ZnEt$_2$ (4 equiv)

CH$_2$Cl$_2$, r.t.

50 → **51**

n = 2: 87% yield, ee = 93%
n = 1: 88% yield, ee = 94%

(S)-BINOL (20 or 40 mol%)
Ti(O*i*-Pr)$_4$ (50 or 100 mol%)

ZnEt$_2$ (2 or 3 equiv)
Cy$_2$NH (5 mol%)
Et$_2$O, r.t.

R = *n*-Bu: 81% yield, ee = 91%
R = *n*-Hept: 92% yield, ee = 88%
R = Ph: 92% yield, ee = 96%
R = 2-Naph: 87% yield, ee = 99%
R = *p*-Tol: 81% yield, ee = 97%
R = *p*-MeOC$_6$H$_4$: 94% yield, ee = 98%
R = *p*-ClC$_6$H$_4$: 83% yield, ee = 96%
R = *m*-ClC$_6$H$_4$: 74% yield, ee = 95%

Scheme 1.46. (S)-BINOL as ligand in addition of phenylacetylene to various aldehydes including enals

51 were then successfully submitted to intramolecular Pauson–Khand reaction, allowing the synthesis of optically active 5,5- and 5,6-fused bicyclic products to be achieved, with retention of enantiomeric purity. In 2013, the same authors applied a closely related methodology to other aldehydes including aromatic ones, which provided the corresponding propargylic alcohols with even higher enantioselectivities of up to 99% ee (Scheme 1.46).[91]

1.3.1.2. *Addition of various terminal alkynes*

Mao *et al.* have extended the scope of the methodology depicted in Scheme 1.41 using resin-supported oxazolidine ligand **40** to terminal alkynes other than phenylacetylene, such as *p*-tolylacetylene, which provided

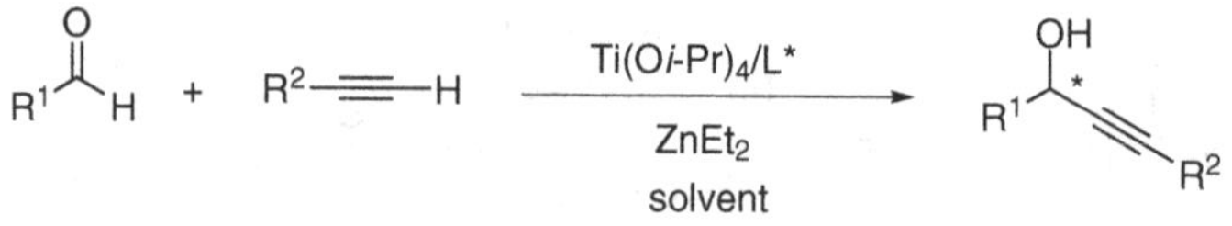

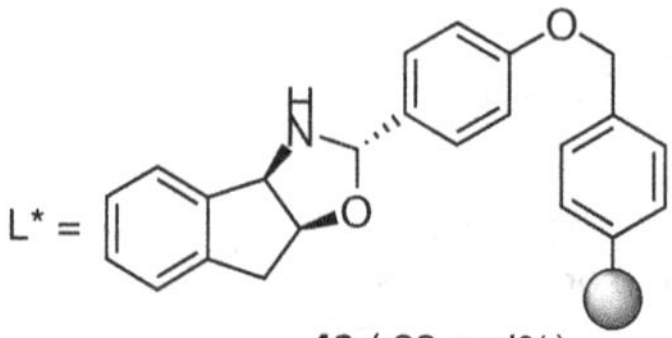

Reaction scheme:

$$R^1\text{CHO} + R^2{-}{\equiv}{-}H \xrightarrow[\substack{ZnEt_2 \\ solvent}]{Ti(O\text{-}i\text{-}Pr)_4/L^*} R^1\text{—CH(OH)—}{\equiv}{-}R^2$$

L* =

40 (28 mol%)
Ti(O*i*-Pr)$_4$ (56 mol%)
ZnEt$_2$ (4 equiv)

THF, r.t.

R^1 = *p*-MeOC$_6$H$_4$, R^2 = *p*-Tol:
81% yield, ee = 92%
R^1 = 2-Naph, R^2 = *p*-Tol:
70% yield, ee = 91%
R^1 = 2-thienyl, R^2 = *p*-Tol:
92% yield, ee = 95%

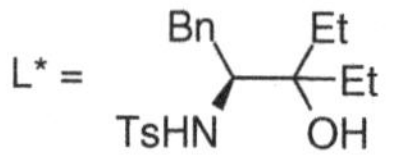

L* =

Bn, Et
TsHN OH

44 (20 mol%)
Ti(O*i*-Pr)$_4$ (20 mol%)
ZnEt$_2$ (3.5 equiv)
DIPEA (50 mol%)
toluene, 42°C

R^1 = Ph, R^2 = TMS: 80% yield, ee = 98%
R^1 = *o*-MeOC$_6$H$_4$, R^2 = TMS: 96% yield, ee = 97%
R^1 = *m*-MeOC$_6$H$_4$, R^2 = TMS: 79% yield, ee = 97%
R^1 = *p*-Tol, R^2 = TMS: 81% yield, ee = 97%
R^1 = 1-Naph, R^2 = TMS: 80% yield, ee >99%
R^1 = 2-Naph, R^2 = TMS: 69% yield, ee = 98%
R^1 = (*E*)-PhCH=CH, R^2 = TMS: 84% yield, ee = 92%
R^1 = 2-thienyl, R^2 = TMS: 79% yield, ee = 98%
R^1 = Ph, R^2 = 1-cyclohexenyl: 85% yield, ee = 91%
R^1 = *p*-Tol, R^2 = 1-cyclohexenyl: 78% yield, ee = 97%
R^1 = *i*-Pr, R^2 = 1-cyclohexenyl: 80% yield, ee = 89%
R^1 = Ph, R^2 = *n*-Pent: 75% yield, ee = 76%
R^1 = Ph, R^2 = CO$_2$Me: 42% yield, ee = 82%

Scheme 1.47. Chiral resin-supported oxazolidine ligand and chiral β-sulfonamide alcohol ligand in addition of various terminal alkynes to aldehydes

excellent enantioselectivities of up to 95% ee for propargylic alcohols arisen from aromatic and heteroaromatic aldehydes, as shown in Scheme 1.47.[83] Meanwhile, some quite challenging alkynes, such as trimethylsilylacety-lene, ethynylcyclohexene, 1-heptyne, and also methyl propiolate were highly enantioselectively added to a range of aromatic, heteroaromatic, α,β-unsaturated, and aliphatic aldehydes by Wang *et al.* on the basis of a novel methodology, in 2009.[87] These reactions were achieved by using a catalytic system based on the combination of a catalytic amount (20 mol%) of a new readily available and inexpensive chiral β-sulfonamide alcohol **44** with a same catalytic amount of Ti(Oi-Pr)$_4$. The process required 3.5 equivalents of ZnEt$_2$ and 0.5 equivalent of a terminal base, such as DIPEA, as an additive. As summarised in Scheme 1.47, a series of chiral propargylic alcohols were produced under these conditions with good yields of up to 96% and remarkable enantioselectivities of up to >99% ee. The role of DIPEA was to facilitate the formation of the alkynylzinc reagents.

These types of reactions have also been induced by BINOL-derived ligands. For example, Pu *et al.* have used (*S*)-BINOL to induce chirality in addition of alkynes to enals **50**, which provided the corresponding chiral propargylic alcohol-based en-type precursors **51** for Pauson–Khand reactions.[93] As shown in Scheme 1.48, the combination of 20 mol% of (*S*)-BINOL with 100 mol% of Ti(Oi-Pr)$_4$ in the presence of four equivalents

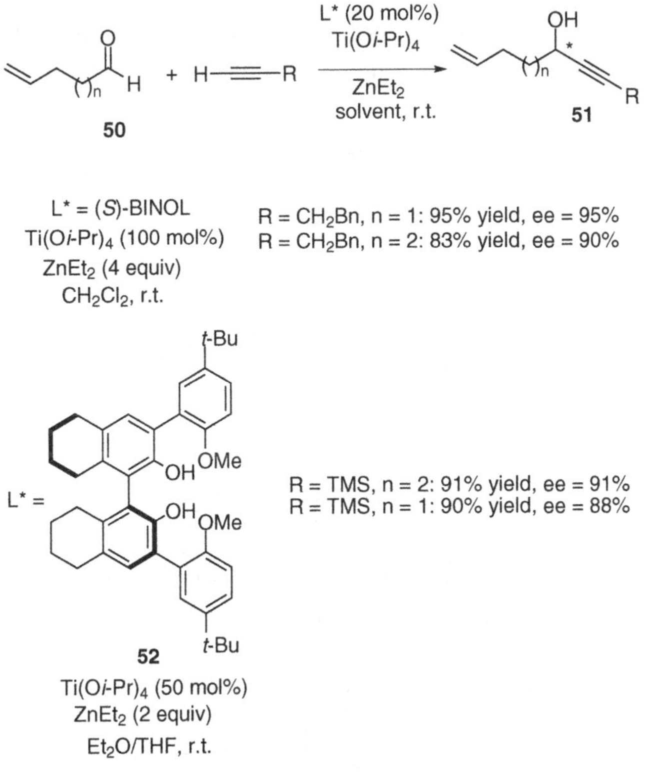

L* (20 mol%)
Ti(Oi-Pr)$_4$

ZnEt$_2$
solvent, r.t.

50 + H━━━R → **51**

L* = (*S*)-BINOL
Ti(Oi-Pr)$_4$ (100 mol%)
ZnEt$_2$ (4 equiv)
CH$_2$Cl$_2$, r.t.

R = CH$_2$Bn, n = 1: 95% yield, ee = 95%
R = CH$_2$Bn, n = 2: 83% yield, ee = 90%

L* =

52

R = TMS, n = 2: 91% yield, ee = 91%
R = TMS, n = 1: 90% yield, ee = 88%

Ti(Oi-Pr)$_4$ (50 mol%)
ZnEt$_2$ (2 equiv)
Et$_2$O/THF, r.t.

Scheme 1.48. (*S*)-BINOL and (*S*)-H$_8$-BINOL-derivative as ligands in addition of alkynes to enals of $ZnEt_2$ in CH_2Cl_2 provided both high yields and enantioselectivities of up to 95% and 95% ee respectively, in the cases of aryl and alkyl alkynes. On the other hand, when trimethylsilyl acetylene was used as alkyne, the best enantioselectivities obtained (91% ee) in this case of substrate were achieved by using (S)-H_8-BINOL-derived ligand **52** at the same catalyst loading (20 mol%) albeit using only 50 mol% of Ti(Oi-Pr)$_4$, and two equivalents of $ZnEt_2$ in a mixed solvent of diethylether/THF (1:5). The formed chiral propargylic ene alcohols **51** were then successfully submitted to intramolecular Pauson–Khand reaction, allowing the synthesis of optically active 5,5- and 5,6-fused bicyclic products to be achieved with retention of enantiomeric purity.

Later, highly enantioselective additions of linear alkyl alkynes to aromatic aldehydes were described by Wang and Yu by using (R)-BINOL at 40 mol% of catalyst loading as chiral ligand.[92] As illustrated in Scheme 1.49, in combination with 100 mol% of Ti(Oi-Pr)$_4$ and 400 mol% of $ZnEt_2$ in toluene, the use of this ligand allowed useful aliphatic chiral propargylic

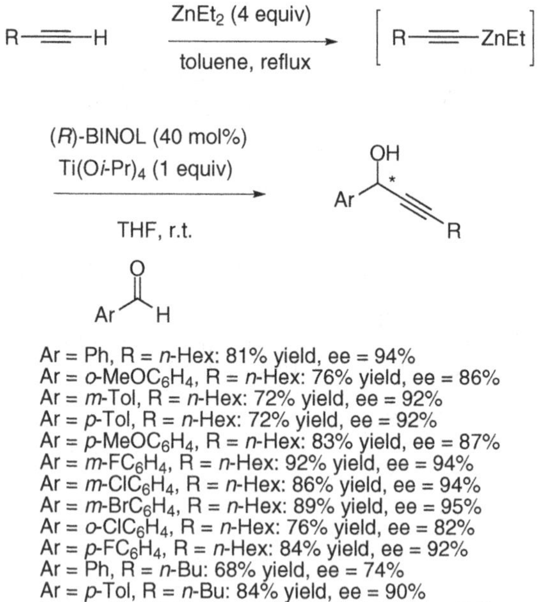

$$R\!-\!\!\equiv\!\!-H \xrightarrow[\text{toluene, reflux}]{ZnEt_2\ (4\ \text{equiv})} \left[R\!-\!\!\equiv\!\!-ZnEt \right]$$

$$\xrightarrow[\text{THF, r.t.}]{\substack{(R)\text{-BINOL (40 mol\%)} \\ \text{Ti(O}i\text{-Pr)}_4\ (1\ \text{equiv})}}$$

Ar = Ph, R = n-Hex: 81% yield, ee = 94%
Ar = o-MeOC$_6$H$_4$, R = n-Hex: 76% yield, ee = 86%
Ar = m-Tol, R = n-Hex: 72% yield, ee = 92%
Ar = p-Tol, R = n-Hex: 72% yield, ee = 92%
Ar = p-MeOC$_6$H$_4$, R = n-Hex: 83% yield, ee = 87%
Ar = m-FC$_6$H$_4$, R = n-Hex: 92% yield, ee = 94%
Ar = m-ClC$_6$H$_4$, R = n-Hex: 86% yield, ee = 94%
Ar = m-BrC$_6$H$_4$, R = n-Hex: 89% yield, ee = 95%
Ar = o-ClC$_6$H$_4$, R = n-Hex: 76% yield, ee = 82%
Ar = p-FC$_6$H$_4$, R = n-Hex: 84% yield, ee = 92%
Ar = Ph, R = n-Bu: 68% yield, ee = 74%
Ar = p-Tol, R = n-Bu: 84% yield, ee = 90%
Ar = m-MeOC$_6$H$_4$, R = n-Bu: 75% yield, ee = 92%
Ar = m-FC$_6$H$_4$, R = n-Bu: 73% yield, ee = 86%

Scheme 1.49. (R)-BINOL as ligand in addition of linear terminal alkynes to aromatic aldehydes alcohols to be synthesised in good to high yields (up to 92%), with high enantioselectivities of up to 94% ee starting from a range of aromatic aldehydes bearing various substituents on different positions.

Earlier, comparable reactions were also performed by Pu *et al.* with (S)-BINOL as chiral ligand in the presence of a stoichiometric quantity of Ti(O*i*-Pr)$_4$.[93] In this case, the use of biscyclohexylamine as an additive (5 mol%) proved to greatly enhance the enantioselectivity of the reaction of linear alkyl alkynes with a variety of aliphatic aldehydes which were found up to 89% ee (Scheme 1.50). In 2013, the authors extended this

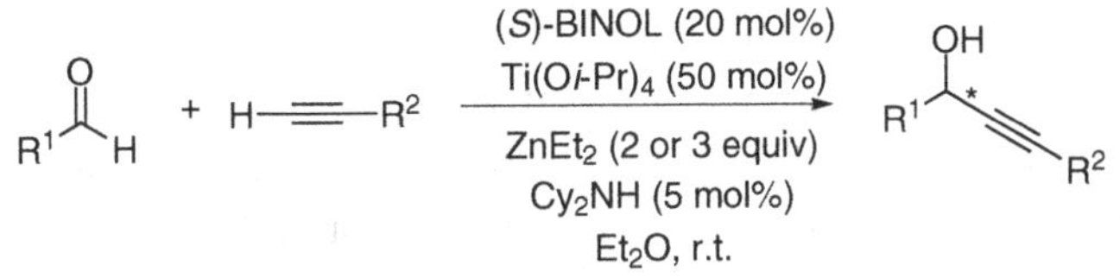

$R^1 = R^2 = n$-Bu: 73% yield, ee = 81%
$R^1 = n$-Bu, $R^2 = n$-Hex: 61% yield, ee = 87%
$R^1 = n$-Bu, $R^2 = (CH_2)_3Cl$: 77% yield, ee = 88%
$R^1 = n$-Hept, $R^2 = n$-Bu: 76% yield, ee = 85%
$R^1 = n$-Hept, $R^2 = n$-Hex: 67% yield, ee = 85%
$R^1 = (CH_2)_2CH=CH_2$, $R^2 = (CH_2)_3Cl$: 70% yield, ee = 89%
$R^1 = (CH_2)_2CH=CH_2$, $R^2 = n$-Bu: 63% yield, ee = 87%
$R^1 = (CH_2)_2CH=CH_2$, $R^2 = n$-Hex: 59% yield, ee = 83%
$R^1 = R^2 = (CH_2)_3Cl$: 60% yield, ee = 88%
$R^1 = (CH_2)_3Cl$, $R^2 = n$-Hex: 57% yield, ee = 77%
$R^1 = BnCH_2$, $R^2 = n$-Bu: 71% yield, ee = 83%
$R^1 = BnCH_2$, $R^2 = n$-Hex: 74% yield, ee = 84%
$R^1 = BnCH_2$, $R^2 = (CH_2)_3Cl$: 65% yield, ee = 89%
$R^1 = Et$, $R^2 = CH_2Bn$: 99% yield, ee = 84%
$R^1 = p$-Tol, $R^2 = CH_2Bn$: 80% yield, ee = 99%
$R^1 = 2$-Naph, $R^2 = CH_2Bn$: 82% yield, ee >99%
$R^1 = p$-ClC$_6$H$_4$, $R^2 = CH_2Bn$: 83% yield, ee >99%
$R^1 = Ph$, $R^2 = n$-Bu: 99% yield, ee = 99%
$R^1 = p$-Tol, $R^2 = n$-Bu: 93% yield, ee = 97%
$R^1 = m$-Tol, $R^2 = n$-Bu: 61% yield, ee = 98%
$R^1 = p$-ClC$_6$H$_4$, $R^2 = n$-Bu: 68% yield, ee = 97%
$R^1 = (E)$-PhCH=CH, $R^2 = n$-Bu: 68% yield, ee = 92%
$R^1 = Ph$, $R^2 = TMS$: 86% yield, ee = 98%
$R^1 = p$-Tol, $R^2 = TMS$: 81% yield, ee = 97%
$R^1 = p$-MeOC$_6$H$_4$, $R^2 = TMS$: 72% yield, ee = 97%
$R^1 = 2$-Naph, $R^2 = TMS$: 79% yield, ee = 97%

Scheme 1.50. (S)-BINOL as ligand in addition of various alkynes to aldehydes with (Cy)$_2$NH methodology to other aldehydes including aromatic ones, which provided the corresponding propargylic alcohols with even higher enantioselectivities of up to >99% ee (Scheme 1.50).[94]

γ-Hydroxy-α,β-acetylenic esters containing three different functional groups constitute very important precursors in the synthesis of highly functionalised organic molecules. Although great efforts have been made in asymmetric alkynylations, it must be recognised that little attention has been paid to the enantioselective reactions of alkynoates to aldehydes. This can be attributed to the higher sensitivity and dissimilar reactivity of alkynoates in comparison with simple alkyl and aryl alkynes. The asymmetric reaction of alkynoates to aldehydes was first reported by Pu *et al.* in 2006.[94] In this work, the reaction between methyl propiolate and aromatic aldehydes was carried out in the presence of diethylzinc, hexamethylphosphoramide (HMPA), and titanium tetraisopropoxide, along with (S)-BINOL as chiral ligand, providing the corresponding chiral propargylic alcohols in high enantioselectivities. In 2009, Mao and Guo described the synthesis of chiral γ-hydroxy-α,β-acetylenic esters on the basis of enantioselective titanium-catalysed addition of alkynoates to aromatic aldehydes, induced by 20 mol% of easily available chiral oxazolidine **53**.[95] An advantage of this process was the use of a catalytic amount (40 mol%) of Ti(O*i*-Pr)$_4$, combined with two equivalents of HMPA, three equivalents of ZnMe$_2$, along with 10 mol% of dimethoxy polyethylene glycol (DIMPEG) as an additive in toluene. As shown in Scheme 1.51, this practical catalytic system allowed a range of chiral γ-hydroxy-α,β-acetylenic esters to be achieved in good yields, with enantioselectivities of up to 84% ee. However, since HMPA is a strong carcinogen, other catalytic systems were investigated avoiding the use of HMPA, such as that reported by Hui *et al.* in 2009.[96] The latter involved the association of 30 mol% of chiral β-hydroxy amide **48** derived from L-tyrosine with the same catalytic amount of Ti(O*i*-Pr)$_4$, in the presence of three equivalents of ZnEt$_2$ in DME. Under these conditions, aliphatic as well aromatic aldehydes in addition to *trans*-cinnamaldehyde reacted with methyl propiolate to give the corresponding chiral alcohols in good to high yields, and enantioselectivities of up to 94% ee, demonstrating the broad generality of this catalytic system for aliphatic and aromatic aldehydes (Scheme 1.51). The best enantioselectivities were reached in the cases of less sterically hindered aliphatic aldehydes and aromatic aldehydes *para*-substituted by

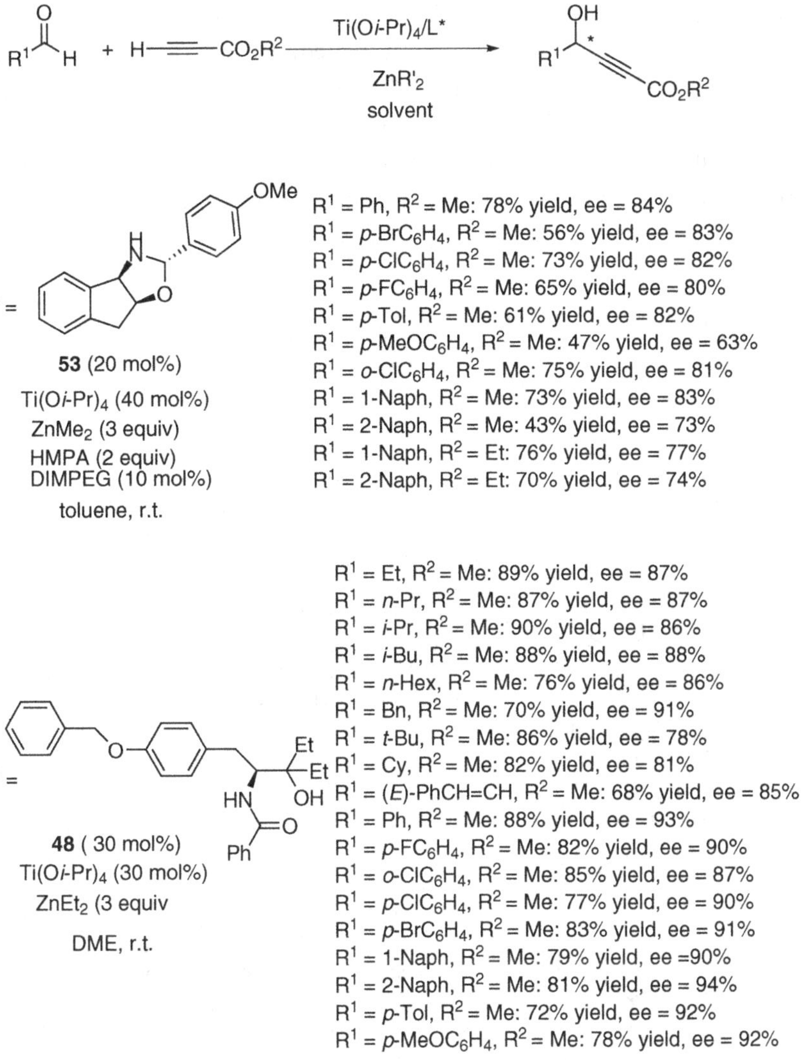

$$R^1\text{CHO} + H\text{—}{\equiv}\text{—}CO_2R^2 \xrightarrow[\substack{ZnR'_2 \\ \text{solvent}}]{Ti(Oi\text{-}Pr)_4/L^*} R^1\text{—}C(OH)\text{—}{\equiv}\text{—}CO_2R^2$$

L* =

53 (20 mol%)
Ti(Oi-Pr)₄ (40 mol%)
ZnMe₂ (3 equiv)
HMPA (2 equiv)
DIMPEG (10 mol%)
toluene, r.t.

R^1 = Ph, R^2 = Me: 78% yield, ee = 84%
R^1 = p-BrC₆H₄, R^2 = Me: 56% yield, ee = 83%
R^1 = p-ClC₆H₄, R^2 = Me: 73% yield, ee = 82%
R^1 = p-FC₆H₄, R^2 = Me: 65% yield, ee = 80%
R^1 = p-Tol, R^2 = Me: 61% yield, ee = 82%
R^1 = p-MeOC₆H₄, R^2 = Me: 47% yield, ee = 63%
R^1 = o-ClC₆H₄, R^2 = Me: 75% yield, ee = 81%
R^1 = 1-Naph, R^2 = Me: 73% yield, ee = 83%
R^1 = 2-Naph, R^2 = Me: 43% yield, ee = 73%
R^1 = 1-Naph, R^2 = Et: 76% yield, ee = 77%
R^1 = 2-Naph, R^2 = Et: 70% yield, ee = 74%

L* =

48 (30 mol%)
Ti(Oi-Pr)₄ (30 mol%)
ZnEt₂ (3 equiv
DME, r.t.

R^1 = Et, R^2 = Me: 89% yield, ee = 87%
R^1 = n-Pr, R^2 = Me: 87% yield, ee = 87%
R^1 = i-Pr, R^2 = Me: 90% yield, ee = 86%
R^1 = i-Bu, R^2 = Me: 88% yield, ee = 88%
R^1 = n-Hex, R^2 = Me: 76% yield, ee = 86%
R^1 = Bn, R^2 = Me: 70% yield, ee = 91%
R^1 = t-Bu, R^2 = Me: 86% yield, ee = 78%
R^1 = Cy, R^2 = Me: 82% yield, ee = 81%
R^1 = (E)-PhCH=CH, R^2 = Me: 68% yield, ee = 85%
R^1 = Ph, R^2 = Me: 88% yield, ee = 93%
R^1 = p-FC₆H₄, R^2 = Me: 82% yield, ee = 90%
R^1 = o-ClC₆H₄, R^2 = Me: 85% yield, ee = 87%
R^1 = p-ClC₆H₄, R^2 = Me: 77% yield, ee = 90%
R^1 = p-BrC₆H₄, R^2 = Me: 83% yield, ee = 91%
R^1 = 1-Naph, R^2 = Me: 79% yield, ee =90%
R^1 = 2-Naph, R^2 = Me: 81% yield, ee = 94%
R^1 = p-Tol, R^2 = Me: 72% yield, ee = 92%
R^1 = p-MeOC₆H₄, R^2 = Me: 78% yield, ee = 92%

Scheme 1.51. Chiral oxazolidine and β-hydroxy amide ligands in addition of alkynoates to aldehydes

electron-withdrawing as well as electron-donating groups. The best enantioselectivity of 94% ee was obtained for 2-naphthaldehyde.

On the other hand, Pu *et al.* developed BINOL-derived ligands to induce this type of reaction. For example, novel H₈-BINOL-based ligand **52** employed at 20 mol% of catalyst loading was found to very efficiently catalyse the alkyl propiolate addition to aliphatic aldehydes, using only a catalytic amount of $Ti(Oi\text{-}Pr)_4$ with two equivalents of $ZnEt_2$ in THF.[97] As shown in Scheme 1.52, this remarkable novel catalytic system provided general excellent enantioselectivities of up to 97% ee for a wide variety of aliphatic aldehydes under mild conditions. Later, the same authors reported a methodology for the methyl propiolate addition to aromatic aldehydes,

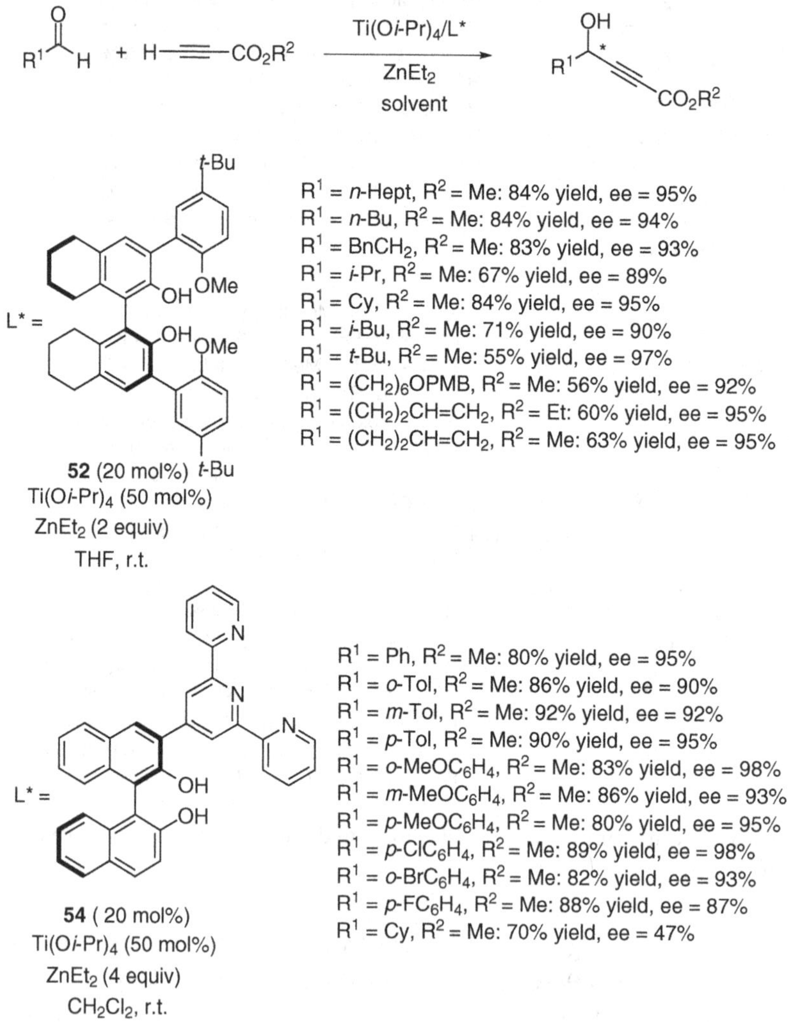

$R^1 = n\text{-Hept}, R^2 = Me$: 84% yield, ee = 95%
$R^1 = n\text{-Bu}, R^2 = Me$: 84% yield, ee = 94%
$R^1 = BnCH_2, R^2 = Me$: 83% yield, ee = 93%
$R^1 = i\text{-Pr}, R^2 = Me$: 67% yield, ee = 89%
$R^1 = Cy, R^2 = Me$: 84% yield, ee = 95%
$R^1 = i\text{-Bu}, R^2 = Me$: 71% yield, ee = 90%
$R^1 = t\text{-Bu}, R^2 = Me$: 55% yield, ee = 97%
$R^1 = (CH_2)_6OPMB, R^2 = Me$: 56% yield, ee = 92%
$R^1 = (CH_2)_2CH=CH_2, R^2 = Et$: 60% yield, ee = 95%
$R^1 = (CH_2)_2CH=CH_2, R^2 = Me$: 63% yield, ee = 95%

52 (20 mol%)
$Ti(Oi\text{-}Pr)_4$ (50 mol%)
$ZnEt_2$ (2 equiv)
THF, r.t.

$R^1 = Ph, R^2 = Me$: 80% yield, ee = 95%
$R^1 = o\text{-Tol}, R^2 = Me$: 86% yield, ee = 90%
$R^1 = m\text{-Tol}, R^2 = Me$: 92% yield, ee = 92%
$R^1 = p\text{-Tol}, R^2 = Me$: 90% yield, ee = 95%
$R^1 = o\text{-MeOC}_6H_4, R^2 = Me$: 83% yield, ee = 98%
$R^1 = m\text{-MeOC}_6H_4, R^2 = Me$: 86% yield, ee = 93%
$R^1 = p\text{-MeOC}_6H_4, R^2 = Me$: 80% yield, ee = 95%
$R^1 = p\text{-ClC}_6H_4, R^2 = Me$: 89% yield, ee = 98%
$R^1 = o\text{-BrC}_6H_4, R^2 = Me$: 82% yield, ee = 93%
$R^1 = p\text{-FC}_6H_4, R^2 = Me$: 88% yield, ee = 87%
$R^1 = Cy, R^2 = Me$: 70% yield, ee = 47%

54 (20 mol%)
$Ti(Oi\text{-}Pr)_4$ (50 mol%)
$ZnEt_2$ (4 equiv)
CH_2Cl_2, r.t.

Scheme 1.52. Chiral H_8-BINOL-derived ligand and chiral BINOL-terpyridine ligand in addition of alkynoates to aldehydes employing novel C_1-symmetric BINOL-terpyridine ligand **54** employed at 20 mol% of catalyst loading, which also used a catalytic amount of Ti(O*i*-Pr)$_4$ (50 mol%) in the presence of four equivalents of ZnEt$_2$ in CH$_2$Cl$_2$.[98] As shown in Scheme 1.52, general high enantioselectivities of up to 98% ee were obtained for a range of aromatic aldehydes in combination with high yields of up to 92%. It was shown that applying the methodology to aliphatic aldehydes gave poor enantioselectivities (47% ee).

1.3.1.3. Addition of 1,3-diynes and 1,3-enynes

Enantioselective additions of diynes have proved more challenging than those of simple alkynes. In 2011, Pu *et al.* reported a highly enantioselective titanium-promoted addition of various 1,3-diynes to a wide variety of aldehydes.[99] This remarkable process employed (*S*)-BINOL as chiral ligand in combination with Ti(O*i*-Pr)$_4$ in the presence of two or three equivalents of ZnEt$_2$, along with Cy$_2$NH as an additive (5 mol%) in diethyl ether as solvent. When reacting aromatic aldehydes, only a catalytic amount (25 mol%) of Ti(O*i*-Pr)$_4$ was sufficient to provide the corresponding chiral diynols in very high enantioselectivities of up to 94% ee, and very good yields of up to 98%. On the other hand, the reaction of aliphatic aldehydes required a stoichiometric amount of Ti(O*i*-Pr)$_4$ to provide the corresponding products in comparable enantioselectivities of up to 95% ee (Scheme 1.53). Furthermore, the addition of diynes to enals gave the corresponding enediynols in generally very high enantioselectivities (88–95% ee). It must be noted that this novel methodology represented the most generally enantioselective catalyst system for the asymmetric diyne addition to aldehydes. The formed products constituted starting materials to synthesise chiral polycyclic products, such as 5,5,7- and 5,5,8-fused tricyclic products, through Pauson–Khand reaction.

In addition, in 2013 Pu *et al.* reported the first highly enantioselective addition of a conjugated enyne to linear aliphatic aldehydes in the presence of a chiral catalyst.[94] This process involved (*S*)-BINOL as chiral ligand, used at 40 mol% of catalyst loading along with one equivalent of Ti(O*i*-Pr)$_4$, three equivalents of ZnEt$_2$, and 5 mol% of Cy$_2$NH as an additive in diethyl ether. It afforded the corresponding chiral enynols in remarkable enantioselectivities of up to 98% ee and high yields, as shown in Scheme 1.54. α,β-unsaturated aldehydes proved compatible to the

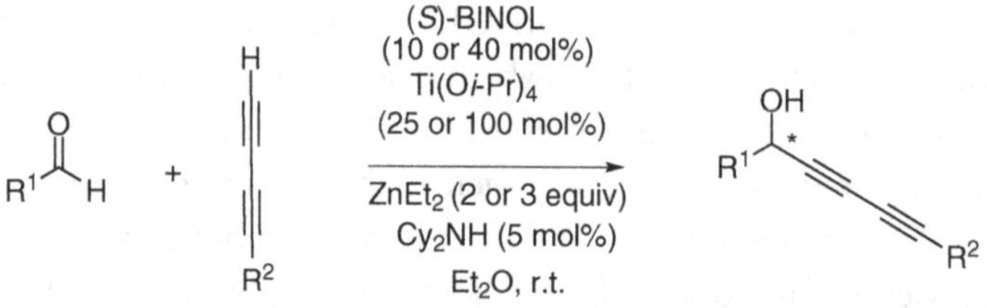

$R^1 = Ph, R^2 = CH_2Bn$: 95% yield, ee = 94%
$R^1 = o\text{-MeOC}_6H_4, R^2 = CH_2Bn$: 81% yield, ee = 91%
$R^1 = p\text{-MeOC}_6H_4, R^2 = CH_2Bn$: 94% yield, ee = 94%
$R^1 = p\text{-ClC}_6H_4, R^2 = CH_2Bn$: 94% yield, ee = 92%
$R^1 = o\text{-Tol}, R^2 = CH_2Bn$: 98% yield, ee = 92%
$R^1 = 2\text{-furyl}, R^2 = CH_2Bn$: 89% yield, ee = 85%
$R^1 = n\text{-Bu}, R^2 = CH_2Bn$: 92% yield, ee = 92%
$R^1 = Cy, R^2 = CH_2Bn$: 91% yield, ee = 91%
$R^1 = i\text{-Bu}, R^2 = CH_2Bn$: 90% yield, ee = 87%
$R^1 = (E)\text{-MeCH=CH}, R^2 = CH_2Bn$: 99% yield, ee = 92%
$R^1 = Ph, R^2 = TIPS$: 98% yield, ee = 91%
$R^1 = Ph, R^2 = n\text{-Hex}$: 87% yield, ee = 94%
$R^1 = Ph, R^2 = CH_2OTBS$: 95% yield, ee = 94%
$R^1 = Ph, R^2 = (CH_2)_3OAc$: 78% yield, ee = 86%
$R^1 = Ph, R^2 = (CH_2)_3Cl$: 95% yield, ee = 91%
$R^1 = Ph, R^2 = (CH_2)_3OTBS$: 97% yield, ee = 92%
$R^1 = Cy, R^2 = TIPS$: 99% yield, ee = 89%
$R^1 = (CH_2)_2CH=CH_2, R^2 = TIPS$: 94% yield, ee = 95%
$R^1 = Cy, R^2 = Ph$: 98% yield, ee = 92%
$R^1 = (CH_2)_3CH=CH_2, R^2 = Ph$: 73% yield, ee = 91%
$R^1 = Cy, R^2 = CH_2OTBS$: 98% yield, ee = 91%
$R^1 = Cy, R^2 = (CH_2)_3OTBS$: 99% yield, ee = 92%

Scheme 1.53. (S)-BINOL as ligand in addition of 1,3-diynes to aldehydes

reaction conditions, since the corresponding propargylic alcohols were obtained in comparable results as other aliphatic aldehydes (90–98% ee). The formed chiral enynols were further converted into trienynes, which were submitted to Pauson–Khand and Diels–Alder reactions to achieve important chiral multicyclic products.

1.3.2. *Ketones as Electrophiles*

The alkynylation of ketones is much less developed than that of aldehydes. It was first reported by Tan and co-workers in 1999, and later by Jiang and

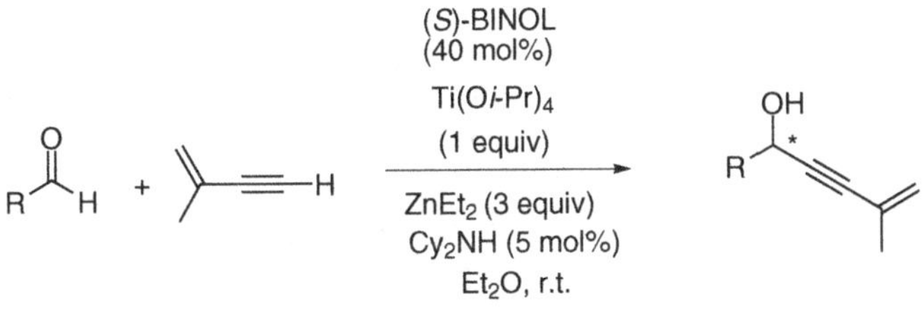

$$
\underset{R}{\overset{O}{\|}}\overset{}{\underset{H}{\bigwedge}} \; + \; \overset{}{\underset{}{\bigvee}}\!\!=\!\!=\!\!-H \quad \xrightarrow[\substack{ZnEt_2 \,(3\ equiv) \\ Cy_2NH \,(5\ mol\%) \\ Et_2O,\ r.t.}]{\substack{(S)\text{-BINOL} \\ (40\ mol\%) \\ Ti(O\text{-}Pr)_4 \\ (1\ equiv)}} \quad \underset{R}{\overset{OH}{\bigwedge}}\!\!\!\overset{*}{=}\!\!\!\overset{}{\underset{}{\bigvee}}
$$

R = *n*-Hept: 85% yield, ee = 90%
R = (CH₂)₂CH=CH₂: 75% yield, ee = 90%
R = Cy: 81% yield, ee = 89%
R = (CH₂)₂C≡CTMS: 83% yield, ee = 88%
R = C(Me)₂CH₂CH=CH₂: 85% yield, ee = 89%
R = Ph: 84% yield, ee = 96%
R = *p*-Tol: 92% yield, ee = 97%
R = 2-Naph: 96% yield, ee = 98%
R = 1-Naph: 78% yield, ee = 97%
R = *p*-ClC₆H₄: 93% yield, ee = 91%
R = *m*-ClC₆H₄: 83% yield, ee = 97%
R = *o*-EtC₆H₄: 94% yield, ee = 92%
R = (*E*)-PhCH=CH: 81% yield, ee = 90%
R = (*Z*)-PhCH=CMe: 85% yield, ee = 98%

Scheme 1.54. (*S*)-BINOL as ligand in addition of 3-methyl-3-buten-1-yne to aldehydes

co-workers, in 2002.[100] The first catalytic enantioselective alkynylation of ketones using chiral titanium catalysts was reported by Cozzi and Alesi, in 2004.[101] This work involved the direct addition of an alkynyltitanium triisopropoxide to ketones, performed in the presence of catalytic amounts of BINOL as chiral ligand, to provide the corresponding chiral propargylic alcohols in good to high enantioselectivities of up to 88% ee. In 2009, Wang *et al.* reported the synthesis of novel chiral hydroxysulfonamide ligands, which were further investigated as promotors in the asymmetric addition of phenylacetylene to aromatic ketones.[102] Among a range of simple L-amino acids tested, chiral hydroxysulfonamide **55**, employed at 20 mol% of catalyst loading in combination with only 40 mol% of Ti(O*i*-Pr)$_4$ in the presence of two equivalents of ZnEt$_2$ in CH$_2$Cl$_2$, was found the most efficient ligand, providing the corresponding tertiary chiral propargylic alcohols in enantioselectivities of up to 83% ee along with good yields (43–81%), as shown in Scheme 1.55.

Trifluoromethyl ketones constitute a class of particularly challenging substrates for the asymmetric titanium-catalysed zinc alkynylide addition, because of the presence of the strongly electron-withdrawing fluorine atoms. Indeed, the activating trifluoromethyl group renders the ketone

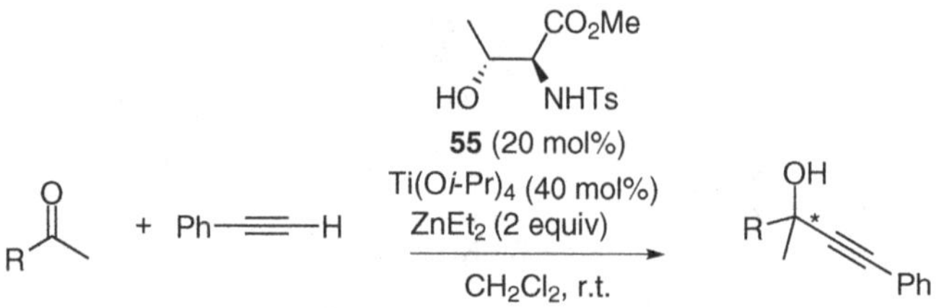

$$\underset{R}{\overset{O}{\|}}\diagdown + Ph\!\!-\!\!\equiv\!\!-H \quad \xrightarrow[\text{CH}_2\text{Cl}_2,\ \text{r.t.}]{\substack{\text{CO}_2\text{Me} \\ \text{HO} \quad \text{NHTs} \\ \textbf{55}\ (20\ \text{mol}\%) \\ \text{Ti}(Oi\text{-Pr})_4\ (40\ \text{mol}\%) \\ \text{ZnEt}_2\ (2\ \text{equiv})}} \quad R\overset{OH}{\underset{*}{\diagup}}\!\!\!\diagdown\!\!\!\equiv\!\!\!-Ph$$

R = Ph: 63% yield, ee = 81%
R = *o*-FC$_6$H$_4$: 78% yield, ee = 82%
R = *m*-BrC$_6$H$_4$: 43% yield, ee = 80%
R = *m*-Tol: 62% yield, ee = 83%
R = *m*-MeOC$_6$H$_4$: 50% yield, ee = 83%
R = *p*-FC$_6$H$_4$: 68% yield, ee = 77%
R = *p*-ClC$_6$H$_4$: 55% yield, ee = 81%
R = *p*-MeOC$_6$H$_4$: 70% yield, ee = 57%
R = 2-Naph: 59% yield, ee = 82%
R = *i*-Bu: 81% yield, ee = 47%
Ar = Bn: 77% yield, ee = 60%

Scheme 1.55. Chiral hydroxysulfonamide ligand in addition of phenylacetylene to ketones

functionality highly reactive and, consequently, has a detrimental effect on the control of the facial selectivity. In 2011, Ma *et al.* reported the first effective method for catalyzing the asymmetric addition of alkynes to trifluoromethyl ketones.[103] This novel methodology involved chiral cinchona alkaloids **56** and **57** as ligands (20 mol%), and had the advantage of using only a catalytic amount of Ti(O*i*-Pr)$_4$. In the presence of two equivalents of ZnEt$_2$ and BaF$_2$ as an additive in CH$_2$Cl$_2$, the reaction of various aromatic alkynes and ketones, including electron-neutral, electron-withdrawing, and electron-donating groups afforded the corresponding chiral tertiary alcohols in high enantioselectivities of up to 91% ee, as shown in Scheme 1.56. It was noteworthy that even aliphatic alkynes gave the products in good to high yields, and enantioselectivities of up to 94% ee. Additionally, the reaction worked with (*E*)-1,1,1-trifluoro-4-phenylbut-en-2-one to afford the corresponding 1,2-adduct in good yield (67%) and enantioselectivity (66% ee). In this work, the authors demonstrated the remarkable effect of the metal fluoride additive, which was proven to be essential for effective asymmetric induction. The major advantage of this process was that both enantiomers of trifluoromethylated propargylic tertiary alcohols could be assessed in high yields, and enantioselectivities according to the ligands used.

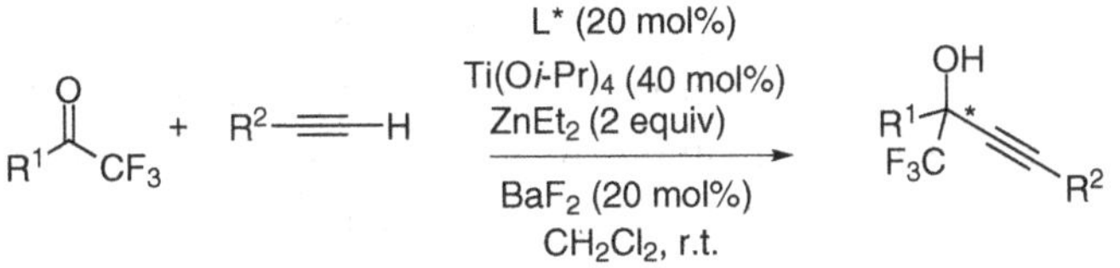

$$R^1 \overset{O}{\underset{}{\parallel}} CF_3 \quad + \quad R^2 \text{—}\!\!\equiv\!\!\text{—} H \quad \xrightarrow[\substack{BaF_2 \ (20 \ mol\%) \\ CH_2Cl_2, \ r.t.}]{\substack{L^* \ (20 \ mol\%) \\ Ti(Oi\text{-}Pr)_4 \ (40 \ mol\%) \\ ZnEt_2 \ (2 \ equiv)}} \quad R^1 \overset{OH}{\underset{F_3C}{\overset{*}{\diagdown}}} \!\!\!\!\equiv\!\!\! R^2$$

$R^1 = R^2 = $ Ph: 89% yield, ee = 91%
$R^1 = p\text{-}FC_6H_4$, $R^2 = $ Ph: 95% yield, ee = 90%
$R^1 = p\text{-}ClC_6H_4$, $R^2 = $ Ph: 98% yield, ee = 88%
$R^1 = p\text{-}Tol$, $R^2 = $ Ph: 82% yield, ee = 85%
$R^1 = p\text{-}MeOC_6H_4$, $R^2 = $ Ph: 81% yield, ee = 86%
$R^1 = p\text{-}PhC_6H_4$, $R^2 = $ Ph: 94% yield, ee = 86%
$R^1 = $ 2-Naph, $R^2 = $ Ph: 76% yield, ee = 89%
$R^1 = $ Ph, $R^2 = p\text{-}Tol$: 98% yield, ee = 84%
$R^1 = p\text{-}ClC_6H_4$, $R^2 = p\text{-}Tol$: 98% yield, ee = 85%
$R^1 = p\text{-}MeOC_6H_4$, $R^2 = p\text{-}Tol$: 86% yield, ee = 87%
$R^1 = $ Ph, $R^2 = c\text{-}Pr$: 96% yield, ee = 65%
$R^1 = $ Ph, $R^2 = n\text{-}Hex$: 75% yield, ee = 94%
$R^1 = (E)\text{-}PhCH\!=\!CH$, $R^2 = $ Ph: 67% yield, ee = 66%

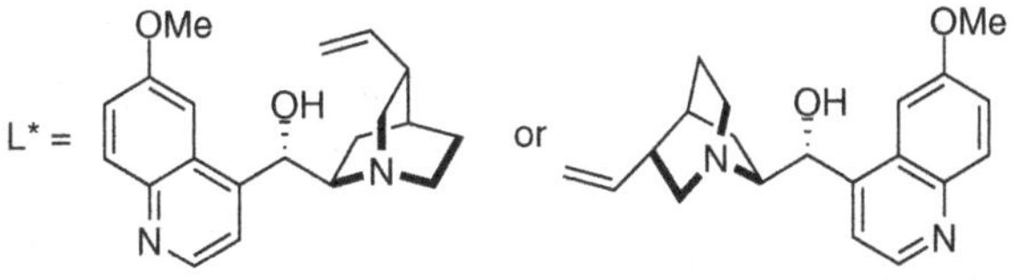

L* =

56   or   **57**

Scheme 1.56. Chiral cinchona alkaloids as ligands in addition of alkynes to trifluoromethyl ketones

1.4. Titanium-Promoted Allylation and Vinylation Reactions of Carbonyl Compounds

1.4.1. Allylations

The condensation of allyl nucleophiles to carbonyl compounds in the presence of titanium catalysts provides corresponding homoallylic alcohols, which are widely applicable in organic synthesis. Enantioselective allylation reactions are particularly interesting reactions, since in addition to creating novel stereogenic centres, an extra double bond is added to the final chiral product that can be further modified to give various functions.

While the first allylation was reported by Hosomi and Sakurai in 1976, using TiCl$_4$ as achiral promoter,[104] the enantioselective version was reported in 1982 by Hayashi and Kumada.[105] In this work, chiral allyltitanium reagents were prepared from the reaction of allyl Grignard or lithium reagents with chiral titanium complexes, allowing by condensation to aldehydes the corresponding chiral homoallylic alcohols to be achieved with high enantioselectivities of up to 88% ee. Ever since, a number of chiral catalysts have been developed, including chiral titanium complexes.[8] As an example, Belokon *et al.* have reported the use of chiral hexadentate Schiff bases, such as **58**, as titanium ligands in asymmetric allylation of aromatic aldehydes with allyltributyltin.[106] The process involved 10 mol% of this chiral ligand in combination with 20 mol% of Ti(O*i*-Pr)$_4$ as catalyst system. The authors demonstrated an unusual effect on the effectiveness of the allylation of aldehydes when TMSCl is used as an additive. It increased both the yield and the enantioselectivity of the reaction. They proposed that the silylation step, regenerating the initial catalytic species, was the rate limiting step of the catalytic sequence. As shown in Scheme 1.57, the best result was obtained for the addition of allyltributyltin to *p*-nitrobenzaldehyde, providing the corresponding chiral allylic alcohol in 94% yield and enantioselectivity of 74% ee. Despite its moderate enantioselectivity, this reaction could be brought to completion after only one hour at room temperature. The authors have proposed dinuclear titanium catalyst **59** as active catalyst of the reaction.

In 2011, Venkateswarlu *et al.* reported a concise total synthesis of cytotoxic *anti*-(3*S*,5*S*)-1-(4-hydroxyphenyl)-7-phenylheptane-3,5-diol **60**, achieved in six steps with 25% overall yield on the basis of titanium-catalysed enantioselective allylation of commercially available 3-phenylpropanal.[107] This reaction was induced by (*R*)-BINOL used at 20 mol% of catalyst loading, providing the corresponding chiral allylic alcohol **61** in 83% yield, and excellent enantioselectivity of 97% ee, as shown in Scheme 1.58. The active catalyst **62** was *in situ* generated from the chiral ligand, a catalytic amount (15 mol%) of Ti(O*i*-Pr)$_4$, and 10 mol% of Ag$_2$O in CH$_2$Cl$_2$ according to Maruoka's method.[108] The allylic chiral alcohol obtained was further converted into the required 1,3-diol **60**.

The enantioselective addition of allylmetal derivatives to ketones using chiral titanium complexes has been less developed than the related addition

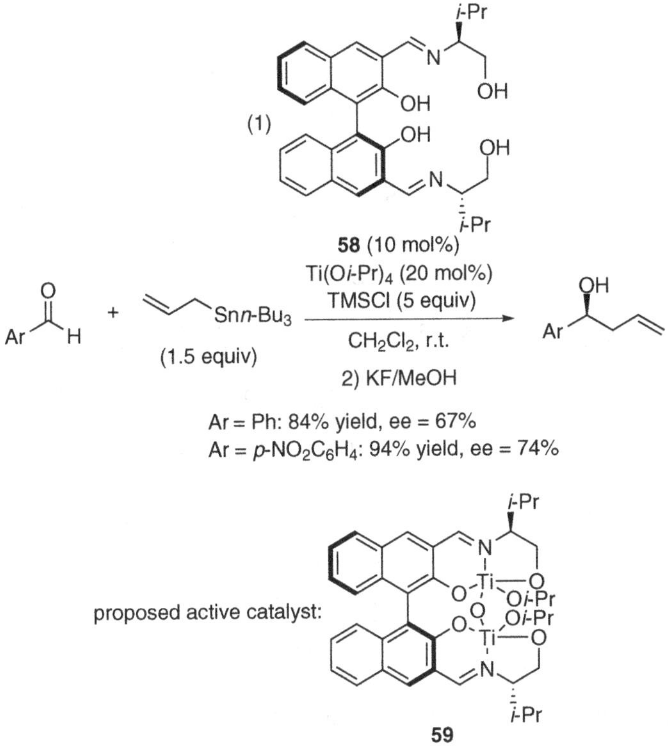

58 (10 mol%)

Ti(O*i*-Pr)$_4$ (20 mol%)

$$Ar \overset{O}{\underset{H}{\bigwedge}} + \underset{(1.5 \text{ equiv})}{\diagdown\diagdown\diagup \text{Sn}n\text{-Bu}_3} \xrightarrow[\substack{\text{CH}_2\text{Cl}_2, \text{ r.t.} \\ \text{2) KF/MeOH}}]{\text{TMSCl (5 equiv)}} Ar \overset{OH}{\underset{}{\diagup}} \diagdown\diagup$$

Ar = Ph: 84% yield, ee = 67%

Ar = *p*-NO$_2$C$_6$H$_4$: 94% yield, ee = 74%

proposed active catalyst:

59

Scheme 1.57. Chiral hexadentate Schiff base as ligand in allylation of aromatic aldehydes

to aldehydes. The first catalytic enantioselective process was published by Tagliavini, in 1999.[109] In this work, enantioselectivities of up to 65% ee were obtained for the addition of tetraallyltin to various ketones in the presence of dichlorotitanium diisopropoxide and BINOL. In 2014, Pericas and Walsh reported the covalent immobilisation of (*R*)-BINOL to polystyrene by employing a 1,2,3-triazole linker to heterogenise a titanium-BINOlate catalyst and its application in the enantioselective allylation of ketones.[110] By using a simple synthetic route, enantiopure 6-ethynyl-BINOL was synthesised and anchored to an azidomethylpolystyrene resin through a copper-catalysed alkyne-azide cycloaddition reaction. The polystyrene-supported BINOL ligand **63**, derived from 30 mol% of both (*R*)-BINOL and Ti(O*i*-Pr)$_4$, was converted into its diisopropoxytitanium derivative *in situ*, and used as a heterogeneous catalyst in the asymmetric allylation of a

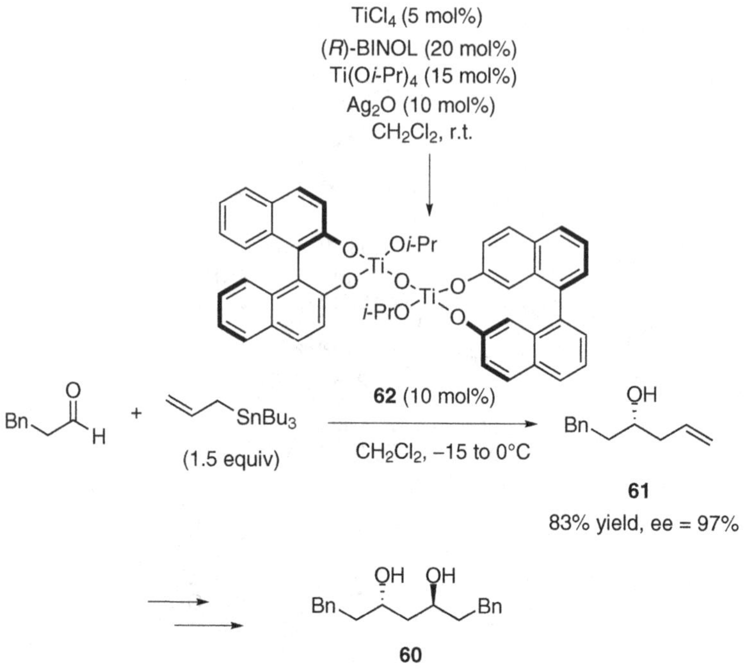

Scheme 1.58. (R)-BINOL as ligand in allylation of 3-phenylpropanal

variety of ketones with allyltributyl tin in CH_2Cl_2 at room temperature, to provide the corresponding chiral tertiary allylic alcohols in good to high yields of up to 98% and high enantioselectivities of up 95% ee, as shown in Scheme 1.59. With the most common substrates, such as substituted acetophenones, the very high enantioselectivities observed were comparable to those recorded with the corresponding homogeneous catalytic system.[111] It must be noted that cyclic ketones, α,β-unsaturated ketones, and heteroaromatic ketones, such as 2-acetylfuran, also afforded the corresponding alcohols in good enantioselectivities (70–88% ee). Furthermore, the authors demonstrated the reusability of the ligand, since both yields and enantioselectivities were preserved after three consecutive reaction cycles. It is noteworthy that this work constituted one of the very few examples of catalytic creation of quaternary centres involving the use of a heterogenised catalytic species.

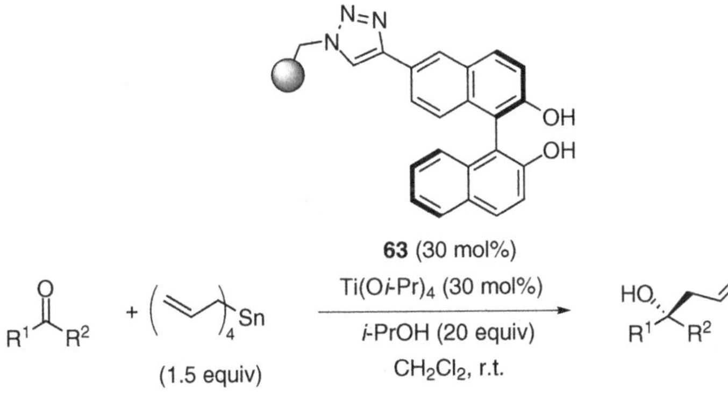

63 (30 mol%)

Ti(O*i*-Pr)$_4$ (30 mol%)

i-PrOH (20 equiv)

CH$_2$Cl$_2$, r.t.

(1.5 equiv)

R^1 = *m*-Tol, R^2 = Me: 96% yield, ee = 95%
R^1 = *m*-CF$_3$C$_6$H$_4$, R^2 = Me: 91% yield, ee = 94%
R^1 = *p*-MeOC$_6$H$_4$, R^2 = Me: 80% yield, ee = 70%
R^1 = *m*-BrC$_6$H$_4$, R^2 = Me: 95% yield, ee = 93%
R^1 = Ph, R^2 = (CH$_2$)$_2$Cl: 67% yield, ee = 73%
R^1 = 2-furyl, R^2 = Me: 70% yield, ee = 86%
R^1 = (*E*)-PhCH=CH, R^2 = Me: 85% yield, ee = 77%
R^1 = BnCH$_2$, R^2 = Me: 98% yield, ee = 76%

Scheme 1.59. (*R*)-BINOL-derived polystyrene-supported ligand in allylation of ketones

1.4.2. *Vinylations*

Highly enantioselective catalysts for the asymmetric vinylation of aldehydes have been developed by the groups of Oppolzer,[112] Wipf,[113] and others.[114] Although chiral allylic alcohols with various substitution patterns are in demand, advances in this area have mostly been focused on the synthesis of β-substituted *E*-allylic alcohols. In contrast, very few methods have been developed for the synthesis of α-substituted allylic alcohols. In this context, in 2013 Harada *et al.* reported a general one pot method for the highly enantioselective synthesis of these products, starting from alkynes and aldehydes, and proceeding through *in situ* generated vinylaluminum reagents.[115] This process involved a catalytic amount of chiral ligand (*R*)-DPP-H$_8$-BINOL in combination with three equivalents of Ti(O*i*-Pr)$_4$ as catalytic system. The use of Me$_2$AlH was essential in the preliminary nickel-catalysed hydroalumination step to generate vinyl aluminum reagents **64** which could not reduce the aldehyde starting material (Scheme 1.60). Remarkable enantioselectivities of up to 94% ee were achieved at

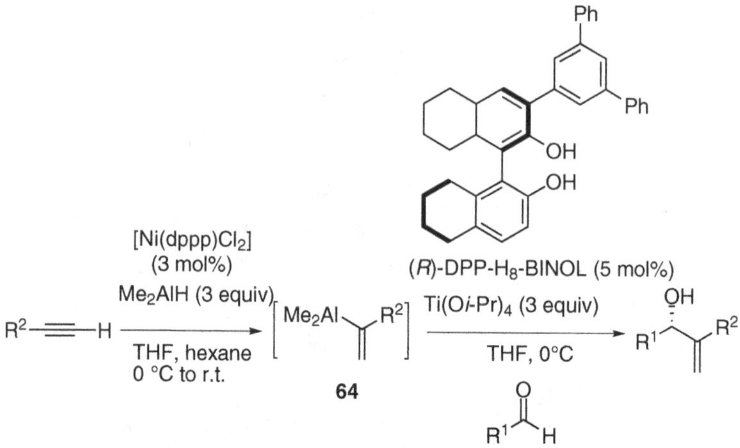

$R^1 = p\text{-}ClC_6H_4, R^2 = CH_2Bn$: 75% yield, ee = 93%
$R^1 = p\text{-}NCC_6H_4, R^2 = CH_2Bn$: 61% yield, ee = 91%
$R^1 = p\text{-}CO_2MeC_6H_4, R^2 = CH_2Bn$: 65% yield, ee = 91%
$R^1 = m\text{-}MeOC_6H_4, R^2 = CH_2Bn$: 74% yield, ee = 92%
$R^1 = 2\text{-thienyl}, R^2 = CH_2Bn$: 69% yield, ee = 91%
$R^1 = Cy, R^2 = CH_2Bn$: 62% yield, ee = 68%
$R^1 = p\text{-}ClC_6H_4, R^2 = Bn$: 69% yield, ee = 91%
$R^1 = p\text{-}BrC_6H_4, R^2 = Bn$: 70% yield, ee = 92%
$R^1 = p\text{-}ClC_6H_4, R^2 = i\text{-}Pent$: 62% yield, ee = 91%
$R^1 = p\text{-}ClC_6H_4, R^2 = Cy$: 33% yield, ee = 90%
$R^1 = p\text{-}ClC_6H_4, R^2 = c\text{-}Pr$: 63% yield, ee = 94%
$R^1 = CH_2Bn, R^2 = c\text{-}Pr$: 68% yield, ee = 71%
$R^1 = p\text{-}ClC_6H_4, R^2 = (CH_2)_8CH{=}CH_2$: 62% yield, ee = 91%
$R^1 = 1\text{-Naph}, R^2 = (CH_2)_8CH{=}CH_2$: 78% yield, ee = 92%
$R^1 = p\text{-}ClC_6H_4, R^2 = (CH_2)_3Cl$: 72% yield, ee = 92%
$R^1 = PhCH{=}CH, R^2 = (CH_2)_3Cl$: 51% yield, ee = 84%
$R^1 = p\text{-}ClC_6H_4, R^2 = (CH_2)_4OTBDPS$: 70% yield, ee = 93%
$R^1 = p\text{-}ClC_6H_4, R^2 = (CH_2)_4OTBS$: 60% yield, ee = 93%
$R^1 = p\text{-}NCC_6H_4, R^2 = (CH_2)_4OTBS$: 60% yield, ee = 93%

Scheme 1.60. (R)-DPP-H$_8$-BINOL as ligand in vinylation of aldehydes

a low catalyst loading of 5 mol% in the subsequent addition reaction to aldehydes. It should be noted that aromatic, heteroaromatic, as well as aliphatic aldehydes were compatible with the reaction, although aliphatic aldehydes generally provided lower enantioselectivities (68–88% ee).

The catalytic asymmetric vinylation of ketones has attracted considerable attention among many research groups due to its more challenging transformations, which are owing to the significant difference in reactivity between aldehydes and ketones. Despite the success of asymmetric vinylation of aldehydes, vinyl additions to the inert ketones remain challenging. To counterbalance the reduced reactivity of ketones, a more reactive vinylating agent was needed. An important discovery in the asymmetric vinylation of ketones was reported by Walsh in 2004, using a protocol in which the reaction coupled hydrozirconation/transmetalation to zinc with the catalyst, to furnish the chiral tertiary alcohols.[116] In this work, the catalyst was prepared from a chiral bis(sulfonamide) ligand and Ti(Oi-Pr)$_4$, and employed diarylzinc as the nucleophiles, to afford chiral tertiary vinylic alcohols in good to excellent enantioselectivities for a range of ketones. In 2009, Gau and Biradar decided to prepare organoaluminum compounds from alkynes and DIBAL-H to be used in the asymmetric addition to ketones, due to their high reactivity, and the greater Lewis acidity of the aluminum centre.[117] The catalytic system of the reaction was based on a catalytic amount (10 mol%) of (S)-BINOL combined with three equivalents of Ti(Oi-Pr)$_4$. It allowed the synthesis of diversified chiral

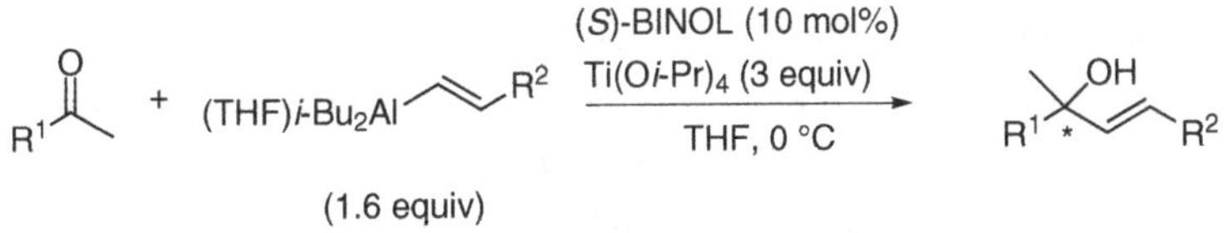

$$
\begin{array}{c}
\underset{R^1}{\overset{O}{\parallel}} + \text{(THF)}i\text{-Bu}_2\text{Al} \diagdown R^2 \xrightarrow[\text{THF, 0 °C}]{\substack{(S)\text{-BINOL (10 mol\%)} \\ \text{Ti(O}i\text{-Pr)}_4 \text{ (3 equiv)}}} \underset{R^1}{\overset{OH}{\diagup}} {}_* \diagup R^2 \\
\text{(1.6 equiv)}
\end{array}
$$

R^1 = Ph, R^2 = n-Bu: 91% yield, ee = 92%
R^1 = m-MeOC$_6$H$_4$, R^2 = n-Bu: 80% yield, ee = 81%
R^1 = p-MeOC$_6$H$_4$, R^2 = n-Bu: 93% yield, ee = 87%
R^1 = o-Tol, R^2 = n-Bu: 87% yield, ee = 98%
R^1 = p-Tol, R^2 = n-Bu: 92% yield, ee = 92%
R^1 = p-ClC$_6$H$_4$, R^2 = n-Bu: 90% yield, ee = 97%
R^1 = p-CF$_3$C$_6$H$_4$, R^2 = n-Bu: 94% yield, ee = 96%
R^1 = p-BrC$_6$H$_4$, R^2 = n-Bu: 89% yield, ee = 91%
R^1 = p-NO$_2$C$_6$H$_4$, R^2 = n-Bu: 92% yield, ee = 95%
R^1 = 1-Naph, R^2 = n-Bu: 82% yield, ee = 87%
R^1 = 2-Naph, R^2 = n-Bu: 92% yield, ee = 92%
R^1 = (E)-PhCH=CH, R^2 = n-Bu: 92% yield, ee = 82%
R^1 = p-NO$_2$C$_6$H$_4$, R^2 = Cl(CH$_2$)$_4$: 93% yield, ee = 96%
R^1 = p-ClC$_6$H$_4$, R^2 = Bn: 94% yield, ee = 95%
R^1 = p-ClC$_6$H$_4$, R^2 = Cy: 90% yield, ee = 85%
R^1 = p-ClC$_6$H$_4$, R^2 = 1-cyclohexenyl: 89% yield, ee = 88%

Scheme 1.61. (S)-BINOL as ligand in vinylation of ketones tertiary allylic alcohols from 1-hexyne and aromatic ketones bearing either electron-donating or electron-withdrawing substituents on the aromatic ring to be achieved in excellent enantioselectivities of up to 98% ee, in combination with general high yields (Scheme 1.61). The scope of this remarkable process was extended to an α,β-unsaturated ketone, such as *trans*-cinnamaldehyde, which gave the corresponding product in 92% yield and enantioselectivity of 82% ee. More importantly, additions of vinylaluminum reagents derived from a variety of alkynes other than 1-hexyne also produced allylic alcohols in excellent enantioselectivities of up to 96% ee (Scheme 1.61).

1.5. Conclusions

This chapter illustrated the impressive amount of enantioselective titanium-promoted alkylation, arylation, alkynylation, allylation, and vinylation reactions of carbonyl compounds achieved in the last seven years, spanning from classical reactions, such as enantioselective 1,2-nucleophilic additions to aldehydes, to those of low reactive organozinc reagents with poor electrophilic ketones. It is only recently that the first highly efficient enantioselective alkylations of aldehydes with trialkylboranes and organolithium reagents have been developed. Moreover, the enantioselective addition of organometallic alkynyl derivatives to carbonyl compounds is today the most expedient route toward chiral propargylic alcohols which constitute strategic building blocks for the enantioselective synthesis of a range of complex important molecules. In particular, highly enantioselective titanium-promoted alkynylations of aldehydes with special alkynes, such as diynes and enynes, have been successfully developed recently among other reactions. All these methodologies have a strategically synthetic advantage to form a new C–C bond, a new functionality (alcohol), with concomitant creation of a stereogenic centre in a single transformation. They arose from the extraordinary ability of chiral titanium catalysts to control stereochemistry, which can be attributed to their rich coordination chemistry and facile modification of titanium Lewis acid centre by structurally modular ligands. Among the Lewis acidic metal complexes, titanium(IV) is the central metal of choice, because of its high Lewis acidity and relatively short metal–ligand bond lengths, in addition to its high abundance, low cost, and low toxicity.

In spite of the impressive number of publications, however, challenges remain in the context of enantioselective nucleophilic additions to carbonyl compounds, such as a better understanding of the role of the active titanium catalysts, and achieving higher turnover numbers of the catalytic cycles. Titanium-mediated organic reactions described in this chapter are indeed very powerful, but their utility often remains hampered by the need to use super-stoichiometric amounts of titanium, particularly in the cases of alkylation and arylation reactions. Even if titanium is not toxic and inexpensive, this remains wasteful and problematic. In this context, the use of sub-stoichiometric amounts of titanium or preformed titanium catalysts used in catalytic amounts have already allowed major advances to be achieved and will have to be more developed in the near future. The ever-growing need for environmentally friendly catalytic processes has prompted organic chemists to focus on more abundant first-row transition metals, such as titanium, to develop new catalytic systems to perform reactions, such as C–C bond formations. Therefore, a bright future is undeniable for more sustainable and enantioselective titanium-promoted transformations.

References

(1) (a) R. Noyori, in *Asymmetric Catalysts in Organic Synthesis*, Wiley, New York, **1994**. (b) *Transition Metals for Organic Synthesis*, M. Beller, C Bolm, eds., Wiley–VCH, Weinheim, **1998**, Vols. I and II. (c) *Comprehensive Asymmetric Catalysis*, E. N. Jacobsen, A. Pfaltz, H. Yamamoto, eds., Springer, Berlin, **1999**. (d) *Catalytic Asymmetric Synthesis*, 2nd ed., I. Ojima, ed.; Wiley–VCH, New York, **2000**. (e) G. Poli, G. Giambastiani, A. Heumann, *Tetrahedron* **2000**, *56*, 5959–5989. (f) E. Negishi, in *Handbook of Organopalladium Chemistry for Organic Synthesis*, John Wiley & Sons, Inc., Hoboken, NJ, **2002**, *2*, 1689–1705. (g) A. de Meijere, P. von Zezschwitz, H. Nüske, B. Stulgies, *J. Organomet. Chem.* **2002**, *653*, 129–140. (h) *Transition Metals for Organic Synthesis*, 2nd ed., M. Beller, C. Bolm, eds., Wiley–VCH, Weinheim, **2004**. (i) L. F. Tietze, I. Hiriyakkanavar, H. P. Bell, *Chem. Rev.* **2004**, *104*, 3453–3516. (j) D. J. Ramon, M. Yus, *Chem. Rev.* **2006**, *106*, 2126–2208.

(2) (a) Y. Yu, K. Ding, G. Chen, in *Acid Catalysis in Modern Organic Synthesis*; H. Yamamoto, K. Ishihara, eds., Wiley–VCH, Weinheim, **2008**; Chapter 14, p. 721. (b) F. Chen, X. Feng, Y. Jiang, *Arkivoc* **2003**, *ii*, 21–31. (c) J. Cossy, S. Bouz, F. Pradaux, C. Willis, V. Bellosta, *Synlett* **2002**, 1595–1606. (d) K. Mikami, M. Terada in *Lewis Acids in Organic Synthesis*, H. Yamamoto, ed., Wiley–VCH, Weinheim, **2000**, Vol. 2, Chapter 16, p. 799. (e) K. Mikami, M. Terada in *Lewis Acid Reagents*, H. Yamamoto, ed.; Wiley–VCH, Weinheim, **1999**, Vol. 1, Chapter 6, pp. 93–136. (f) D. Ramon,

M. Yus, *Rec. Res. Dev. Org. Chem.* **1998**, *2*, 489–523. (g) A. H. Hoveyda, J. P. Morken, *Angew. Chem., Int. Ed. Engl.* **1996**, *35*, 1262–1284. (h) B. M. Trost, In *Stereocontrolled Organic Synthesis*, B. M. Trost, ed., Blackwell, Oxford, **1994**, p. 17.

(3) (a) M. Yoshioka, T. Kawakita, M. Ohno, *Tetrahedron Lett.* **1989**, *30*, 1657–1660. (b) H. Takahashi, T. Kawakita, M. Yoshioka, S. Kobayashi, M. Ohno, *Tetrahedron Lett.* **1989**, *30*, 7095–7098.

(4) For reviews on (asymmetric) additions of organozinc reagents to carbonyl compounds, see: (a) C. M. Binder, B. Singaram, *Org. Prep. Proc. Int.* **2011**, *43*, 139–208. (b) K. Soai, T. Shibata in *Comprehensive Asymmetric Catalysis Supplement I*; E. N. Jacobsen, A. Pfaltz, eds. Springer, Berlin, **2004**; p 95. (c) L. Pu, H.-B. Yu, *Chem. Rev.* **2001**, *101*, 757–824. (d) K. Soai in *Comprehensive Asymmetric Catalysis II*; E. N. Jacobsen, A. Pfaltz, H. Yamamoto, eds., Springer, Berlin, **1999**; p. 911.

(5) For reviews on (asymmetric) allylations of carbonyl compounds, see: (a) M. Yus, J. C. Gonzales-Gomez, F. Foubelo, *Chem. Rev.* **2011**, *111*, 7774–7854. (b) S. E. Denmark, J. Fu, *Chem. Rev.* **2003**, *103*, 2763–2793. (c) Y. Yamamoto, N. Asao, *Chem. Rev.* **1993**, *93*, 2207–2293. (d) R. O. Duthaler, A. Hafner, *Chem. Rev.* **1992**, *92*, 807–832.

(6) E. Frankland, *Ann. Chem. Pharm.* **1849**, *71*, 171–213.

(7) V. C. R Grignard, *Hebd. Sceances Acad. Sci.* **1900**, *130*, 1322–1324.

(8) (a) W. Schlenk, J. Holtz, *Chem. Ber.* **1917**, *50*, 262–274. (b) K. Ziegler, H. Colonius, *J. Liebigs Ann. Chem.* **1930**, *479*, 135–149. (c) G. Wittig, U. Pockels, H. Droge, *Chem. Ber.* **1938**, *71*, 1903–1912. (d) H. Gilman, A. L. Jacoby, *J. Org. Chem.* **1938**, *3*, 108–119. (e) H. Gilman, W. Langham, A. L. Jacoby, *J. Am. Chem. Soc.* **1939**, *61*, 106–109.

(9) M. Betti, E. Lucchi, *Chem. Abstr.* **1940**, *34*, 2354.

(10) H. L. Cohen, G. F. Wright, *J. Org. Chem.* **1953**, *18*, 432–446.

(11) B. Weber, D. Seebach, *Tetrahedron* **1994**, *50*, 7473–7484.

(12) (a) N. Oguni, T. Omi, *Tetrahedron Lett.* **1984**, *25*, 2823–2824. (b) N. Oguni, T. Omi,, Y. Yamamoto, A. Nakamura, *Chem. Lett.* **1983**, 841–842.

(13) M. Kitamura, S. Suga, K. Kawai, R. Noyori, *J. Am. Chem. Soc.* **1986**, *108*, 6071–6072.

(14) P. Knochel, *Chemtracts* **1995**, *8*, 205–221.

(15) B. Schmidt, D. Seebach, *Angew. Chem., Int. Ed. Engl.* **1991**, *30*, 99–101.

(16) K. Soai, S. Niwa, *Chem. Rev.* **1992**, *92*, 833–856.

(17) K.-H. Wu, H.-M. Gau, *Organometallics* **2004**, *23*, 580–588.

(18) H. Takahashi, T. Kawakita, M. Ohno, M. Yoshioka, S. Kobayashi, *Tetrahedron* **1992**, *48*, 5691–5700.

(19) M. Mori, T. Nakai, *Tetrahedron Lett.* **1997**, *38*, 6233–6236.

(20) S. N. MacMillan, K. T. Ludforth, J. M. Tanski, *Tetrahedron: Asymmetry* **2008**, *19*, 543–548.

(21) K. Pathak, I. Ahmad, S. H. R. Abdi, R. I. Kureshy, N.-U. H. Khan, R. V. Jasra, *J. Mol. Catal. A* **2008**, *280*, 106–114.

(22) B. Liu, C. Fang, Z.-B. Dong, J.-S. Li, *Appl. Organomet. Chem.* **2008**, *22*, 55–58.

(23) B. Liu, Z.-B. Dong, C. Fang, H.-B. Song, J.-S. Li, *Chirality* **2008**, *20*, 828–832.

(24) Z.-B. Dong, B. Liu, C. Fang, J.-S. Li, *J. Organomet. Chem.* **2008**, *693*, 17–22.

(25) Z.-G. Zhang, Z.-B. Dong, J.-S. Li, *Chirality* **2010**, *22*, 820–826.

(26) S. Gou, Z. M. A. Judeh, *Tetrahedron Lett.* **2009**, *50*, 281–283.

(27) X. Liu, P. Wang, Y. Yang, P. Wang, Q. Yang, *Chem. Asian J.* **2010**, *5*, 1232–1239.

(28) A. R. Abreu, M. M. Pereira, J. C. Bayon, *Tetrahedron* **2010**, *66*, 743–749.

(29) A. R. Abreu, M. Lourenço, D. Peral, M. T. S. Rosado, M. E. S. Eusébio, O. Palacios, J. C. Bayon, M. M. Pereira, *J. Mol. Catal. A* **2010**, *325*, 91–97.

(30) L. Ma, M. M. Wanderley, W. Lin, *ACS Catal.* **2011**, *1*, 691–697.

(31) S. Beckendorf, O. G. Mancheno, *Synthesis* **2012**, *44*, 2162–2172.

(32) Y. Kinoshita, S. Kanehira, Y. Hayashi, T. Harada, *Chem. Eur. J.* **2013**, *19*, 3311–3314.

(33) Y. S. Sokeirik, A. Hoshina, M. Omote, K. Sato, A. Tarui, I. Kumadaki, A. Ando, *Chem. Asian J.* **2008**, *3*, 1850–1856.

(34) H. Pellissier, *Tetrahedron* **2008**, *64*, 10279–10317.

(35) M. A. Dean, S. R. Hitchcock, *Tetrahedron: Asymmetry* **2008**, *19*, 2563–2567.

(36) C. M. Jones, H. Li, A. J. Hickman, L. D. Hughs, S. J. Sobelman, A. R. Johnson, *Tetrahedron: Asymmetry* **2012**, *23*, 501–507.

(37) A. Venegas, L. Rivas, G. Huelgas, C. A. de Parrodi, D. Madrigal, G. Aguirre, M. Parra-Hake, D. Chavez, R. Somanathan, *Tetrahedron: Asymmetry* **2010**, *21*, 2944–2948.

(38) A. Viso, R. F. de la Pradilla, M. Urena, *Tetrahedron* **2009**, *65*, 3757–3766.

(39) T. Watanabe, I. Kurata, C. Hayashi, M. Igarashi, R. Sawa, Y. Takahashi, Y. Akamatsu, *Bioorg. Med. Chem. Lett.* **2010**, *20*, 5843–5846.

(40) T. Bauer, S. Smolinski, *Appl. Catal. A* **2010**, *375*, 247–251.

(41) G. Bian, H. Huang, H. Zong, Song, L. *Chirality* **2012**, *24*, 825–832.

(42) (a) M. Yus, D. J. Ramon, O. Rrieto, *Tetrahedron: Asymmetry* **2003**, *14*, 1103–1114.

(b) M. Yus, D. J. Ramon, O. Prieto, *Tetrahedron: Asymmetry* **2002**, *13*, 2291–2293.

(c) D. J. Ramon, M. Yus, *Tetrahedron Lett.* **1998**, *39*, 1239–1242.

(43) A. Hui, J. Zhang, H. Sun, Z. Wang, *Arkivoc* **2008**, *ii*, 25–32.

(44) G. Huelgas, L. K. LaRochelle, L. Rivas, Y. Luchinina, R. A. Toscano, P. J. Carroll, P. J. Walsh, C. A. de Parrodi, *Tetrahedron* **2011**, *67*, 4467–4474.

(45) S. Gonzalez-Lopez, M. Yus, D. J. Ramon, *Tetrahedron: Asymmetry* **2012**, *23*, 611–615.

(46) (a) V. J. Forrat, D. J. Ramon, M. Yus, *Tetrahedron: Asymmetry* **2008**, *19*, 537–541.

(b) V. J. Forrat, D. J. Ramon, M. Yus, *Tetrahedron: Asymmetry* **2009**, *20*, 65–67.

(47) A. S. C. Chan, F.-Y. Zhang, C.-W. Yip, *J. Am. Chem. Soc.* **1997**, *119*, 4080–4081.

(48) D. Seebach, A. K. Beck, R. Imwinkelried, S. Roggo, A. Wonnacott, *Helv. Chim. Acta* **1987**, *70*, 954–974.

(49) C. Garcia, P. J. Walsh, *Org. Lett.* **2003**, *5*, 3641–3644.

(50) S.-H. Hsieh, C.-A. Chen, D.-W. Chuang, M.-C. Yang, H.-T. Yang, H.-M. Gau, *Chirality* **2008**, *20*, 924–929.

(51) S. Zhou, D.-W. Chuang, S.-J. Chang, H.-M. Gau, *Tetrahedron: Asymmetry* **2009**, *20*, 1407–1412.

(52) S. Zhou, K.-H. Wu, C.-A. Chen, H.-M. Gau, *J. Org. Chem.* **2009**, *74*, 3500–3505.

(53) E. Fernandez-Mateos, B. Macia, M. Yus, *Tetrahedron: Asymmetry* **2012**, *23*, 789–794.

(54) Q. Chen, J. Ou, G.-S. Chen, T.-H. Liu, B. Xie, H.-B. Liang, L.-Q. Xie, Y.-X. Chen, *Chirality* **2014**, *26*, 268–271.

(55) Chen, C.-A. Wu, K.-H. Gau, H.-M. *Adv. Synth. Catal.* **2008**, *350*, 1626–1634.

(56) K.-H. Wu, D.-W. Chuang, C.-A. Chen, H.-M. Gau, *Chem. Commun.* **2008**, 2343–2345.

(57) D. B. Biradar, S. Zhou, H.-M. Gau, *Org. Lett.* **2009**, *11*, 3386–3389.

(58) (a) B. Weber, D. Seebach, *Angew. Chem., Int. Ed. Engl.* **1992**, *31*, 84–86. (b) B. Weber, D. Seebach, *Tetrahedron* **1994**, *50*, 6117–6128.

(59) (a) Y. Muramatsu, T. Harada, *Angew. Chem., Int. Ed.* **2008**, *47*, 1088–1090. (b) Y. Muramatsu, S. Kanehira, M. Tanigawa, Miyawaki, T. Harada, *Bull. Chem. Soc. Jpn.* **2010**, *83*, 19–32.

(60) (a) Y. Muramatsu, T. Harada, *Chem. Eur. J.* **2008**, *14*, 10560–10563.

(61) D. Itakura, T. Harada, *Synlett* **2011**, *19*, 2875–2879.

(62) X.-Y. Fan, Y.-X. Yang, F.-F. Zhuo, S.-L. Yu, X. Li, Q.-P. Guo, Z.-X. Du, C.-S. Da, *Chem. Eur. J.* **2010**, *16*, 7988–7991.

(63) (a) Y. Liu, C.-S. Da, S.-L. Yu, X.-G. Yin, J.-R.Wang, X.-Y. Fan, W.-P. Li, R. Wang, *J. Org. Chem.* **2010**, *75*, 6869–6878. (b) C.-S. Da, J.-R. Wang, X.-G. Yin, X.-Y. Fan, Y. Liu,, S.-L. Yu, *Org. Lett.* **2009**, *11*, 5578–5581.

(64) E. Fernandez-Mateos, B. Macia, D. J. Ramon, M. Yus, *Eur. J. Org. Chem.* **2011**, 6851–6855.

(65) E. Fernandez-Mateos, B. Macia, M. Yus, *Adv. Synth. Catal.* **2013**, *355*, 1249–1254.

(66) Q. Li, H.-M. Gau, *Chirality* **2011**, *23*, 929–939.

(67) K.-H. Wu, S. Zhou, C.-A. Chen, M.-C. Yang, R.-T. Chiang, C.-R. Chen, H.-M. Gau, *Chem. Commun.* **2011**, *47*, 11668–11670.

(68) A. Uenishi, Y. Nakagawa, H. Osumi, T. Harada, *Chem. Eur. J.* **2013**, *19*, 4896–4905.

(69) (a) M. Srebnik, *Tetrahedron Lett.* **1991**, *32*, 2449–2452. (b) W. Oppolzer, R. Radinov, N. *J. Am. Chem. Soc.* **1993**, *115*, 1593–1594. (c) F. Langer, L. Schwink, A. Devasagayaraj, P.-Y. Chavant, P. Knochel, *J. Org. Chem.* **1996**, *61*, 8229–8243. (d) S.-J. Jeon, H. Li, C. Garcia, L. K. La Rochelle, P. J. Walsh, *J. Org. Chem.* **2005**, *70*, 448–455.

(70) T. Ukon, T. Harada, *Eur. J. Org. Chem.* **2008**, 4405–4407.

(71) T. Shono, T. Harada, *Org. Lett.* **2010**, *12*, 5270–5273.

(72) R. Kumar, H. Kawasaki, T. Harada, *Org. Lett.* **2013**, *15*, 4198–4201.

(73) C. Najera, M. Yus, *Curr. Org. Chem.* **2003**, *7*, 867–926.

(74) D. Seebach, H. Dörr, B. Bastani, V. Ehrig, *Angew. Chem., Int. Ed. Engl.* **1969**, *8*, 982–983.

(75) B. Lecachey, C. Fressigné, H. Oulyadi, A. Harrison-Marchand, J. Maddaluno, *Chem. Commun.* **2011**, *47*, 9915–9917.

(76) E. Fernandez-Mateos, B. Macia, M. Yus, *Eur. J. Org. Chem.* **2012**, 3732–3736.

(77) For initial examples of enantioselective catalytic alkynylation of aldehydes, see: (a) K. Niwa, K. Soai, *J. Chem. Soc. Perkin Trans. I* **1990**, 937–942. (b) M. Ishizaki, O. Hoshino, *Tetrahedron: Asymmetry* **1994**, *5*, 1901–1904.

(78) (a) G. Lu, X.-S. Li, W.-L. Chan, A. S. C. Chan, *Chem. Commun.* **2002**, 172–173. (b) L. Pu, D. Moore, *Org. Lett.* **2002**, *4*, 4143–4146.

(79) Z. Xu, J. Mao, Y. Zhang, *Org. Biomol. Chem.* **2008**, *6*, 1288–1292.

(80) J. Mao, Z. Bao, J.Guo, S. Ji, *Tetrahedron* **2008**, *64*, 9901–9905.

(81) L. Wu, L. Zheng, L. Zong, J. Xu, Y. Cheng, *Tetrahedron* **2008**, *64*, 2651–2657.

(82) S. Gou, Z. Ye, Z. Huang, X. Ma, *Appl. Organometal. Chem.* **2010**, *24*, 374–379.

(83) T. Bauer, S. Smolinski, P. Gawel, J. Jurczak, *Tetrahedron Lett.* **2011**, *52*, 4882–4884.

(84) L. Qiu, Q. Wang, L. Lin, X. Liu, X. Jiang, Q. Zhao, G. Hu, R. Wang, *Chirality* **2009**, *21*, 316–323.

(85) M. Catir, M. Cakici, S. Karabuga, S. Ulukanli, E. Sahin, H. Kilic, *Tetrahedron: Asymmetry* **2009**, *20*, 2845–2853.

(86) Kelgtermans, L. Dobrzanska, L. Van Meervelt, W. Dehaen, *Tetrahedron* **2011**, *67*, 3685–3689.

(87) X.-P. Hui, C. Yin, Z.-C. Chen, L.-N. Huang, P.-F. Xu, G.-F. Fan, *Tetrahedron* **2008**, *64*, 2553–2558.

(88) Y.-M. Li, Y.-Q. Tang, X.-P. Hui, L.-N. Huang, P.-F. Xu, *Tetrahedron* **2009**, *65*, 3611–3614.

(89) X.-P. Hui, L.-N. Huang, Y.-M. Li, R.-L. Wang, P.-F Xu, *Chirality* **2010**, *22*, 347–354.

(90) M. Turlington, Y. Yue, X.-Q. Yu, L. Pu, *J. Org. Chem.* **2010**, *75*, 6941–6952.

(91) W. Chen, J.-H. Tay, J. Ying, X.-Q. Yu, L. Pu, *J. Org. Chem.* **2013**, *78*, 2256–2265.

(92) X. Du, Q. Wang, X. He, R.-G. Peng, X. Zhang, X.-Q. Yu, *Tetrahedron: Asymmetry* **2011**, *22*, 1142–1146.

(93) Y. Du, M. Turlington, X. Zhou, L. Pu, *Tetrahedron Lett.* **2010**, *51*, 5024–5027.

(94) G. Gao, Q. Wang, X.-Q. Yu, R.-G. Xie, L. Pu, *Angew. Chem., Int. Ed.* **2006**, *45*, 122–125.

(95) J. Mao, J. Guo, *Synlett* **2009**, *14*, 2295–2300.

(96) T. Xu, C. Liang, Y. Cai, J. Li, Y.-M. Li, X.-P. Hui, *Tetrahedron: Asymmetry* **2009**, *20*, 2733–2736.

(97) M. Turlington, A. DeBerardinis, L. Pu, *Org. Lett.* **2009**, *11*, 2441–2444.

(98) X. Chen, W. Chen, L. Wang, X.-Q. Yu, D.-S. Huang,, Pu, L. *Tetrahedron* **2010**, *66*, 1990–1993.

(99) M. Turlington, Y. Du, S. G. Ostrum, V. Santosh, K. Wren, T. Lin, M. Sabat, L. Pu, *J. Am. Chem. Soc.* **2011**, *133*, 11780–11794.

(100) (a) L. S. Tan, C. Y. Chen, R. D. Tillyer, E. J. J. Grabowski, P. J. Reider, *Angew. Chem., Int. Ed.* **1999**, *38*, 711–713. (b) B. Jiang, Z. L. Chen, X. X. Tang, *Org. Lett.* **2002**, *4*, 3451–3453.

(101) P. G. Cozzi, S. Alesi *Chem. Commun.* **2004**, 2448–2449.

(102) Y.-F. Zhou, Z.-J. Han, L. Qiu, J.-Y. Liang, F.-B. Ren, R. Wang, *Chirality* **2009**, *21*, 473–479.

(103) G.-W. Zhang, W. Meng, H. Ma, J. Nie, W.-Q. Zhang, J.-A. Ma, *Angew. Chem., Int. Ed.* **2011**, *50*, 3538–3542.

(104) A. Hosomi, Sakurai, H. *Tetrahedron Lett.* **1976**, *17*, 1295–1298.

(105) T. Hayashi, M. Konishi, H. Ito, M. Kumada, *J. Am. Chem. Soc.* **1982**, *104*, 4962–4963.

(106) Y. N. Belokon, D. Chusov, D. A. Borkin, L. V. Yashkina, P. Bolotov, T. Skrupskaya, M. North, *Tetrahedron: Asymmetry* **2008**, *19*, 459–466.

(107) R. S. Puruskotham, K. Ashabetha, R. D. Kumar, B. B. Chinna, Y. Venkateswarlu, *Synthesis* **2011**, *19*, 3180–3184.

(108) H. Hanawa, T. Hashimoto, K. Maruoka, *J. Am. Chem. Soc.* **2003**, *125*, 1708–1709.

(109) S. Casolari, D. D'addario, E. Tagliavini, *Org. Lett.* **1999**, *1*, 1061–1063.

(110) J. Yadav, G. R. Stanton, X. Fan, J. R. Robinson, E. J. Schelter, P. J. Walsh, M. Pericas, A. *Chem. Eur. J.* **2014**, *20*, 7122–7127.

(111) K. M. Waltz, J. Gavenonis, P. J. Walsh, *Angew. Chem., Int. Ed.* **2002**, *114*, 3849–3852.

(112) W. Oppolzer, R. N. Radinov, *Helv. Chim. Acta* **1992**, *75*, 170–173.

(113) (a) P. Wipf, W. Xu, H. Takahashi, H. Jahn, P. D. G. Coish, *Pure Appl. Chem.* **1997**, *69*, 639–644. (b) Wipf, P. Nunes, R. L. *Tetrahedron* **2004**, *60*, 1269–1279.

(114) S. Bräse, S. Dahmen, S. Höfener, F. Lauterwasser, M. Kreis, R. E. Ziegert, *Synlett* **2004**, 2647–2669.

(115) R. Kumar, H. Kawasaki, T. Harada, *Chem. Eur. J.* **2013**, *19*, 17707–17710.

(116) (a) H. Li, P. J. Walsh, *J. Am. Chem. Soc.* **2004**, *126*, 6538–6539. (b) H. Li, P. J. Walsh, *J. Am. Chem. Soc.* **2005**, *127*, 8355–8361.

(117) D. B. Biradar, H.-M. Gau, *Org. Lett.* **2009**, *11*, 499–502.

Chapter 2

Enantioselective Titanium-Catalysed Cyanation Reactions of Carbonyl Compounds and Derivatives

2.1. Introduction

The catalysis of organic reactions by metals still constitutes one of the most useful and powerful tools in organic synthesis.[1] By the very fact of the lower costs of titanium catalysts in comparison with other transition metals, and their nontoxicity — allowing their use for medical purposes (prostheses) — enantioselective titanium-mediated transformations have received ever-growing attention during the last decades that leading to exciting and fruitful research.[1j,2] This interest might also be related to the fact that titanium complexes are of high abundance, exhibit a remarkably diverse chemical reactivity, and constitute ones of the most useful Lewis acids in asymmetric catalysis.[3] This usefulness is particularly highlighted in the area of enantioselective cyanation reactions of carbonyl compounds since this field is largely dominated by the use of titanium-based chiral catalysts, especially for aldehyde substrates.

The last twenty years have seen an explosion of interest in the asymmetric synthesis of cyanohydrins, driven by the ease with which chiral cyanohydrins can be converted into other important bifunctional products. The cyanation methodologies have a strategically synthetic advantage in forming a new carbon to carbon (C–C) bond, new functionalities

85

(alcohol and cyanide) with concomitant creation of a stereogenic centre in a single transformation. The asymmetric reactions of various cyanide sources with carbonyl compounds to yield the corresponding chiral cyanohydrins can be catalysed by enzymes,[4] chiral Lewis acid catalysts,[5] or chiral organocatalysts.[6] They constitute very highly versatile synthetic transformations, related to the fact that these chiral products constitute important precursors for numerous valuable organic intermediates and chiral starting materials for the synthesis of a number of important natural products and bioactive molecules.[5a–5c,7] Indeed, chiral cyanohydrins are well-known natural products and versatile synthetic intermediates for many pharmaceuticals and agrochemicals. They are present as glycosides in over 3000 plants, bacteria, fungi, and insects as antifeedants forming part of the self-defense system.[8] Cyanohydrins also serve as sources of nitrogen for the biosynthesis of amino acids.[9] Furthermore, they provide opportunities for other transformations, because both the hydroxyl and the nitrile groups can be altered into a large number of functionalities under conditions preserving high enantiopurity. For example, chiral cyanohydrins can be readily transformed into a range of key products, such as α-hydroxy acids, α-amino acids, α-hydroxy aldehydes, α-hydroxy ketones, β-amino alcohols, among others.

The cyanohydrin synthesis was first reported by Winkler in 1832 using hydrogen cyanide as the cyanide source,[10] while the first asymmetric addition of hydrogen cyanide to aldehydes was reported by Rosenthaler in 1908, using an oxynitrilase enzyme as catalyst.[11] All early works were carried out with this cyanide source, but the volatility and extreme toxicity of this reagent caused obvious difficulties, and numerous alternative cyanide-based reagents have so far been developed. The reversibility of the cyanohydrin synthesis causes a potential problem for the asymmetric cyanohydrin synthesis, since the products are prone to racemise by elimination and re-addition of hydrogen cyanide. In this context, reagents, such as trimethylsilyl cyanide or acyl cyanides, have a major advantage in comparison with hydrogen cyanide, since they produce *O*-protected cyanohydrins, which are not prone to racemisation by this mechanism.

The asymmetric addition of cyanide to imines (Strecker-type reaction) can also be catalysed by chiral titanium complexes. It provides a robust, direct, and economically viable access to various naturally and non-naturally occurring α-amino acids after subsequent hydrolysis of the formed chiral amino nitriles. This chapter provides a comprehensive overview of the major developments in enantioselective titanium-catalysed cyanation reactions of carbonyl compounds and derivatives reported since the beginning of 2008, when this general field was reviewed by Yu and co-workers in a book chapter dealing with titanium Lewis acids,[2a] and earlier by Yus *et al.* in 2006.[12] (In addition, two reports dealing with the general field of enantioselective synthesis of cyanohydrins were independently published in 2008 by North and by Khan *et al.*,[5bc] followed by a microreview reported by Feng, in 2010.)[7b]

The chapter has been divided into three parts. The first part deals with enantioselective titanium-catalysed cyanation reactions of aldehydes, the second part collects less developed enantioselective titanium-catalysed cyanation reactions of ketones, while the third part includes enantioselective titanium-catalysed Strecker-type reactions.

2.2. Aldehydes as Electrophiles

2.2.1. *Cyanations of Aldehydes with Trimethylsilyl Cyanide*

2.2.1.1. *Salen ligands*

The first example of enantioselective cyanation of aldehydes using titanium complexes was reported in 1986 by Reetz *et al.*, employing BINOL as chiral ligand.[13] One year later, Narasaka *et al.* investigated the reaction of aromatic and aliphatic aldehydes with trimethylsilyl cyanide catalysed by a chiral titanium catalyst *in situ* generated from $TiCl_2(Oi\text{-}Pr)_2$ and TADDOL,[14] providing the corresponding cyanohydrins in enantioselectivities of up to 96% *ee*.[15] Undoubtedly, catalytic systems based on titanium complexes have been the most extensively studied in this field. Most of the catalysts were *in situ* prepared by treatment of the ligand with titanium tetraisopropylate, which has important consequences for the nature of the cyanating agent, since trimethylsilyl cyanide, for example, reacts with liberated isopropanol to form hydrogen cyanide *in situ*. Since the early work of Reetz using BINOL as ligand,[13] a wide range of other chiral

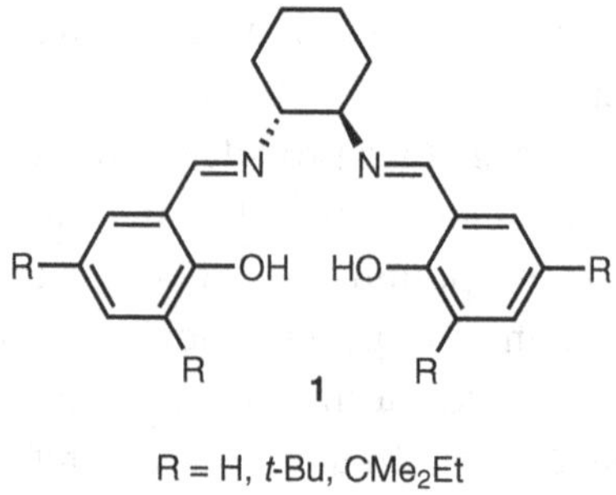

R = H, *t*-Bu, CMe$_2$Et

Figure 2.1. Most used chiral salen ligands

ligands have been investigated with the most used being chiral diimines derived from salicylaldehyde of type **1** (Figure 2.1).

In the last few years, a number of preformed chiral titanium catalysts have been successfully applied to asymmetric cyanations of aldehydes, in addition to *in situ* generated chiral titanium catalysts. For example, North *et al.* investigated novel preformed chiral C_1- and C_2-symmetric salen titanium complexes in the asymmetric addition of trimethylsilyl cyanide to benzaldehyde. The C_2-symmetrical ligands were prepared from a range of acyclic commercially available chiral diamines, and then complexed to titanium tetraisopropylate to give chiral dichloride complexes **2a–e**.[16] When induced by catalysts **2a–e** employed at 0.1–10 mol% of catalyst loadings, the corresponding cyanohydrin trimethylsilyl ethers were obtained in moderate to quantitative yields (34–100%) in combination with moderate enantioselectivities of up to 52% *ee*, as shown in Scheme 2.1.

The most notable feature of this reaction was the inversion of the enantioselectivity in comparison with that observed by using usual salen complexes derived from (*R,R*)-1,2-diaminocyclohexane of type **1**.[17] Indeed, this was the first time that salen ligands derived from an (*R,R*)-diamine have been found to catalyse the asymmetric addition of cyanide to the *Si* face of benzaldehyde. The magnitude of the asymmetric induction using catalysts **2b–d** was very low (up to 30% ee), whilst complex **2a** ($R^1 = R^2 = Ph$) gave the best result (52% ee). The efficiency of complexes **2a–e** was compared with that of the corresponding vanadium complexes, which provided the products in better enantioselectivities and yields. Both the magnitude and the sense of the asymmetric induction could be

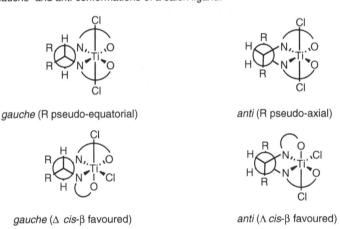

with 0.1 mol% of **2a** ($R^1 = R^2 = Ph$): 64% yield, ee = 52% (*S*)
with 2 mol% of **2b** ($R^1 = R^2 = Me$): 34% yield, ee = 26% (*R*)
with 10 mol% of **2c** ($R^1 = R^2 = n\text{-Pr}$): >99% yield, ee = 30% (*R*)
with 2 mol% of **2d** ($R^1 = R^2 = i\text{-Pr}$): 44% yield, ee = 2% (*R*)
with 1 mol% of **2e** ($R^1 = R^2 = t\text{-Bu}$): 62% yield, ee = 50% (*R*)
with 2 mol% of **3** ($R^1 = i\text{-Pr}$, $R^2 = H$): 61% yield, ee = 54% (*S*)

Gauche- and *anti*-conformations of a salen ligand:

gauche (R pseudo-equatorial)   *anti* (R pseudo-axial)

gauche (Δ *cis*-β favoured)   *anti* (Λ *cis*-β favoured)

Stereochemistry inducing transition state for asymmetric cyanohydrin synthesis using titanium salen complexes:

Scheme 2.1. Titanium-based C_1- and C_2-symmetrical salen catalysts in asymmetric cyanation of benzaldehyde explained on the basis of the diamine unit within the salen ligand, which could adopt either a *gauche* or an *anti*-conformation (Scheme 2.1). Cyclic amines, such as 1,2-diaminocyclohexanes of type **1**, can adopt the *gauche*-conformation, but for acyclic diamines the *anti*-conformation is generally thermodynamically preferred, as it minimizes steric interactions between the R groups; and between the R groups and the imine hydrogens. Whilst a salen ligand normally prefers to be planar, it is well known that it can adopt a *cis*-β configuration, and this is necessary to allow formation of the established bimetallic transition state for asymmetric cyanohydrin synthesis in which cyanide is transferred intramolecularly to the coordinated aldehyde (Scheme 2.1).

When applied to octahedral complexes, with a salen ligand in the *cis*-β configuration, the consequence of changing the conformation of the diamine unit from *gauche* to *anti* is to change the preferred stereochemistry of the titanium salen unit from Δ to Λ. This is due to the 'twist' of the diamine unit, which determines the overall 'twist' of the salen ligand (Scheme 2.1). Thus, for complexes **2b–e** it appears that the salen ligand preferred to adopt the *anti*-conformation, resulting in a Λ-configuration around the metal centres, thus inverting the enantioselectivity of the cyanohydrin synthesis compared to complex **2a**. In another work, these authors converted amino acids, such as (*S*)-alanine, (*S*)-valine, (*S*)-phenylalanine, and (*R*)-phenylglycine, into C_1-symmetrical salen ligands which were complexed to titanium tetraisopropylate to give the corresponding chiral dichloride complexes. These catalysts were investigated in the same cyanation of benzaldehyde.[18] Among them, only complex **3** gave reasonable levels of yield (61%) and asymmetric induction of 54% ee, as shown in Scheme 2.1. Again, the authors have compared the efficiency of these titanium catalysts with that of the corresponding vanadium ones and observed generally lower yields and enantioselectivities.

Later, Sun *et al.* described the synthesis of other tetradentate ligands derived from salen.[19] These chiral ligands were prepared by reacting Grignard reagents with the C=N of salen ligands, to construct a C–N bond. The corresponding titanium complexes were further obtained by the reaction of equimolar amounts of these ligands, with titanium tetraisopropylate in dichloromethane at reflux. These complexes were subsequently investigated as catalysts in the asymmetric trimethylsilyl cyanation

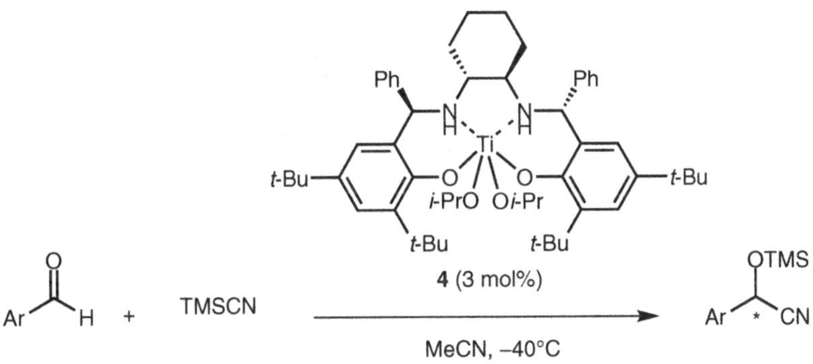

Ar = Ph: 99% yield, ee = 81%
Ar = p-MeOC$_6$H$_4$: 99% yield, ee = 92%
Ar = o-MeOC$_6$H$_4$: 96% yield, ee = 33%
Ar = p-ClC$_6$H$_4$: 50% yield, ee = 66%
Ar = o-ClC$_6$H$_4$: 93% yield, ee = 30%
Ar = o-BrC$_6$H$_4$: 98% yield, ee = 29%
Ar = 2,6-Cl$_2$C$_6$H$_3$: 97% yield, ee = 34%
Ar = p-Tol: 84% yield, ee = 90%
Ar = o-Tol: 50% yield, ee = 37%

Scheme 2.2. Titanium catalyst of tetradentate ligand derived from salen in asymmetric cyanation of aryl aldehydes

of aldehydes. Among them, catalyst **4** used at 3 mol% of catalyst loading provided the best enantioselectivities of up to 92% ee in combination with good to excellent yields of up to 99% (Scheme 2.2). The results showed that general excellent enantioselectivities were obtained with electron-rich aromatic aldehydes, while electron-deficient aromatic aldehydes gave lower enantiomeric excesses. Moreover, it was evident that the steric properties of the aromatic aldehyde had a great effect on the enantioselectivity of the reaction. For example, the *ortho*-substituted aromatic aldehydes provided lower enantioselectivities than the *para*-substituted ones. On the other hand, no reaction occurred with aliphatic aldehydes as substrates.

In 2011, the same authors reported the synthesis of a new enantiopure salen ligand **5**, bearing a diphenylphosphine oxide on the 3-position of one aromatic ring, and its use, in combination with titanium tetraisopropylate, as an *in situ* generated monometallic bifunctional catalyst for

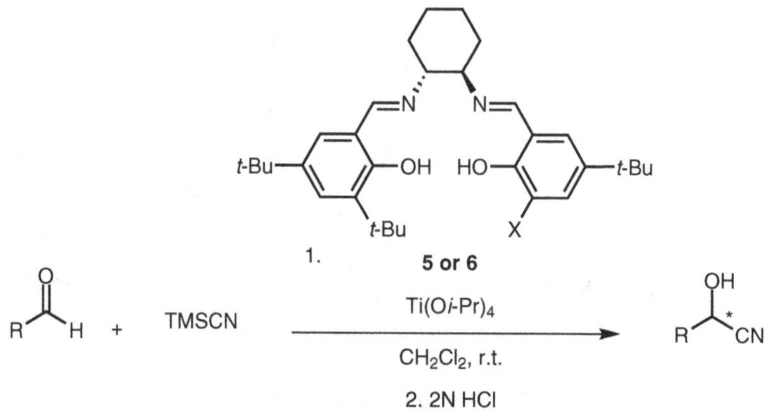

with 1 mol% of ligand **5** (X = $CH_2P(O)Ph_2$)
in combination with 1 mol% of Ti(O*i*-Pr)$_4$:
R = Ph: 98% yield, ee = 70%
R = *p*-Tol: 99% yield, ee = 69%
R = *p*-MeOC$_6$H$_4$: 93% yield, ee = 69%
R = *o*-MeOC$_6$H$_4$: 99% yield, ee = 60%
R = 2-furyl: 97% yield, ee = 52%
R = *i*-Pr: 95% yield, ee = 26%
with 0.055 mol% of ligand **6** (X = CH_2NEt_2)
in combination with 0.05 mol% of Ti(O*i*-Pr)$_4$:
R = Ph: 99% yield, ee = 81%
R = *p*-Tol: 99% yield, ee = 80%
R = *p*-MeOC$_6$H$_4$: 26% yield, ee = 88%
R = *o*-MeOC$_6$H$_4$: 57% yield, ee = 50%
R = *p*-ClC$_6$H$_4$: 99% yield, ee = 68%
R = 2-furyl: 92% yield, ee = 58%

Scheme 2.3. *In situ* generated unsymmetric titanium salen catalysts in asymmetric cyanation of aromatic and aliphatic aldehydes

asymmetric cyanosilylation reaction of aldehydes.[20] As shown in Scheme 2.3, this catalyst system exhibited, at 1 mol% of catalyst loading, an excellent activity with yields of 93–99%, albeit in combination with moderate enantioselectivities of 26–70% ee for final cyanohydrins derived from subsequent acidic hydrolysis of the corresponding cyanohydrin trimethylsilyl ethers. Remarkably, the reaction of trimethylsilyl cyanide with *p*-nitrobenzaldehyde under these conditions was complete within ten minutes. The authors have proposed that the appended phosphine oxide on the ligand played a role of Lewis base to activate trimethylsilyl cyanide, and cooperatively catalysed the cyanosilylation reaction with the central Ti(IV) ion as Lewis acid.

Almost at the same time, Lu and co-workers reported the synthesis of related unsymmetric salen ligands bearing a diethylamino group instead of diphenylphosphine oxide group, attached on the ligand framework.[21] Used at a remarkable catalyst loading of only 0.055 mol%, ligand 6 combined with only 0.05 mol% of titanium tetraisopropylate allowed a range of chiral cyanohydrins to be achieved after subsequent de-protection by treatment with aqueous HCl from the reaction of trimethylsilyl cyanide with the corresponding aldehydes (Scheme 2.3). Substituted benzaldehydes with an electron-donating group exhibited significantly low reactivity in comparison with benzaldehyde, which provided an almost quantitative yield in cyanohydrin. On the contrary, *p*-nitrobenzaldehyde and *p*-chlorobenzaldehyde showed excellent reactivity in the process with trimethylsilyl cyanide, and were completely transformed into the corresponding products within one hour. The chiral cyanohydrins were obtained in moderate to high enantioselectivities of up to 88% ee, as shown in Scheme 2.3. It must be noted that the lowest enantioselectivities were observed in the cases of using aliphatic aldehydes as substrates (26% ee for isobutyraldehyde).

In 2011, pyrrolidine-based chiral salen titanium complex 7, derived from L-tartaric acid, was applied by Sun *et al.* in combination with achiral *N*-oxide 8 as catalyst system to induce chirality in the addition of trimethylsilyl cyanide to aldehydes.[22] The corresponding cyanohydrin trimethylsilyl ethers were obtained in excellent yields of up to 98% and moderate to high enantioselectivities of up to 90% ee by using 1 mol% of this catalyst system. The best result (90% ee) was reached for benzaldehyde as substrate. As shown in Scheme 2.4, high enantiomeric excesses were obtained with electron-rich aromatic aldehydes, while electron-deficient aromatic aldehydes gave lower enantioselectivities. Moreover, it was shown that the steric properties of the aldehydes had a great effect on the enantioselectivity values of the products, since the *ortho*-substituted aromatic aldehydes gave lower enantioselectivities than the *meta*-substituted and *para*-substituted aromatic aldehydes. In terms of yield, *para*-substituted

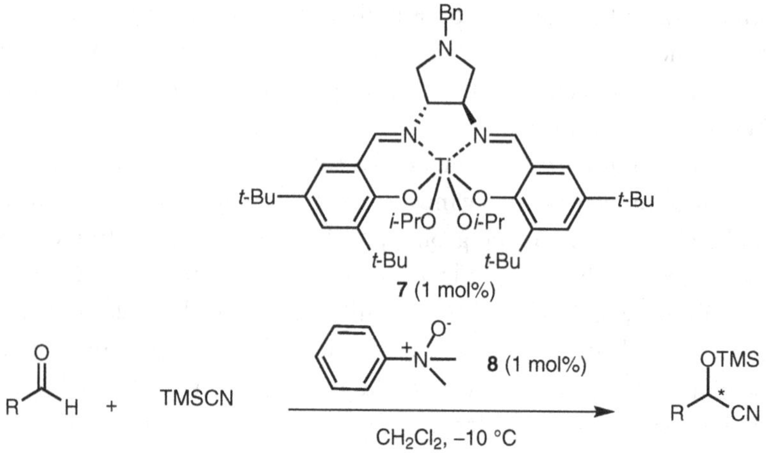

R = Ph: 96% yield, ee = 90%
R = *p*-Tol: 60% yield, ee = 86%
R = *m*-Tol: 98% yield, ee = 87%
R = *o*-Tol: 93% yield, ee = 67%
R = *p*-MeOC$_6$H$_4$: 56% yield, ee = 87%
R = *m*-MeOC$_6$H$_4$: 98% yield, ee = 84%
R = *o*-MeOC$_6$H$_4$: 85% yield, ee = 82%
R = *p*-ClC$_6$H$_4$: 95% yield, ee = 89%
R = *o*-ClC$_6$H$_4$: 96% yield, ee = 51%
R = 2,6-Cl$_2$C$_6$H$_3$: 97% yield, ee = 39%
R = 2-furyl: 80% yield, ee = 69%
R = (*E*)-PhCH=CH: 70% yield, ee = 61%
R = Me: 88% yield, ee = 48%
R = *n*-Hex: 90% yield, ee = 78%

Scheme 2.4. Pyrrolidine titanium salen catalyst in asymmetric cyanation of aliphatic and aromatic aldehydes

aromatic aldehydes were superior to *meta*- and *ortho*-substituted ones. The scope of the process was then extended to aliphatic aldehydes, giving high yields and moderate to good enantioselectivities of up to 78% ee. Cinnamaldehyde also proved to be compatible with the reaction, providing the corresponding protected cyanohydrin in 70% yield and 61% ee. The authors demonstrated that the presence of *N*-oxide **8** was indispensable to achieve asymmetric induction in the process.

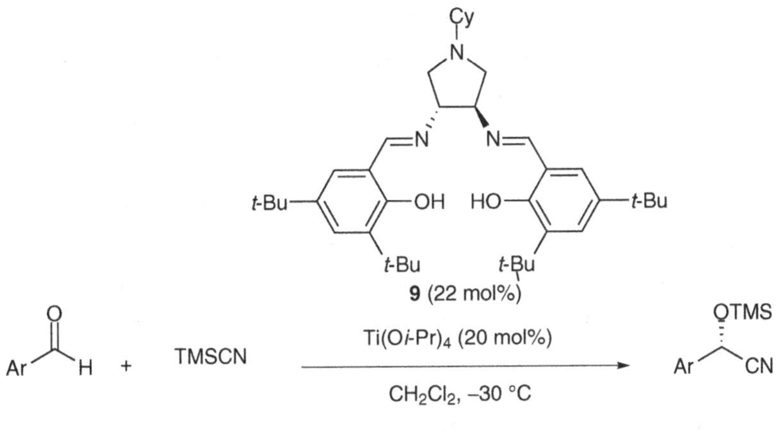

Ar = Ph: >99% conversion, ee = 82%
Ar = *p*-Tol: 95% conversion, ee = 86%
Ar = *p*-ClC$_6$H$_4$: 93% conversion, ee = 86%
Ar = *p*-BrC$_6$H$_4$: 88% conversion, ee = 81%
Ar = *o*-ClC$_6$H$_4$: 94% conversion, ee = 57%
Ar = *m*-Tol: 97% conversion, ee = 92%
Ar = *o*-Tol: 97% conversion, ee = 76%

Scheme 2.5. *In situ* generated pyrrolidine titanium salen catalyst in asymmetric cyanation of aryl aldehydes

Closely related pyrrolidine-based chiral salen ligand **9** was previously synthesised and evaluated by Serra *et al.* in the cyanation of aryl aldehydes with trimethylsilyl cyanide.[23] The active titanium catalyst was generated *in situ*, by reacting this chiral ligand with titanium tetraisopropylate, and was found to be very active, since benzaldehyde provided the corresponding *O*-trimethylsilylmandelonitrile in complete conversion, along with enantioselectivity of 82% ee. The scope of this methodology was later extended to a range of aryl aldehydes, allowing the corresponding cyanohydrin trimethylsilyl ethers to be achieved in good to high enantioselectivities of up to 92% ee, as shown in Scheme 2.5.[24] Notably, no direct relationship was observed between the electronic properties of the substrate and the enantioselectivity of the products. In this work, the different activity and selectivity of various pyrrolidine-based chiral salen catalysts studied were found to be dependent on the *N*-substituent of the pyrrolidine ligand.

Therefore, ligand *N*-cyclohexylpyrrolidine **9** gave better enantioselectivities than the corresponding *N*-benzylpyrrolidine catalyst (88 and 82% ee, respectively, for benzaldehyde as substrate) while *N*-phenylpyrrolidine catalyst gave a low enantioselectivity (15% ee for benzaldehyde as substrate).

A feature of all the most effective catalysts for asymmetric cyanohydrin synthesis is the need for cooperative catalysis by two functional groups that activate the aldehyde and cyanide, respectively. This also preorganizes the two reactants into the correct orientation, to accelerate the reaction and enhance the enantioselectivity. This is achieved by using bimetallic titanium chiral salen complexes, for which the bimetallic nature of the complex is critical as both titanium ions have a role in the catalysis, activating the aldehyde and cyanide, respectively.[25] Highly efficient green conditions have been reported by North and Omedes-Pujol by using chiral dititanium catalyst **10** in the asymmetric synthesis of cyanohydrin trimethylsilyl ethers.[26] As shown in Scheme 2.6, a catalyst loading as low as 0.1 mol% was sufficient to allow the addition of trimethylsilyl cyanide to a range of aliphatic and aromatic aldehydes to be achieved at room temperature in generally high yields of up to 98% and good to excellent enantioselectivities of up to 97% ee. With the aim of rendering the reaction conditions even more green, the authors carried out the reactions in propylene carbonate instead of dichloromethane as solvent, however, in each case of substrates studied, significantly lowered enantioselectivities were observed (10–61% ee), and for aromatic aldehydes the conversion was also reduced substantially (16–47%).

To maximise the amount of bimetallic complex present in solution, ligands in which two salen ligands were covalently linked together were designed and synthesised by Ding *et al.* in 2010.[27] These ligands were then complexed to titanium and the resulting complexes used at a remarkable catalyst loading of only 0.01 mol% to promote the addition of trimethylsilyl cyanide to aldehydes. Among a range of bis(salen)ligands bridged by spacers with diverse length and spatial orientations, ligand **11**, having a *cis*-5-norbornene-*endo*-2,3-dicarboxylate bridge, proved to be the most efficient in terms of both activity and enantioselectivity. As shown in Scheme 2.7, impressive results were achieved, since a range of chiral cyanohydrin trimethylsilyl ethers were obtained in both excellent yields

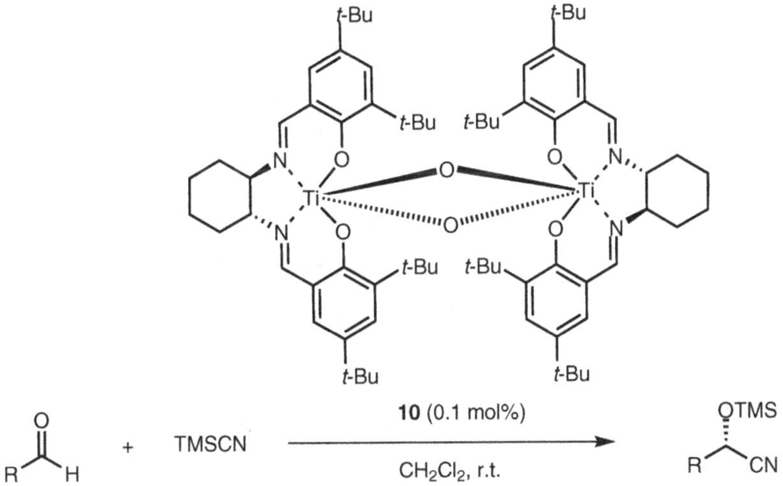

R = Ph: 95% conversion, ee = 80%
R = *p*-FC$_6$H$_4$: 40% conversion, ee = 78%
R = *p*-ClC$_6$H$_4$: 98% conversion, ee = 83%
R = *o*-Tol: 76% conversion, ee = 76%
R = *m*-Tol: 95% conversion, ee = 97%
R = *p*-Tol: 82% conversion, ee = 69%
R = *n*-Oct: 71% conversion, ee = 73%
R = *t*-Bu: 93% conversion, ee = 47%

Scheme 2.6. Bimetallic titanium salen catalyst in asymmetric cyanation of aliphatic and aromatic aldehydes

and enantioselectivities of up to 99% and 97% ee, respectively. It should be noted that the best results were reached in the cases of aromatic, heteroaromatic, as well as α,β-unsaturated aldehydes, while aliphatic aldehydes provided moderate to good enantioselectivities of up to 72% ee in combination with excellent yields. Remarkably, this catalyst was shown extremely efficient since in the addition of trimethylsilyl cyanide to benzaldehyde for example, the catalyst loading could be additionally reduced to as low as 0.0005 mol% at a higher substrate concentration (3M) to afford the corresponding product in 86% yield and 95% ee.

In 2011, Sun *et al.* reported the synthesis of a tethered bimetallic titanium pyrrolidine-based chiral salen complex **12**, which was then further

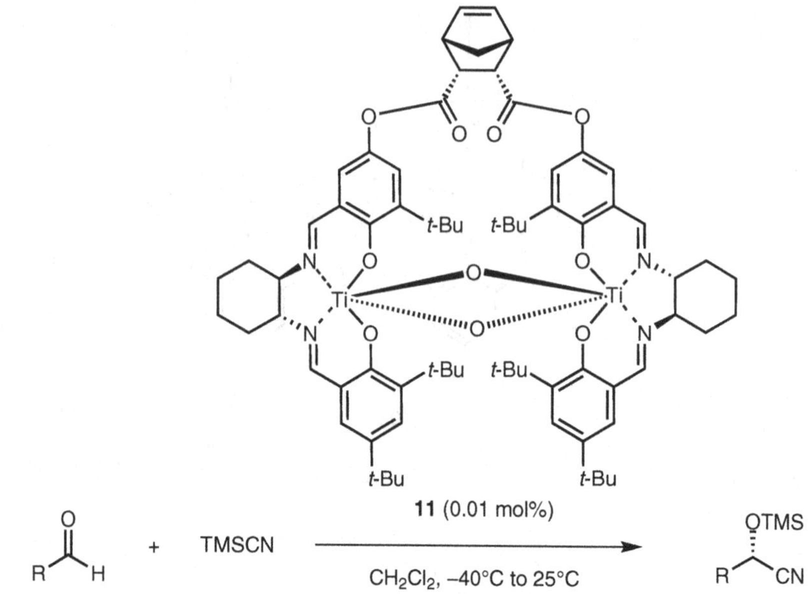

$$\text{R}\overset{\text{O}}{\underset{\text{H}}{\diagdown}} + \text{TMSCN} \xrightarrow[\text{CH}_2\text{Cl}_2,\ -40°\text{C to }25°\text{C}]{\textbf{11}\ (0.01\ \text{mol\%})} \text{R}\overset{\text{OTMS}}{\underset{\text{CN}}{\diagdown}}$$

R = Ph: 99% yield, ee = 97%
R = p-MeOC$_6$H$_4$: 98% yield, ee = 97%
R = p-Tol: 99% yield, ee = 90%
R = p-ClC$_6$H$_4$: 99% yield, ee = 93%
R = p-BrC$_6$H$_4$: 99% yield, ee = 93%
R = p-NO$_2$C$_6$H$_4$: 99% yield, ee = 90%
R = p-FC$_6$H$_4$: 99% yield, ee = 96%
R = m-MeOC$_6$H$_4$: 99% yield, ee = 97%
R = o-MeOC$_6$H$_4$: 99% yield, ee = 92%
R = m-ClC$_6$H$_4$: 99% yield, ee = 96%
R = m-PhOC$_6$H$_4$: 87% yield, ee = 92%
R = (E)-PhCH=CH: 95% yield, ee = 96%
R = (E)-MeCH=CH: 93% yield, ee = 82%
R = CH$_2$Bn: 89% yield, ee = 72%
R = 2-Naph: 99% yield, ee = 96%
R = 1-Naph: 99% yield, ee = 92%
R = 2-furyl: 97% yield, ee = 95%
R = i-Pr: 96% yield, ee = 64%
R = Cy: 99% yield, ee = 67%

Scheme 2.7. Tethered intramolecular bimetallic titanium salen catalyst in asymmetric cyanation of aliphatic and aromatic aldehydes investigated in the asymmetric cyanohydrin synthesis.[28] Used at a catalyst loading of 0.1 mol%, this catalyst provided a range of chiral cyanohydrin trimethylsilyl ethers in excellent yields and good enantioselectivities of up to 86% ee, as shown in Scheme 2.8. The results indicated that the introduction of an electron-withdrawing group on the benzene ring of the aryl aldehyde disfavored the stereocontrol of the reaction. On the other hand, the electron-rich aromatic aldehydes were transformed into the corresponding chiral products in similar or slightly lower enantioselectivities than that of benzaldehyde. The steric properties of the aromatic aldehydes were found to have great effect on the ee values of the products in this reaction. For example, the *ortho*-substituted aromatic aldehydes gave lower enantioselectivities and yields than the *meta*-substituted and *para*-substituted aromatic aldehydes. Aliphatic aldehydes, such as heptaldehyde, also reacted smoothly in a 64% ee.

2.2.1.2. BINOL-derived ligands

Binol-derived chiral ligands have been also successfully investigated for their ability to induce chirality in titanium-catalysed addition of trimethylsilyl cyanide to aldehydes. As an example, Gau and You reported the synthesis of chiral non-C_2-symmetric BINOL ligands monosubstituted in position 3 with various nitrogen-containing heterocycles.[29] Among imidazolyl, methyleneimidazolyl, pyrrole, pyrazole, and triazole substituents, the imidazolyl unit proved to be the most efficient. As shown in Scheme 2.9, the reaction of aliphatic as well as aromatic aldehydes with trimethylsilyl cyanide, promoted by 2–10 mol% of a chiral titanium catalyst *in situ* generated from ligand **13** and titanium tetraisopropylate afforded, after subsequent acidic hydrolysis, the corresponding chiral cyanohydrins in remarkable yields and enantioselectivities of up to 99% and 98% ee, respectively. Indeed, a number of aromatic aldehydes with sterically hindered, electron-poor, electron-neutral, or electron-rich substituents all provided general excellent yields, and enantioselectivities in the range of 91 to 98% ee. Similarly, a variety of aliphatic aldehydes, including linear and branched ones, provided excellent results with enantioselectivities of up to 98% ee. Interestingly, the authors found that the use of freshly distilled trimethylsilyl cyanide dried over calcium hydride

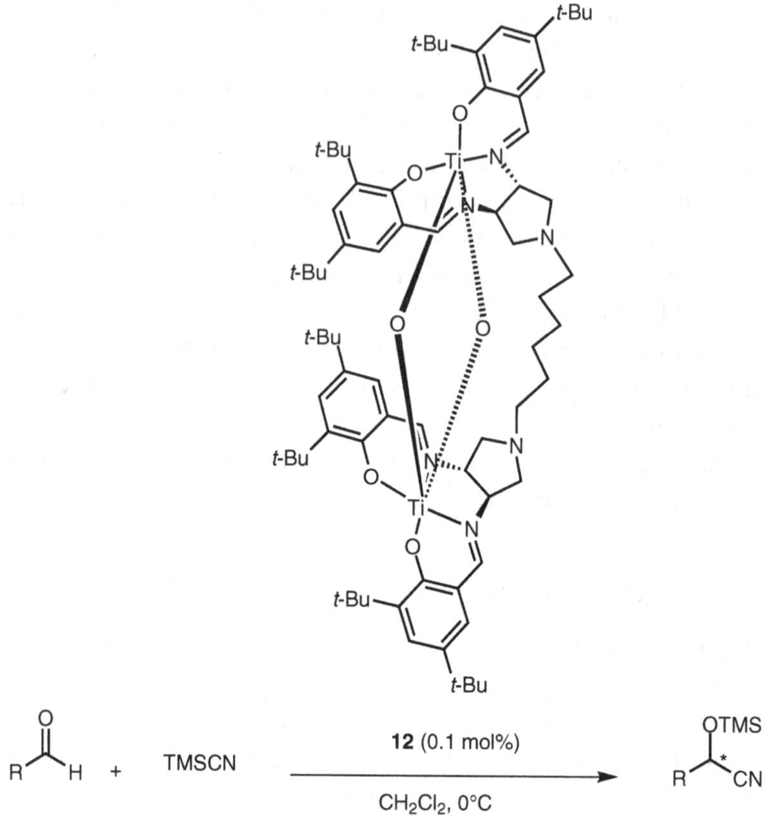

R = Ph: 98% yield, ee = 82%
R = *p*-Tol: 97% yield, ee = 75%
R = *m*-Tol: 90% yield, ee = 86%
R = *o*-Tol: 90% yield, ee = 66%
R = *p*-MeOC$_6$H$_4$: 73% yield, ee = 75%
R = *m*-MeOC$_6$H$_4$: 92% yield, ee = 76%
R = *o*-MeOC$_6$H$_4$: 94% yield, ee = 85%
R = *p*-ClC$_6$H$_4$: 98% yield, ee = 81%
R = *o*-ClC$_6$H$_4$: 95% yield, ee = 57%
R = 2-furyl: 93% yield, ee = 70%
R = (*E*)-PhCH=CH: 66% yield, ee = 12%
R = *n*-Hex: 90% yield, ee = 64%

Scheme 2.8. Bimetallic titanium pyrrolidine salen catalyst in asymmetric cyanation of aliphatic and aromatic aldehydes

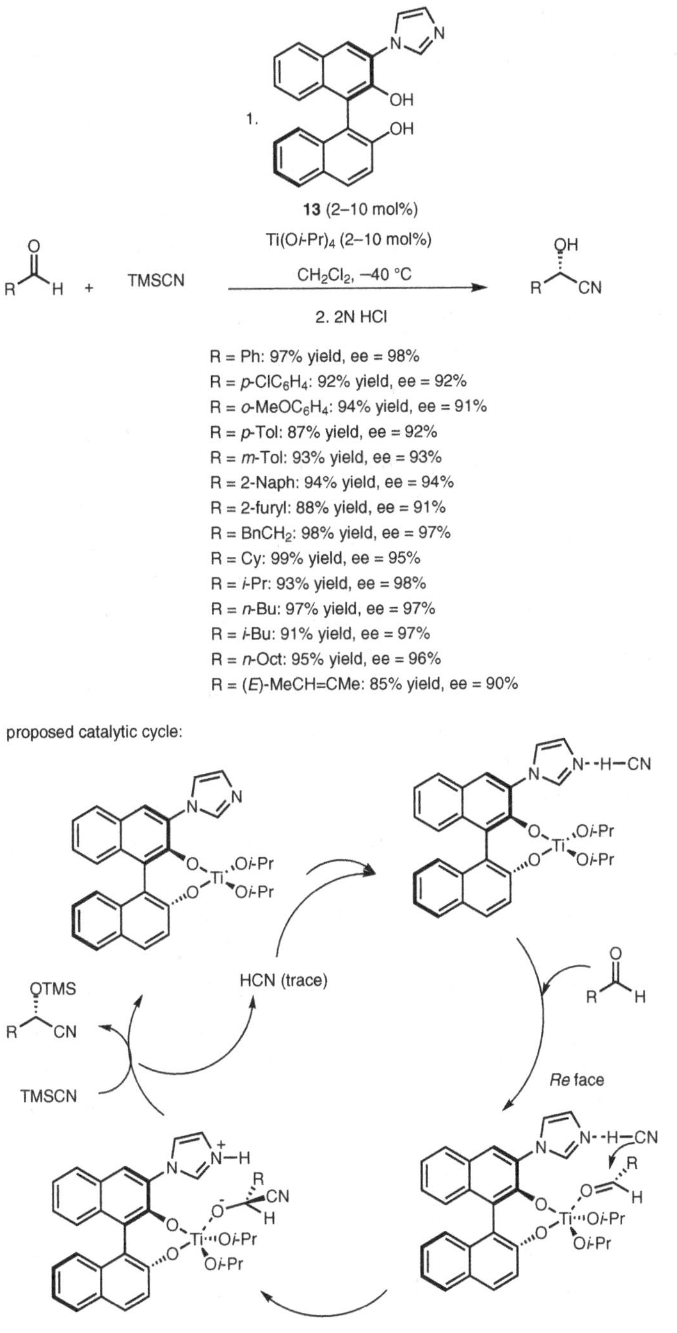
1.
13 (2–10 mol%)
Ti(O*i*-Pr)4 (2–10 mol%)
CH2Cl2, –40 °C
2. 2N HCl
TMSCN
R = Ph: 97% yield, ee = 98%
R = *p*-ClC6H4: 92% yield, ee = 92%
R = *o*-MeOC6H4: 94% yield, ee = 91%
R = *p*-Tol: 87% yield, ee = 92%
R = *m*-Tol: 93% yield, ee = 93%
R = 2-Naph: 94% yield, ee = 94%
R = 2-furyl: 88% yield, ee = 91%
R = BnCH2: 98% yield, ee = 97%
R = Cy: 99% yield, ee = 95%
R = *i*-Pr: 93% yield, ee = 98%
R = *n*-Bu: 97% yield, ee = 97%
R = *i*-Bu: 91% yield, ee = 97%
R = *n*-Oct: 95% yield, ee = 96%
R = (*E*)-MeCH=CMe: 85% yield, ee = 90%
proposed catalytic cycle:
HCN (trace)
TMSCN
Re face

Scheme 2.9. *In situ* generated imidazole BINOL titanium catalyst in asymmetric cyanation of aliphatic and aromatic aldehydes gave a low enantioselectivity, and only a moderate yield of product, compared with direct use of the commercial reagent. In this context, they proposed that hydrogen cyanide was the actual reactive nucleophile in the process, as shown in the proposed catalytic cycle depicted in Scheme 2.9, in which a complex simultaneously bound to the two reaction partners should position the activated aldehyde at a perpendicular site close to the internal base unit, while hydrogen cyanide, which interacted with the 3-N position of the imidazolyl moiety, could transfer cyanide to the aldehyde. This novel general methodology constituted one of the most practical and effective strategies to reach chiral cyanohydrins.

C_2-symmetric BINOL ligands have also been investigated. In 2011, Tang and Zhang reported the use of a chiral titanium complex *in situ* generated from (R)-3,3'-bis(diphenylphosphinoyl)-BINOL 14 and titanium tetraisopropylate, as a novel bifunctional catalyst in trimethylsilyl cyanation of aldehydes.[30] The authors envisioned that the titanium worked as a Lewis acid to activate the carbonyl group of the aldehyde, and the oxygen atom of the phosphine oxide functioned as a Lewis base to activate the silylated nucleophile. By using 10 mol% of ligand 14, cyanohydrins derived from aromatic aldehydes were produced after subsequent acidic treatment in high yields, but with moderate enantioselectivities of up to 51% ee, as shown in Scheme 2.10.

Earlier, Belokon *et al.* described the synthesis of chiral hexadentate ligands based on (R)-BINOL, and chiral amino alcohols to be investigated in asymmetric addition of trimethylsilyl cyanide to benzaldehyde.[31] The optimal ligand 15 was based on (S)-leucinol, and allowed the corresponding cyanohydrin trimethylsilyl ether to be achieved in quantitative yield and enantioselectivity of 86% ee, as shown in Scheme 2.11. The authors have proposed that this ligand formed in the presence of titanium tetraisopropylate the tetranuclear active catalyst 16. (This work could also be situated in the next paragraph dealing with Schiff base ligands.)

2.2.1.3. *Schiff base ligands*

After the pioneering work reported by Hayashi and Oguni[32] on the enantioselective cyanosilylation of aldehydes using chiral Schiff base ligands

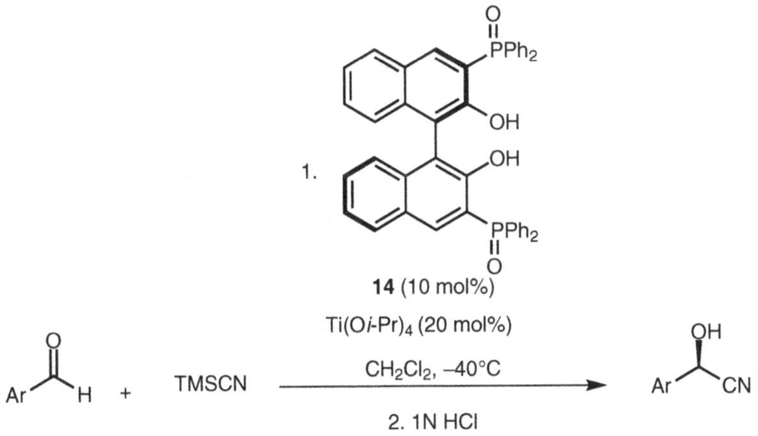

Ar = Ph: 89% yield, ee = 51%
Ar = *p*-ClC$_6$H$_4$: 88% yield, ee = 22%
Ar = *o*-MeOC$_6$H$_4$: 93% yield, ee = 11%
Ar = *p*-MeOC$_6$H$_4$: 85% yield, ee = 19%
Ar = *o*-Tol: 90% yield, ee = 27%
Ar = *m*-Tol: 91% yield, ee = 35%
Ar = 2-Naph: 88% yield, ee = 21%

Scheme 2.10. *In situ* generated (*R*)-3,3′-bis(diphenylphosphinoyl)-BINOL titanium catalyst in asymmetric cyanation of aromatic aldehydes

and titanium tetraisopropylate, various chiral catalytic systems based on Schiff bases have been developed. Extensive studies of these systems, employing a variety of Schiff bases derived from different chiral amino alcohols or diamine compounds, revealed that the enantioselective cyanosilylation of aldehydes was highly dependent on the type of the Schiff base.[33] In 2011, Jaworska *et al.* reported the synthesis of chiral Schiff bases derived from α-pinene[34] Among several ligands bearing various substituents on the phenolic ring, ligand **17** was selected to be the most efficient, providing at 20 mol% of catalyst loading, and cyanohydrins of aliphatic and aromatic aldehydes in moderate enantioselectivities of up to 62% ee (Scheme 2.12). The use of this *in situ* generated catalyst gave much better results when (*E*)-cinnamaldehyde was employed as substrate since an excellent enantioselectivity of >99% ee was achieved.

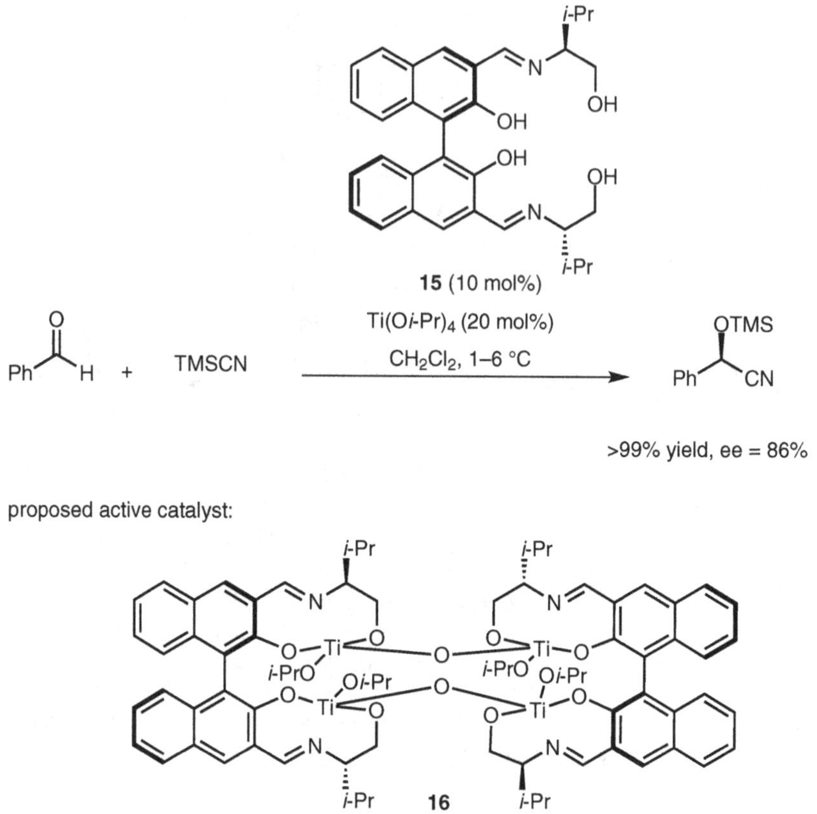

>99% yield, ee = 86%

proposed active catalyst:

Scheme 2.11. *In situ* generated (*S*)-leucinol-derived hexadentate BINOL tetrametallic titanium catalyst in asymmetric cyanation of benzaldehyde

In 2014, Bosiak *et al.* investigated a new class of chiral camphor-derived Schiff bases in comparable reactions.[35] Five tridentate Schiff bases were prepared from (1*R*,2*S*,3*R*,4*S*)-3-amino-1,7,7-trimethylbicyclo[2.2.1] heptan-2-ol and salicylaldehydes. Among them, the *in situ* generated titanium complex of the Schiff base **18**, derived from 2-hydroxy-3-isopropylbenzaldehyde, was the most selective catalyst for cyanosilylation of aliphatic, aromatic, as well as heteroaromatic aldehydes. As shown in Scheme 2.13, the corresponding cyanohydrin trimethylsilyl ethers were obtained in low to excellent enantioselectivities of up to >99% ee in combination with low to high yields. This best enantioselectivity of >99% ee

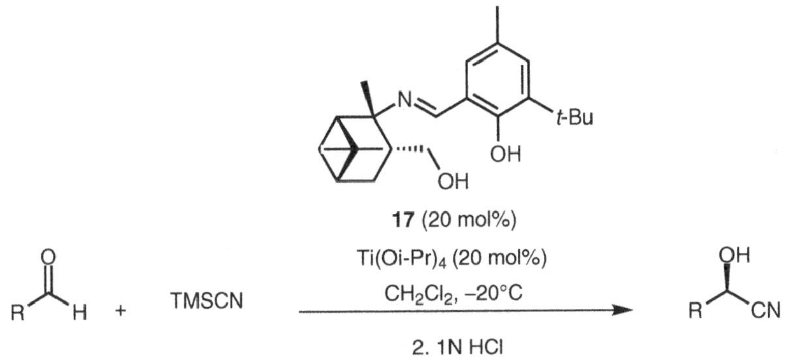

17 (20 mol%)

Ti(Oi-Pr)$_4$ (20 mol%)

CH$_2$Cl$_2$, –20°C

2. 1N HCl

R = Ph: 73% yield, ee = 60%
R = *p*-MeOC$_6$H$_4$: 89% yield, ee = 62%
R = *m*-MeOC$_6$H$_4$: 53% yield, ee = 62%
R = *m*-ClC$_6$H$_4$: 94% yield, ee = 58%
R = Cy: 75% yield, ee = 45%
R = *n*-Pent: 45% yield, ee = 54%
R = (*E*)-PhC=CH: 83% yield, ee >99%
R = 2-furyl: 40% yield, ee = 42%

Scheme 2.12. *In situ* generated α-pinene-derived Schiff base titanium catalyst in asymmetric cyanation of aliphatic and aromatic aldehydes

was reached in the case of (*E*)-cinnamaldehyde as substrate. High enantioselectivities were also obtained for the reaction of furan-2-carbaldehyde (91% ee), and *para*- and *meta*-methoxybenzaldehydes (98 and 97% ee, respectively). Moreover, aliphatic aldehydes provided enantioselectivities of up to 98% ee albeit with generally lower yields.

2.2.1.4. *Other ligands*

Polymer-supported metal catalysts have advantages for use in enantioselective synthesis since the metal complexes can be recovered and reused. In 2008, Moberg *et al.* described the synthesis of a chiral salen polymeric ligand containing a carboxylic acid group attached to macroporous monosized polystyrene-divinylbenzene beads, grafted with (4-hydroxybutyl) vinyl ether groups.[36] The titanium complex of the beads **19** was used in enantioselective cyanation of benzaldehyde with trimethylsilyl cyanide

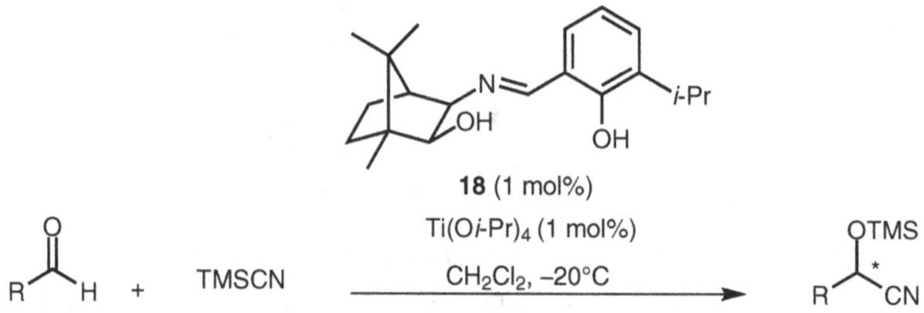

18 (1 mol%)

Ti(O*i*-Pr)$_4$ (1 mol%)

CH$_2$Cl$_2$, –20°C

R = Ph: 86% yield, ee = 81%
R = *o*-MeOC$_6$H$_4$: 17% yield, ee = 13%
R = *p*-MeOC$_6$H$_4$: 91% yield, ee = 98%
R = *m*-MeOC$_6$H$_4$: 5% yield, ee = 97%
R = *p*-Tol: 7% yield, ee = 80%
R = *o*-ClC$_6$H$_4$: 72% yield, ee = 59%
R = *m*-ClC$_6$H$_4$: 87% yield, ee = 65%
R = *p*-ClC$_6$H$_4$: 55% yield, ee = 62%
R = *o*-FC$_6$H$_4$: 24% yield, ee = 15%
R = 2-furyl: 50% yield, ee = 91%
R = Cy: 81% yield, ee = 88%
R = *n*-Pr: 10% yield, ee = 98%
R = (*E*)-PhC=CH: 93% yield, ee >99%
R = *n*-Pent: 49% yield, ee = 49%

Scheme 2.13. *In situ* generated camphor-derived Schiff base titanium catalyst in asymmetric cyanation of aliphatic and aromatic aldehydes

using a microreactor. The use of flow through microreactor technology offers advantages over conventional laboratory techniques. Indeed, reactions performed in a continuous flow often provide higher yields and selectivities than those performed in batch operations. Moreover, the lower consumption of reagents and consumables and decreased production of waste resulting from miniaturisation are of obvious interest for economic, safety, and environmental reasons. When using 6 mol% of titanium catalyst **19** at a flow rate of 0.8 μL/min, the process afforded the corresponding cyanohydrin trimethylsilyl ether in 92% conversion combined with a moderate enantioselectivity of 72% ee, as shown in Scheme 2.14. The catalyst could be reused and the system was stable with the selectivity barely not effected by variations in the flow rate.

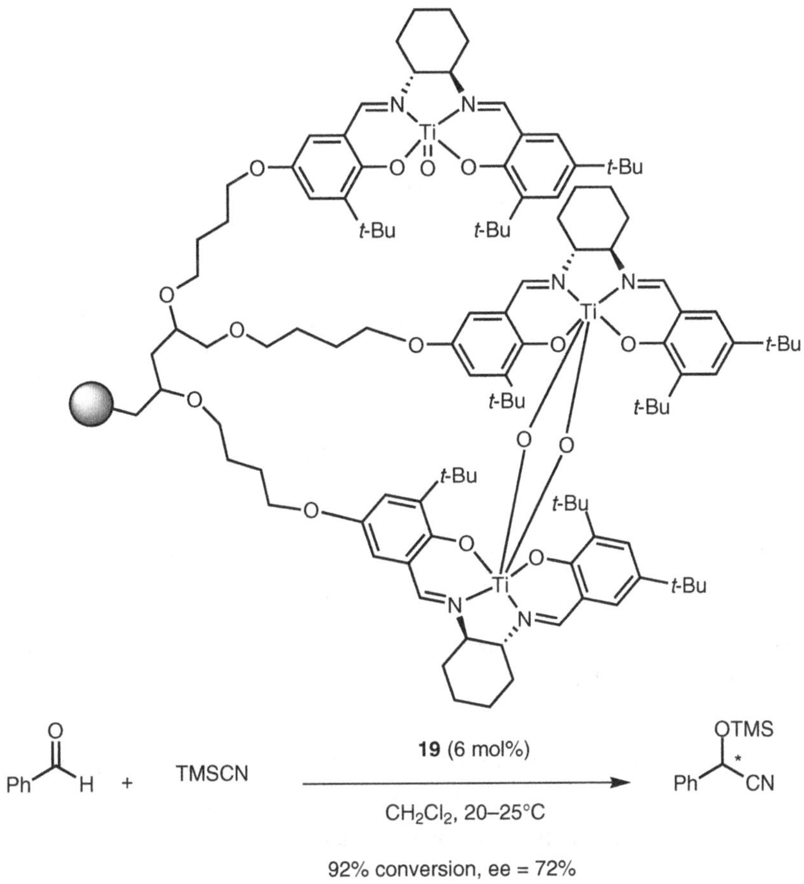

Scheme 2.14. Polymeric titanium salen catalyst in asymmetric cyanation of benzaldehyde

Later, Punniyamurthy and Sakthivel reported the synthesis of polymeric linear titanium catalyst **20**, having salen and 1,4-dioctyloxybenzene as alternate segments.[37] This catalyst was applied to the asymmetric cyanation of a range of aldehydes at 1 mol% of catalyst loading in chloroform, providing the corresponding chiral cyanohydrin trimethylsilyl ethers in general excellent yields combined with moderate to good enantioselectivities of up to 88% ee, as shown in Scheme 2.15. The process tolerated aliphatic aldehydes and presented the advantages of simplified product isolation, easier recovery, and recyclability of the polymer catalyst. (Note

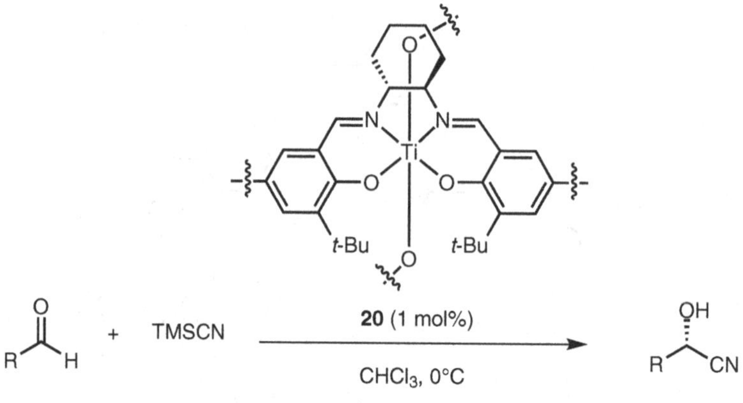

R = *p*-MeOC$_6$H$_4$: 97% yield, ee = 78%
R = *o*-MeOC$_6$H$_4$: 95% yield, ee = 88%
R = *o*-EtOC$_6$H$_4$: 91% yield, ee = 78%
R = *o*-ClOC$_6$H$_4$: 97% yield, ee = 60%
R = *m*-BrC$_6$H$_4$: 90% yield, ee = 75%
R = *p*-Tol: 97% yield, ee = 68%
R = 2-Naph: 90% yield, ee = 74%
R = *i*-Pr: 90% yield, ee = 66%

Scheme 2.15. Polymeric linear titanium salen catalyst in asymmetric cyanation of aliphatic and aromatic aldehydes

that although the two works included in this section involved polymeric catalysts derived from salen, it was decided to situate them here, rather than in the section dealing with salen ligands related to their polymeric nature.)

2.2.2. Cyanations of Aldehydes with Other Sources of Cyanide

2.2.2.1. Salen ligands

Early works in this field were carried out with hydrogen cyanide as the cyanide source, but the volatility and extreme toxicity of this reagent cause obvious difficulties and, in this context, numerous alternative cyanide-based reagents have been developed. So far, the most commonly used cyanide source in cyanohydrin formation is trimethylsilyl cyanide which was first independently reported in 1973 by the groups of Evans[38] and Lidy.[39] Acyl cyanides[40] can also be used as efficient cyanating agents, and have the advantage of producing *O*-protected cyanohydrins which do not revert to carbonyl compounds; thus, the cyanohydrin formation becomes irreversible, and even substrates for which the equilibrium between the carbonyl compound and the cyanohydrin is unfavorable can be prepared in this way. However, the use of acyl cyanides as cyanide sources often requires high catalyst loadings and very low temperature, and additionally, recyclable catalysts remain extremely rare.

A remarkable example was recently reported by Khan *et al.* with the use of only 0.5 mol% of an *in situ* generated chiral macrocyclic titanium salen complex derived from ligand **21** in the asymmetric cyanoethoxy carbonylation of aldehydes.[41] This catalyst, with a flexible polyether linkage, demonstrated excellent performance with ethyl cyanoformate as the cyanide source, providing the corresponding *O*-protected cyanohydrins of aromatic aldehydes, irrespective of electron-donating or -withdrawing groups on the phenyl ring attached to the aromatic aldehydes. It also produced aliphatic aldehydes in generally excellent yields of up to 99% and very high enantioselectivities of up to 95% ee, as shown in Scheme 2.16. The process used DMAP as a Lewis base to activate ethyl cyanoformate. This catalyst was shown to retain its performance at a multigram level and was conveniently recycled for a number of times. On the basis of kinetic studies, the authors proposed the mechanism depicted in Scheme 2.16 in which titanium could act as a Lewis acid to activate the carbonyl group and DMAP as a Lewis base to activate ethyl cyanoformate via the formation of intermediate species **C**. As the kinetics of the reaction were independent of the substrate concentration, it was assumed that the interaction of the substrate with the catalyst was very fast, proceeding via intermediate **A**, and the overall rate of the reaction did not depend on the substrate concentration. As the reaction rate was first-order-dependent on catalyst and ethyl cyanoformate concentrations, intermediate **B** could have been involved in the rate-determining step. The utility of this methodology was demonstrated in the synthesis of the pharmaceutically important chiral drugs (*R*)-proethalol and (*R*)-phenylephedrine.

Among other cyanide-based reagents, the simplest ones are inexpensive metal cyanide salts such as potassium cyanide (KCN). However, they are just precursors of volatile and extremely toxic hydrogen cyanide. In 2014, in order to find an economical way to (*S*)-2-hydroxy-(3-phenoxyphenyl) acetonitrile derivatives, which are components of pyrethroid insecticides,

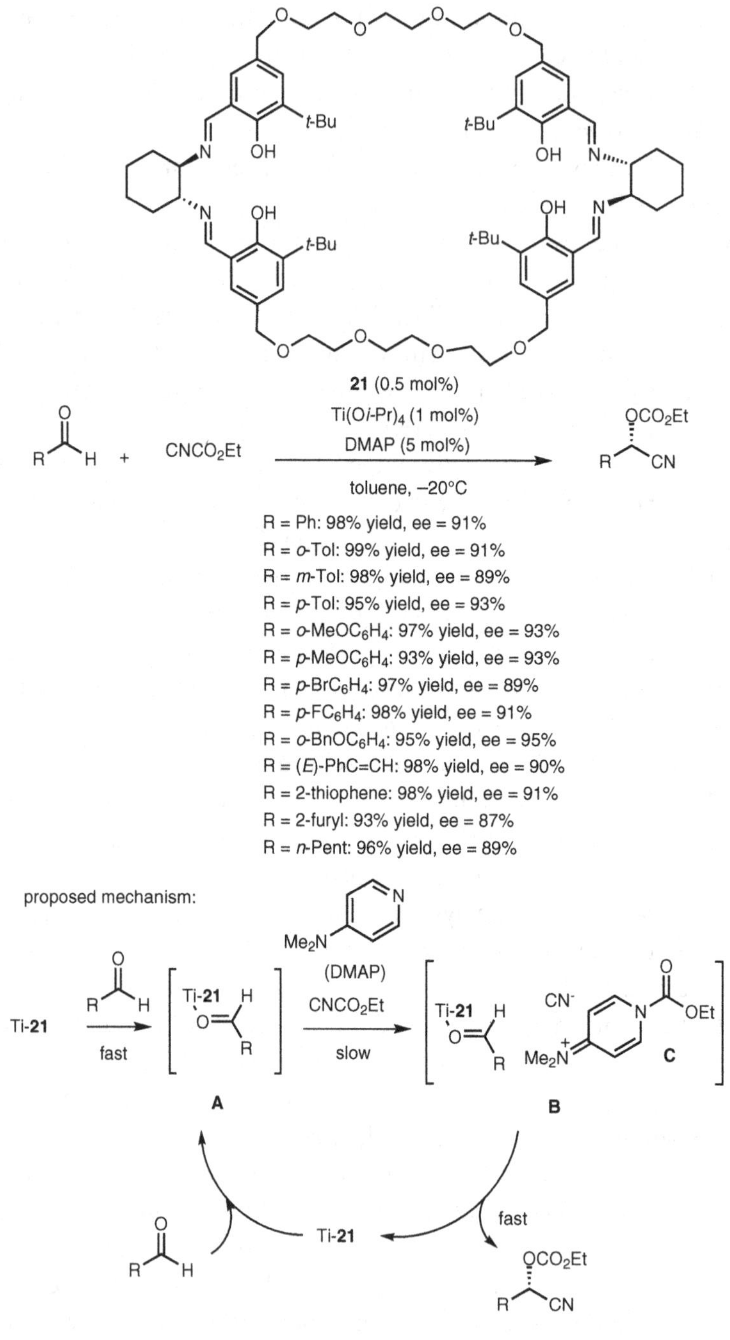
t-Bu
OH
N
21 (0.5 mol%)
Ti(Oi-Pr)4 (1 mol%)
DMAP (5 mol%)
toluene, −20°C
CNCO2Et
OCO2Et
CN
R = Ph: 98% yield, ee = 91%
R = o-Tol: 99% yield, ee = 91%
R = m-Tol: 98% yield, ee = 89%
R = p-Tol: 95% yield, ee = 93%
R = o-MeOC6H4: 97% yield, ee = 93%
R = p-MeOC6H4: 93% yield, ee = 93%
R = p-BrC6H4: 97% yield, ee = 89%
R = p-FC6H4: 98% yield, ee = 91%
R = o-BnOC6H4: 95% yield, ee = 95%
R = (E)-PhC=CH: 98% yield, ee = 90%
R = 2-thiophene: 98% yield, ee = 91%
R = 2-furyl: 93% yield, ee = 87%
R = n-Pent: 96% yield, ee = 89%
proposed mechanism:
Me2N
(DMAP)
Ti-21
fast
slow
A
B
C
OEt
CN-

Scheme 2.16. *In situ* generated macrocyclic titanium salen catalyst in asymmetric cyanation of aliphatic and aromatic aldehydes

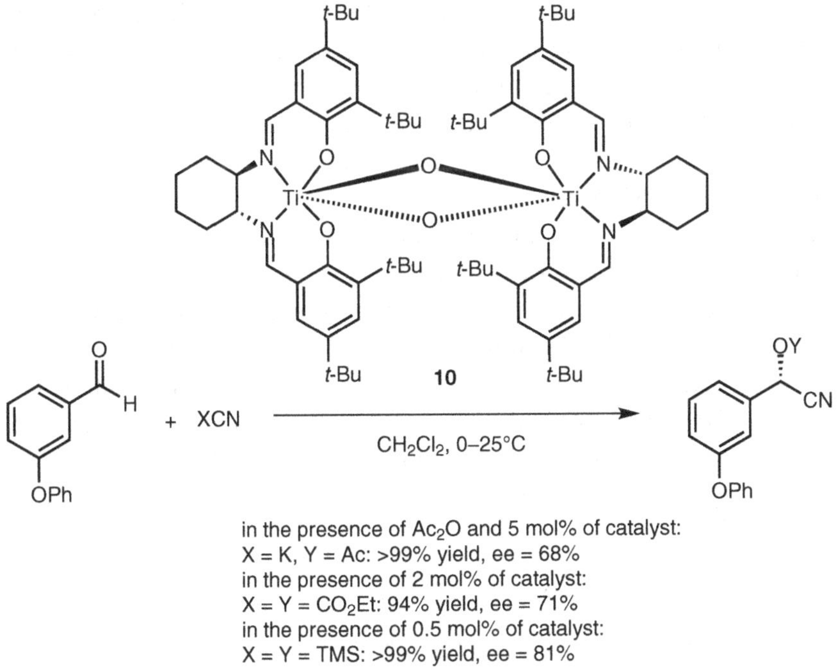

Scheme 2.17. Bimetallic titanium salen catalyst in asymmetric cyanation of *m*-phenoxybenzaldehyde

North and Watson applied this cyanide source (KCN) in the asymmetric cyanation of 3-phenoxybenzaldehyde, catalysed by chiral bimetallic titanium salen catalyst **10**.[42] In the presence of 5 mol% of this catalyst and acetic anhydride, the corresponding cyanohydrin acetate was produced in quantitative yield and moderate enantioselectivity of 68% ee, as shown in Scheme 2.17. The saving in cyanide source cost in this reaction was offset by the need to use 5 mol% of catalyst **10**. In this context, the authors performed the reaction with ethyl cyanoformate as another cyanide source, with the aim of using a lower catalyst loading. It was shown that the process catalysed by only 2 mol% of the same catalyst afforded in the presence of this cyanide source the corresponding 2-ethoxycarbonyl (S)-2-hydroxy-(3-phenoxyphenyl)acetonitrile in 94% yield and slightly better enantioselectivity of 71% ee, as shown in Scheme 2.17. In addition, the authors also investigated trimethylsilyl cyanide as cyanide source, which finally provided the best result since a quantitative yield of the corresponding cyanohydrin trimethylsilyl ether was achieved with a better enantioselectivity of 81% ee when performing the reaction at only 0.5 mol% of catalyst loading (Scheme 2.17). The authors concluded that there was a little cost difference between the production of cyanohydrin trimethylsilyl ether from using trimethylsilyl cyanide and cyanohydrin acetate from using KCN. The synthesis of this acetate was marginally less expensive due to the less expensive cyanide source KCN used, though this was largely offset by the need to use more catalyst to prepare the cyanohydrin acetate than the cyanohydrin silyl ether. In addition, the synthesis of the cyanohydrin acetate was a heterogeneous process, whereas the synthesis of the cyanohydrin silyl ether was a homogeneous reaction. Finally, overall the optimal route for the asymmetric synthesis of pyrethroids entailed the synthesis of the cyanohydrin trimethylsilyl ether.

2.2.2.2. *Other ligands*

Ethyl cyanoformate was also employed as cyanide source by Li *et al.* in the enantioselective cyanoformylation of a range of aromatic aldehydes catalysed by *in situ* generated Schiff base titanium catalysts.[43] The precursor Schiff base ligands were prepared by condensation of cinchona alkaloid-derived 9-amino compounds with salicylaldehyde derivatives. Among them, the optimal ligand was shown to be bifunctional ligand **22**, used at 5 mol% of catalyst loading in ethanol at 0°C. Under these mild reaction conditions, a variety of chiral cyanohydrin ethyl carbonates could be synthesised in general high yields of up to 96%, associated with good enantioselectivities of up to 85% ee, as shown in Scheme 2.18. It was found that the aromatic aldehydes with electron-donating groups gave generally similar results than benzaldehyde. Among them, 2-methoxybenzaldehyde provided the best result (85% ee). On the other hand, the aromatic aldehydes with electron-withdrawing groups gave lower enantioselectivities. The substrate scope was extended to heteroaromatic aldehydes which gave moderate enantioselectivities (73% ee for furfural and 61% ee for pyridylaldehyde) and also to aliphatic aldehydes which provided moderate enantioselectivities (67% ee for 3-phenylpropionaldehyde). The authors have proposed the transition state depicted in Scheme 2.18, in which the

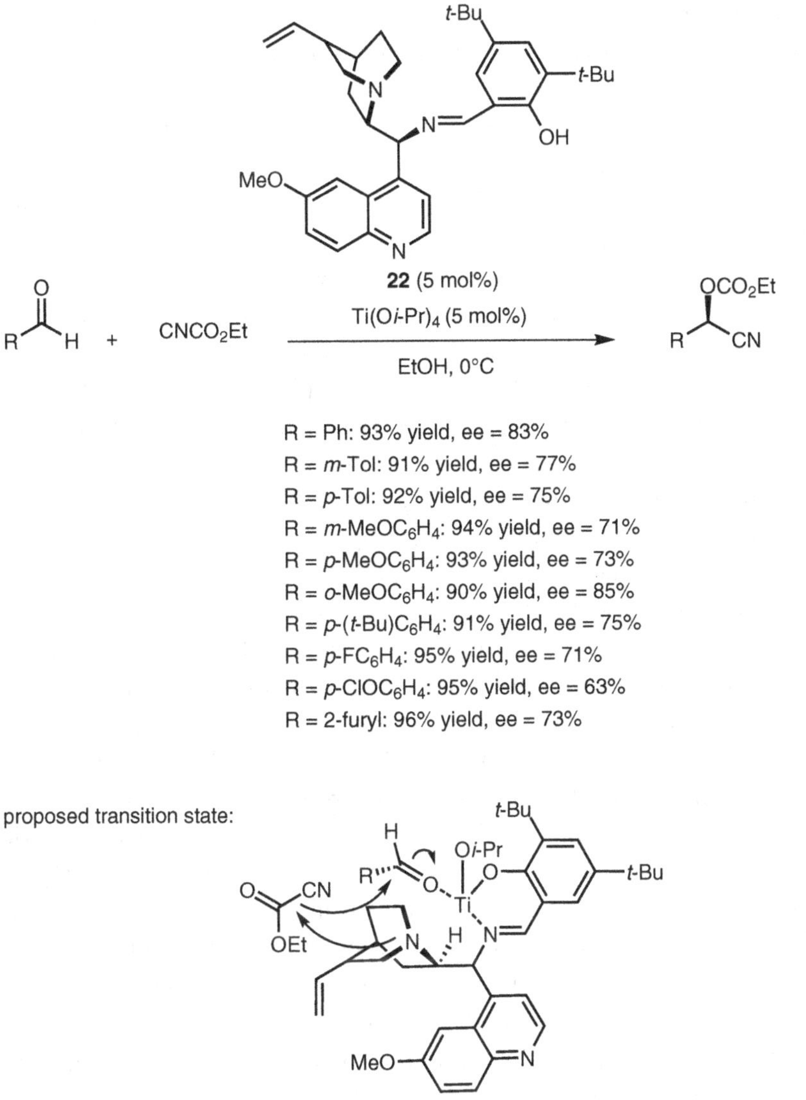

$$R\overset{O}{\underset{H}{\|}} + CNCO_2Et \xrightarrow[\text{EtOH, 0°C}]{\substack{\textbf{22} \text{ (5 mol\%)} \\ Ti(Oi\text{-}Pr)_4 \text{ (5 mol\%)}}} R\overset{OCO_2Et}{\underset{CN}{\|}}$$

R = Ph: 93% yield, ee = 83%
R = *m*-Tol: 91% yield, ee = 77%
R = *p*-Tol: 92% yield, ee = 75%
R = *m*-MeOC$_6$H$_4$: 94% yield, ee = 71%
R = *p*-MeOC$_6$H$_4$: 93% yield, ee = 73%
R = *o*-MeOC$_6$H$_4$: 90% yield, ee = 85%
R = *p*-(*t*-Bu)C$_6$H$_4$: 91% yield, ee = 75%
R = *p*-FC$_6$H$_4$: 95% yield, ee = 71%
R = *p*-ClOC$_6$H$_4$: 95% yield, ee = 63%
R = 2-furyl: 96% yield, ee = 73%

proposed transition state:

Scheme 2.18. *In situ* generated Schiff base titanium catalyst in asymmetric cyanation of aldehydes hydroxyl group of the salicylaldehyde and the nitrogen from the Schiff base double bond, coordinated with titanium tetraisopropylate, could act as a Lewis acid to activate the carbonyl group, and the tertiary nitrogen atom of the cinchona alkaloid could act as a base to activate ethyl cyanoformate. In this proposed model, the activated cyanide attacked the aldehyde through the less hindered *Re* face, and then the corresponding product was afforded.

In 2008, Moberg *et al.* reported the synthesis of a chiral salen polymeric ligand containing a carboxylic acid group, attached to macroporous monosized polystyrene-divinylbenzene beads grafted with (4-hydroxybutyl)vinyl ether groups.[36] The titanium complex of the beads **19** was used for enantioselective cyanations of benzaldehyde with acetyl cyanide in the presence of triethylamine using a microreactor. The corresponding cyanohydrin acetate was obtained in 48% conversion and 70% ee, as shown in Scheme 2.19. The cyanation employing acetyl cyanide was found to be slower than that performed with trimethylsilyl cyanide (Scheme 2.14) and conversions were therefore low, even at low flow rates.

Since 2008, there has been an explosion in the number of multiple-catalyst systems for various organic transformations developed.[44] Among them, a wide number employ two organocatalysts. As an example, a combination of chiral cinchona alkaloid **23** with achiral biphenol **24** was used by Feng *et al.* to *in situ* generate, in the presence of titanium tetraisopropylate, a chiral titanium catalyst, applied to induce chirality in cyanations of aldehydes with ethyl cyanoformate as cyanide source.[45] As shown in Scheme 2.20, the products derived from aromatic and heteroaromatic aldehydes gave remarkable results, with generally excellent yields, and enantioselectivities ranging from 90 to 96% ee. Furthermore, aliphatic aldehydes were tolerated, providing moderate to even higher enantioselectivities, up to 98% ee in some cases of substrates (Scheme 2.20). Surprisingly, steric bulky pivalaldehyde gave the corresponding cyanohydrin with 95% ee and 92% yield. It should be noted that when *i*-PrOH was used as an additive, the reactivity was greatly enhanced, with the enantioselectivity slightly improved. Interestingly, varying the amount of *i*-PrOH (0.25–3 equiv) had no obvious effect on the outcomes. The authors proposed that the reaction proceeded through a dual activation mechanism. As illustrated in Scheme 2.20, the titanium acted as a Lewis acid to activate the aldehyde,

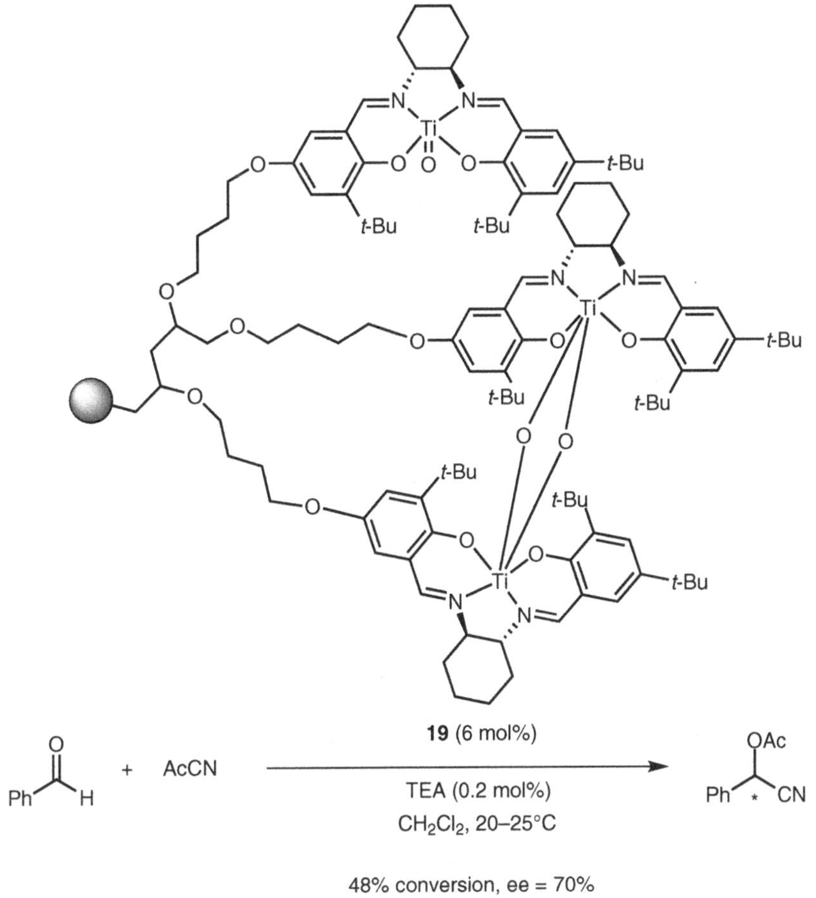

19 (6 mol%)

Ph–CHO + AcCN → TEA (0.2 mol%), CH$_2$Cl$_2$, 20–25°C → Ph-*-(OAc)(CN)

48% conversion, ee = 70%

Scheme 2.19. Polymeric titanium salen catalyst in asymmetric cyanation of benzaldehyde

while the tertiary amine of cinchona alkaloid worked as a Lewis base to activate hydrogen cyanide, which was actually the real cyanide reagent generated *in situ* from isopropanol and ethyl cyanoformate. The active species **D** was formed via dual activation of the substrates. Then, the enantioselective addition of hydrogen cyanide to the aldehyde gave intermediate **E**, which readily afforded the cyanohydrin by intramolecular combination of a proton. Meanwhile, the catalyst was regenerated. Promoted by the catalyst, the cyanohydrin could easily react with ethyl cyanoformate, to give

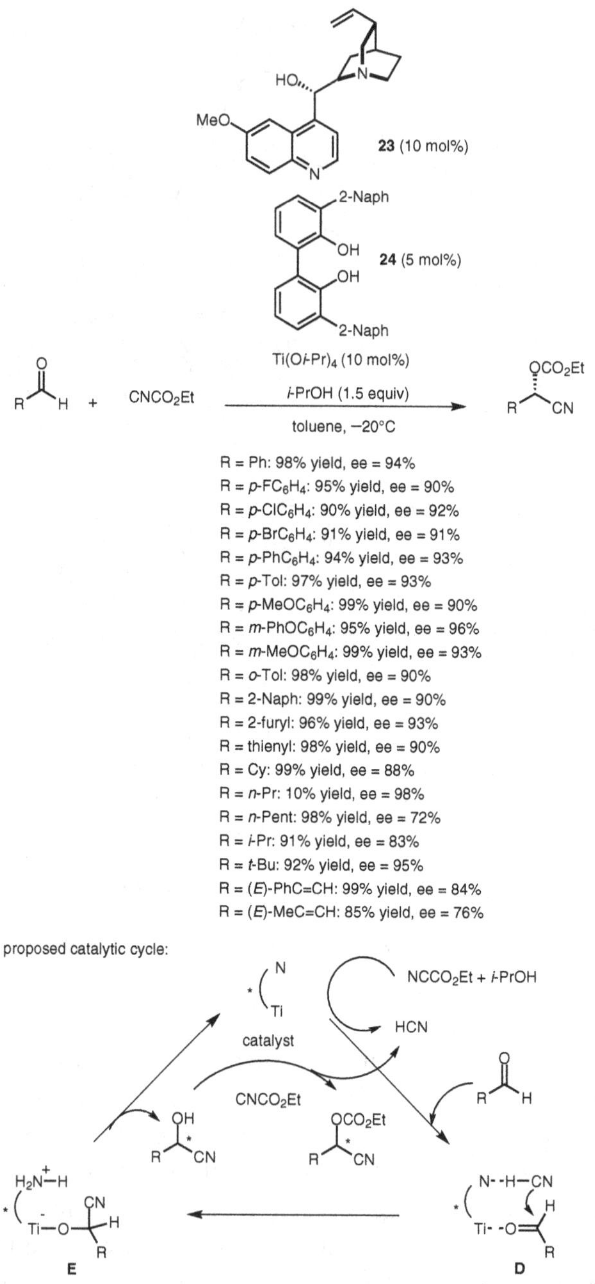

HO
MeO
N
23 (10 mol%)
2-Naph
OH
24 (5 mol%)
OH
2-Naph
Ti(Oi-Pr)4 (10 mol%)
i-PrOH (1.5 equiv)
toluene, −20°C
R H + CNCO2Et
OCO2Et
R CN
R = Ph: 98% yield, ee = 94%
R = p-FC6H4: 95% yield, ee = 90%
R = p-ClC6H4: 90% yield, ee = 92%
R = p-BrC6H4: 91% yield, ee = 91%
R = p-PhC6H4: 94% yield, ee = 93%
R = p-Tol: 97% yield, ee = 93%
R = p-MeOC6H4: 99% yield, ee = 90%
R = m-PhOC6H4: 95% yield, ee = 96%
R = m-MeOC6H4: 99% yield, ee = 93%
R = o-Tol: 98% yield, ee = 90%
R = 2-Naph: 99% yield, ee = 90%
R = 2-furyl: 96% yield, ee = 93%
R = thienyl: 98% yield, ee = 90%
R = Cy: 99% yield, ee = 88%
R = n-Pr: 10% yield, ee = 98%
R = n-Pent: 98% yield, ee = 72%
R = i-Pr: 91% yield, ee = 83%
R = t-Bu: 92% yield, ee = 95%
R = (E)-PhC=CH: 99% yield, ee = 84%
R = (E)-MeC=CH: 85% yield, ee = 76%
proposed catalytic cycle:
N
Ti
catalyst
NCCO2Et + i-PrOH
HCN
CNCO2Et
OH
R CN
OCO2Et
R CN
R H
H2N–H
CN
Ti–O–H
R
E
N- -H–CN
H
Ti- -O=
R
D

Scheme 2.20. *In situ* generated titanium catalyst from a combination of a cinchona alkaloid and a substituted 2,2′-biphenol in asymmetric cyanation of aromatic and aliphatic aldehydes

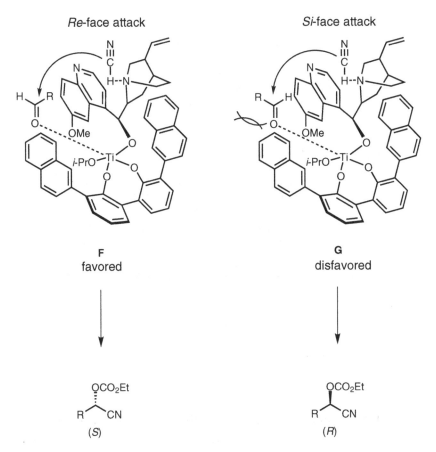

Figure 2.2. Proposed transition states for cyanation of aldehydes catalysed by a combination of chiral cinchona alkaloid **23** and achiral biphenol **24**

the final product and reproduce the recyclable hydrogen cyanide. Then, the next catalytic cycle is started.

To explain the asymmetric induction of the precedent reaction, the authors have proposed two plausible transition states, **F** and **G** (Figure 2.2). As the repulsion between the R group in the substrate and the 2-naphthyl group in the catalyst can occur for transition state **G**, the transition state **F** is more favourable. Moreover, in transition state **F** the *Re* face attack is much more accessible than the *Si* face, because the latter is well shielded by the 2-naphthyl group in the catalyst. As a result, the product with the *S* configuration is selectively produced.

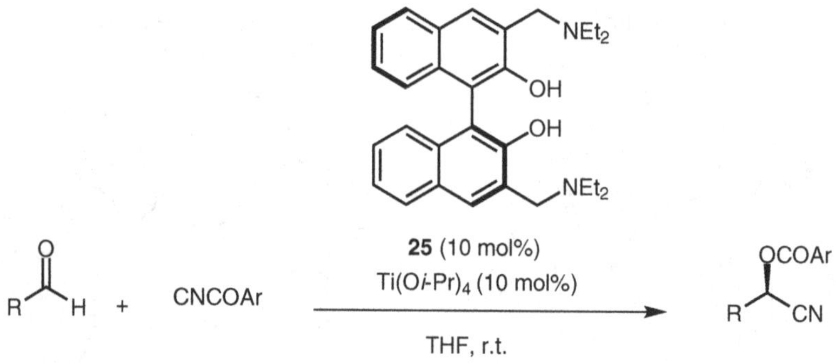

R = Ar = Ph: 91% yield, ee = 68%
R = Ph, Ar = *p*-MeOC$_6$H$_4$: 75% yield, ee = 56%
R = Ph, Ar = *p*-ClC$_6$H$_4$: 87% yield, ee = 58%
R = *m*-PhOC$_6$H$_4$, Ar = Ph: 92% yield, ee = 66%
R = *p*-ClC$_6$H$_4$, Ar = Ph: 85% yield, ee = 58%
R = (*E*)-PhCH=CH, Ar = Ph: 75% yield, ee = 84%
R = Cy, Ar = Ph: 83% yield, ee = 12%
R = BnCH$_2$, Ar = Ph: 93% yield, ee = 66%
R = PhCH$_2$OCH$_2$, Ar = Ph: 85% yield, ee = 38%

Scheme 2.21. *In situ* generated BINOLAM titanium catalyst in asymmetric benzoylcyanation of aliphatic and aromatic aldehydes

In 2011, Saa and Najera investigated the enantioselective cyanobenzoylation of aldehydes with less reactive cyanide derivatives, such as aroyl cyanides, using an *in situ* generated titanium catalyst from chiral BINOLAM ligand **25** and titanium tetraisopropylate in THF at room temperature.[46] As shown in Scheme 2.21, a range of O-aroylcyanohydrins can be achieved in good to high yields of up to 93% associated to moderate enantioselectivities of up to 84% ee. The authors carried out DFT theoretical studies to gain insight into the mechanism of the reaction, which led to them conclude that the reaction conditions were in fact non-Curtin–Hammett-like, as the benzoylation of ligated *i*-PrOH was a very slow reaction, taking place prior to the relevant reaction.

2.3. Ketones as Electrophiles

In comparison with aldehydes, the catalytic asymmetric addition of nucleophilic reagents, including cyanides to ketones, is a more challenging task for synthetic chemists, owing to their low electrophilicity and the reduced propensity of ketone carbonyl to coordinate with Lewis acids. In 2008, Feng *et al.* reported a highly enantioselective cyanosilylation of ketones, catalysed by a combination of a (*S*)-prolinamide derivative **26** as chiral ligand, titanium tetraisopropylate, and an achiral phenolic *N*-oxide **27**.[47] The reaction was performed in THF at −45°C, with trimethylsilyl cyanide as cyanide source. By using 2.5 mol% of chiral ligand **27** and achiral phenolic *N*-oxide **27**, in combination with 10 mol% of titanium tetraisopropylate, a range of tertiary cyanohydrin trimethylsilyl ethers could be synthesised from the corresponding aromatic and aliphatic aldehydes, in excellent yields and high enantioselectivities of up to 96% ee, as shown in Scheme 2.22. The electronic character of the *ortho*- or *para*- substituent of acetophenone was found to be not crucial for the enantioselectivity. Steric bulkier aromatic ketones, such as β-acetonaphthone and α-tetralone, afforded the corresponding cyanohydrin ethers in 96 and 85% ee, respectively. Moreover, an α,β-unsaturated ketone was well tolerated and gave the corresponding 1,2-addition product exclusively in 92% yield and 91% ee. Finally, aliphatic and α,β-saturated ketones afforded the corresponding products in high yields, with good enantioselectivities of up to 89% ee. The authors have proposed the catalytic cycle with a bifunctional pathway illustrated in Scheme 2.22 to explain the results. The tridentate ligand **26** was able to coordinate with titanium tetraisopropylate through an oxygen atom and two amide moieties, meanwhile, the achiral phenolic *N*-oxide **27** was able to assemble with titanium through phenolic group to form **H**. Titanium acted as a Lewis acid to activate the carbonyl group and the *N*-oxide moiety acted as a Lewis base to activate trimethylsilyl cyanide, simultaneously. In transition state **I**, the large steric hindrance between the R group of the ketone, and the two phenyl groups of the amide was able to direct the activated CN attacking the carbonyl group from the less hindered side. Then, the corresponding final *O*-TMS ethers were released with the regeneration of the catalyst.

In 2009, a combination of chiral cinchona alkaloid **28** with achiral biphenol **29** was used in the presence of titanium tetraisopropylate by Feng *et al.* to *in situ* generate a chiral titanium catalyst, which was applied to induce chirality in cyanations of ketones with trimethylsilyl cyanide as cyanide source.[45] The reactivity and enantioselectivity of the process were optimal when using *n*-hexane as solvent, although the catalyst was

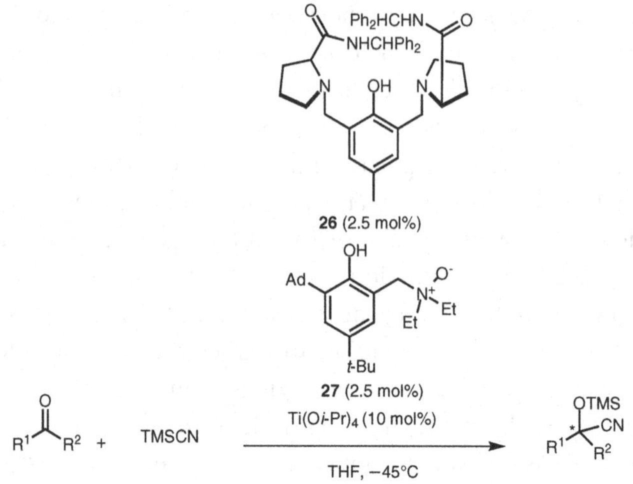

R^1 = Ph, R^2 = Me: 96% yield, ee = 90%
R^1 = *o*-FC$_6$H$_4$, R^2 = Me: 93% yield, ee = 90%
R^1 = *m*-ClC$_6$H$_4$, R^2 = Me: 93% yield, ee = 94%
R^1 = *p*-ClC$_6$H$_4$, R^2 = Me: 90% yield, ee = 89%
R^1 = *p*-Tol, R^2 = Me: 91% yield, ee = 86%
R^1 = 2-thiophene, R^2 = Me: 78% yield, ee = 92%
R^1 = 2-Naph, R^2 = Me: 89% yield, ee = 96%
R^1 = Ph, R^2 = (CH$_2$)$_3$: 80% yield, ee = 85%
R^1 = (*E*)-PhCH=CH, R^2 = Me: 92% yield, ee = 91%
R^1 = BnCH$_2$, R^2 = Me: 92% yield, ee = 89%
R^1 = *i*-Pr, R^2 = Me: 89% yield, ee = 62%
R^1 = *n*-Pent, R^2 = Me: 95% yield, ee = 82%
R^1 = Cy, R^2 = Me: 90% yield, ee = 71%

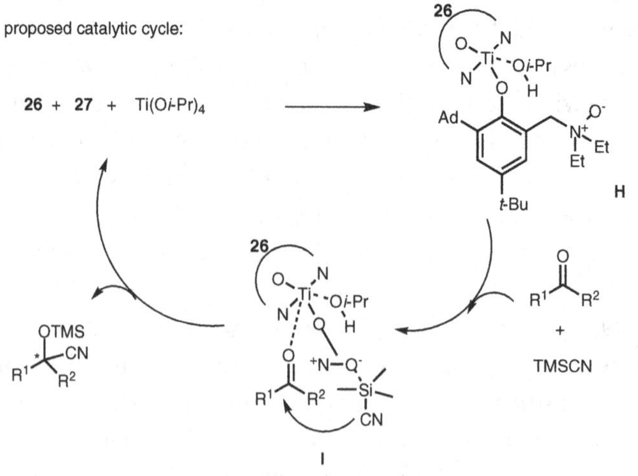

proposed catalytic cycle:

Scheme 2.22. *In situ* generated titanium catalyst from a combination of a (*S*)-prolinamide and a phenolic *N*-oxide in asymmetric cyanation of ketones

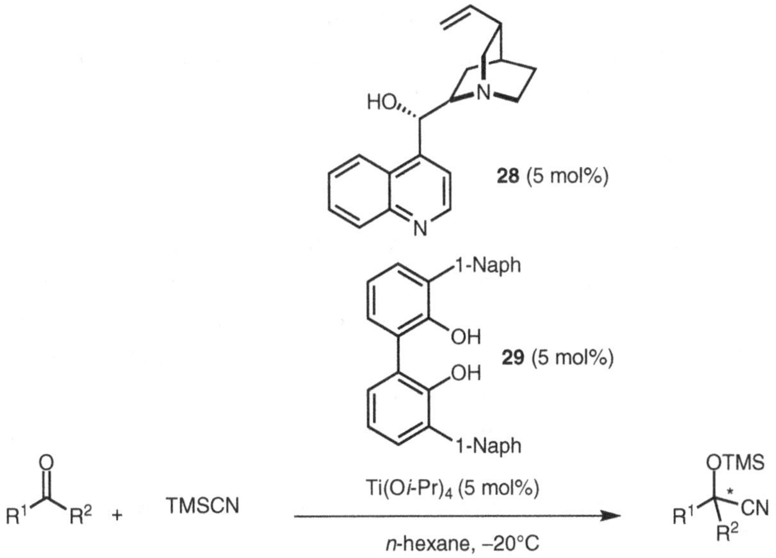

28 (5 mol%)

29 (5 mol%)

1-Naph

OH

OH

1-Naph

$$R^1 \overset{O}{\underset{}{\parallel}} R^2 + TMSCN \xrightarrow[n\text{-hexane, }-20°C]{Ti(O\textit{i}-Pr)_4 \text{ (5 mol\%)}} R^1 \overset{OTMS}{\underset{R^2}{\overset{*}{\parallel}}} CN$$

R^1 = Ph, R^2 = Me: 98% yield, ee = 94%
R^1 = p-FC$_6$H$_4$, R^2 = Me: 99% yield, ee = 92%
R^1 = p-ClC$_6$H$_4$, R^2 = Me: 98% yield, ee = 87%
R^1 = p-BrC$_6$H$_4$, R^2 = Me: 99% yield, ee = 98%
R^1 = p-Tol, R^2 = Me: 95% yield, ee = 90%
R^1 = p-MeOC$_6$H$_4$, R^2 = Me: 83% yield, ee = 94%
R^1 = p-NO$_2$C$_6$H$_4$, R^2 = Me: 92% yield, ee = 88%
R^1 = m-ClC$_6$H$_4$, R^2 = Me: 99% yield, ee = 70%
R^1 = Ph, R^2 = Et: 97% yield, ee = 74%
R^1 = p-FC$_6$H$_4$, R^2 = Me: 99% yield, ee = 92%
R^1 = Ph, R^2 = Et: 97% yield, ee = 74%

Scheme 2.23. *In situ* generated titanium catalyst from a combination of a cinchona alkaloid and a substituted 2,2′-biphenol in asymmetric cyanation of ketones

undissolved in it. In contrast with related reactions of aldehydes (Scheme 2.20), employing *i*-PrOH as additive, had no positive effect on the results. As shown in Scheme 2.23, *para*-substituted acetophenones with either electron-donating or electron-withdrawing groups were suitable substrates (87–98% ees). However, for the *meta*- and *ortho*-substituted substrates, less satisfactory results were obtained (32–70% ees). On the other hand, propiophenone was converted into the corresponding cyanohydrin in excellent yield (97%) albeit moderate enantioselectivity (74% ee). Other ketones, such as aliphatic, heterocyclic, and cyclic ketones, were also tested, but only 13–30% ee values were obtained.

2.4. Strecker-Type Reactions

The well-known Strecker reaction was first documented by Strecker in 1850, constituting one of the earliest one pot multicomponent reactions discovered.[48] By simply mixing acetaldehyde, ammonia, and hydrogen cyanide, the resulting amino nitrile was formed in high yield. Subsequent hydrolysis of this amino nitrile afforded alanine. Since this discovery, the Strecker reaction has gradually attracted more and more attention, since it provides a direct and economical access to various naturally and non-naturally occurring α-amino acids,[49] including enantio-enriched products.[50] Since the first example of a catalytic enantioselective Strecker reaction reported by Lipton *et al.* in 1996,[51] a number of highly efficient catalytic systems have been developed, including chiral titanium complexes. For example, Hoppe *et al.* reported in 2008 on the synthesis of a library of *N*-arenesulfonyl-1,3-oxazolidinyl-substituted biphenyldiols **30a–d**, which were subsequently examined as ligands in enantioselective Strecker reactions.[52] After systematic screening, it was clear that the optimal ligands were the ones with the bulky mesitylene group, and the sterically more-demanding alkyl side-chains in the oxazolidine moiety. By employing 10 mol% of a combination of titanium tetraisopropylate and ligand **30a–d** as catalyst system and isopropanol as an additive, the Strecker reaction of various aldimines proceeded well, to furnish the corresponding products with moderate to good yields and enantioselectivities of up to 89% ee, as shown in Scheme 2.24. Notably, the optimal ligand varied as the substrate changed. Although a ligand that is efficient overall has not been found for this system, the results revealed the advantages of designing a type of fine-tunable ligand with several variable sites that could be modified.

In 2009, Vilaivan *et al.* developed a practical and efficient method for the synthesis of chiral α-arylamino nitriles, based on an enantioselective Strecker reaction catalysed by an *in situ* generated titanium complex of structurally simple amino alcohol ligand **31** at a low catalyst loading of

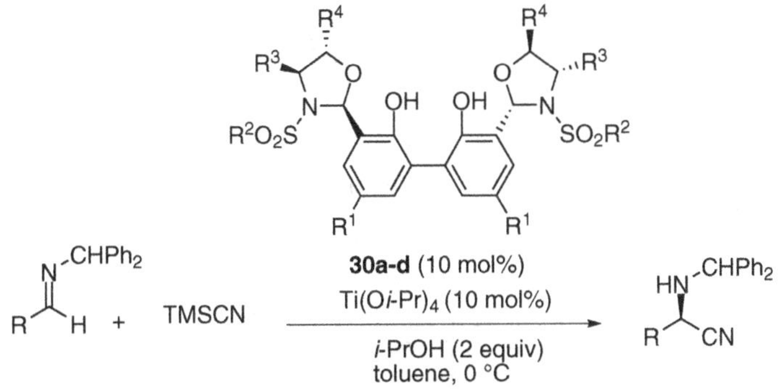

with R^1 = *t*-Bu, R^2 = Mes, R^3 = *i*-Pr, R^4 = H:
R = *p*-ClC_6H_4: 87% yield, ee = 84%
R = *m*-Tol: 83% yield, ee = 72%
R = *o*-Tol: 91% yield, ee = 85%
with R^1 = R^3 = *t*-Bu, R^2 = Mes, R^4 = H:
R = Ph: 94% yield, ee = 89%
with R^1 = H, R^2 = Mes, R^3 = Me, R^4 = H:
R = *p*-Tol: 92% yield, ee = 80%
with R^1 = *t*-Bu, R^2 = Mes, R^3 = *s*-Bu, R^4 = H:
R = *t*-Bu: 86% yield, ee = 63%

Scheme 2.24. *In situ* generated biphenyldiol titanium catalyst in asymmetric Strecker reaction of aldimines

2.5 mol%.[53] As shown in Scheme 2.25, the reaction of a range of aromatic aldimines with trimethylsilyl cyanide afforded the corresponding *N*-benzhydryl α-amino nitriles in excellent yields (89–99%), and remarkable enantioselectivities ranging from 96 to >98% ee. The process did not require the slow addition of the cyanating reagent. The formed Strecker products were subsequently hydrolysed into the corresponding chiral arylglycine derivatives in good yields (60–92%), and moderate to excellent enantioselectivities (85–98% ee).

In 2010, Chai *et al.* found that the catalyst generated from *N*- salicyl-β-amino alcohol **32** and partially hydrolysed titanium alkoxide (PHTA) was a more efficient catalyst for asymmetric Strecker reactions.[54] PHTA was prepared by hydrolising Ti(O*n*-Bu)$_4$ using the residual H_2O in toluene with stirring for 18 hours. Under optimised reaction conditions, various aldimines could be converted into the corresponding Strecker products at

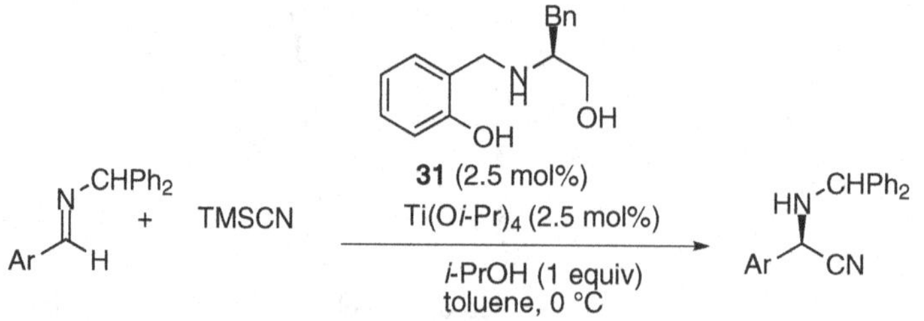

Ar = Ph: 95% yield, ee = 98%
Ar = o-Tol: 98% yield, ee = 98%
Ar = p-Tol: 97% yield, ee >98%
Ar = o-BrC$_6$H$_4$: 97% yield, ee >98%
Ar = p-BrC$_6$H$_4$: 98% yield, ee = 96%
Ar = o-ClC$_6$H$_4$: 98% yield, ee >98%
Ar = p-ClC$_6$H$_4$: 94% yield, ee = 98%
Ar = m-FC$_6$H$_4$: 99% yield, ee = 96%
Ar = 1-Naph: 91% yield, ee = 98%

Scheme 2.25. *In situ* generated amino alcohol titanium catalyst in asymmetric Strecker reaction of aldimines

room temperature in a short time, typically 15–60 minutes. Significantly, several kinds of N-protecting groups, such as Ph$_2$CH, Bn, and Boc, proved to be compatible with the catalyst system, giving both excellent yields and enantioselectivities of up to 99% and 98% ee, respectively, as shown in Scheme 2.26. Later, Chai *et al.* reported similar results, using hydrogen cyanide as the cyanide reagent in the presence of catalytic amounts of trimethylsilyl cyanide (10–25 mol%).[55] The hydrogen cyanide had to be slowly added over a period of one hour after the addition of trimethylsilyl cyanide. A preliminary mechanistic study indicated that trimethylsilyl cyanide was in fact the actual cyanide reagent responsible for the enantioselective delivery of the cyanide ion to the aldimine in the catalytic cycle. Also, hydrogen cyanide presumably acted as the proton source to assist the release of the product, while at the same time recovering the trimethylsilyl cyanide for the next cycle.

In 2012, the same authors reported the synthesis of a robust heterogeneous self-supported chiral titanium cluster catalyst **33**, and its application in the enantioselective Strecker reaction of aldimines.[56] As shown in Scheme 2.27, a range of chiral N-protected amino nitriles were synthesised in excellent yields, with enantioselectivities of up to 99% and 98% ee,

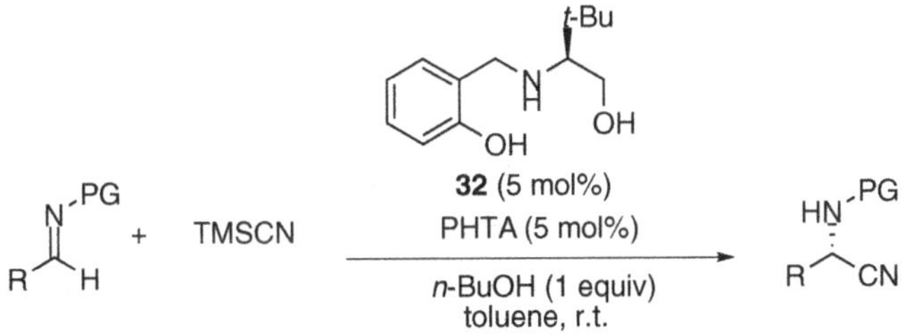

32 (5 mol%)
PHTA (5 mol%)

n-BuOH (1 equiv)
toluene, r.t.

R = Ph, PG = CHPh$_2$: 99% yield, ee = 96%
R = *o*-ClC$_6$H$_4$, PG = CHPh$_2$: 95% yield, ee = 98%
R = *p*-Tol, PG = CHPh$_2$: 96% yield, ee = 97%
R = *m*-MeOC$_6$H$_4$, PG = CHPh$_2$: 93% yield, ee = 93%
R = (*E*)-PhCH=CH, PG = CHPh$_2$: 96% yield, ee = 97%
R = 3-pyridinyl, PG = CHPh$_2$: 81% yield, ee = 94%
R = Ph, PG = Boc: 99% yield, ee = 98%
R = *p*-Tol, PG = Boc: 88% yield, ee = 96%
R = *p*-MeOC$_6$H$_4$, PG = Boc: 89% yield, ee = 95%
R = 2-furyl, PG = Boc: 87% yield, ee = 90%
R = Ph, PG = Bn: 99% yield, ee = 87%
R = *o*-ClC$_6$H$_4$, PG = Bn: 94% yield, ee = 89%
R = *p*-Tol, PG = Bn: 95% yield, ee = 87%

Scheme 2.26. *In situ* generated amino alcohol titanium catalyst in asymmetric Strecker reaction of aldimines

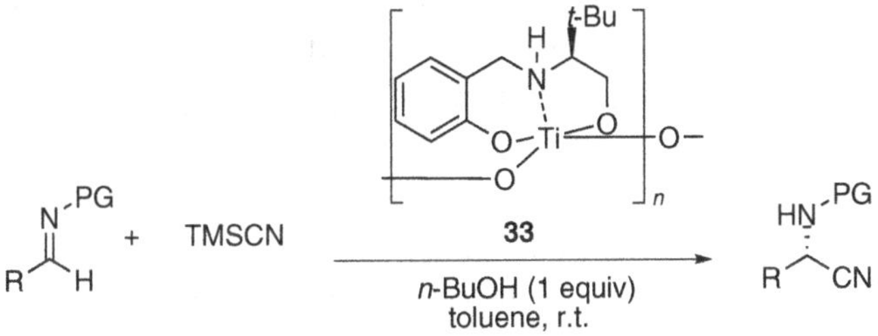

33

n-BuOH (1 equiv)
toluene, r.t.

R = Ph, PG = CHPh$_2$: 99% yield, ee = 98%
R = Ph, PG = Bn: 99% yield, ee = 86%
R = *o*-BrC$_6$H$_4$, PG = CHPh$_2$: 98% yield, ee = 98%
R = *p*-Tol, PG = CHPh$_2$: 99% yield, ee = 97%
R = *o*-BrC$_6$H$_4$, PG = Bn: 97% yield, ee = 88%
R = *p*-Tol, PG = Bn: 99% yield, ee = 87%
R = *p*-CF$_3$C$_6$H$_4$, PG = Bn: 98% yield, ee = 88%
R = (*E*)-PhCH=CH, PG = CHPh$_2$: 96% yield, ee = 94%

Scheme 2.27. Self-supported titanium cluster catalyst in asymmetric Strecker reaction of aldimines respectively, by reaction of the corresponding aldimines with trimethylsilyl cyanide at room temperature. The catalyst was successfully recycled more than ten times without any significant loss in the catalytic performance. This constituted the first heterogeneous catalyst that showed high activity and enantioselectivity at room temperature for the cyanation of imines. Moreover, the robustness, high performance, and recyclability of the catalyst enabled it to be used in a packed-bed reactor to carry out the cyanation under continuous flow. Under these conditions, remarkable enantioselectivities of up to 97% ee and quantitative conversions were achieved.

2.5. Conclusions

This chapter collects the advances in enantioselective titanium-catalysed cyanation reactions of carbonyl compounds and derivatives that have been reported since 2008. In spite of the synthetic usefulness of the resultant chiral cyanohydrins, asymmetric cyanohydrin synthesis has known a very long development over the last century, with most progress being achieved over the last two decades. In the cases of aldehyde substrates, a range of chiral titanium catalysts are now available, providing high levels of enantioselectivity for any substrate, with different cyanide sources, at low catalyst loadings, and without requiring low temperatures and long reaction times. For example, remarkable results (99% yield, 97% ee) were reported in 2010 by using a chiral tethered bimetallic titanium salen complex at only 0.01 mol% of catalyst loading, which could be reduced to as low as 0.0005 mol% while still providing exceptional results (86% yield, 95% ee). In the context of cyanation of aldehydes with trimethylsilyl cyanide, non-C_2-symmetric BINOL chiral ligands, monosubstituted in position 3 with nitrogen-containing heterocycles, were also demonstrated to be remarkably efficient. Even if the field of asymmetric cyanation reactions of carbonyl compounds is still largely dominated by the use of trimethylsilyl cyanide as the cyanide source, two excellent works using ethyl cyanoformate as cyanide source reported in the last seven years have to be highlighted. The first was based on the use of an *in situ* generated macrocyclic titanium salen catalyst with a flexible polyether linkage used at a low catalyst loading of 0.5 mol%, providing both excellent yields and enantioselectivities (99% yield, 95% ee). The second involved the use of a combination of two ligands, for instance a chiral cinchona alkaloid and an achiral biphenol, which provided up to 99% yield and 98% ee.

The use of ketones as substrates remains, however, less well developed in comparison with aldehydes. In the last seven years, two interesting examples with enantioselectivities of up to 98% ee have been described, both based on combinations of two ligands and employing trimethylsilyl cyanide as the cyanide source. Despite the important number of works on asymmetric titanium-catalysed cyanations of carbonyl compounds in the last few years, many challenges remain. First, a better understanding of the reaction mechanism is needed, as well as further application of known catalyst systems. Second, the cyanation of ketones is still to be extended as well demonstrated in this chapter. Finally, one of the most important factors for taking the catalytic cyanation reaction to industry is the choice of the source of cyanide. More than 90% of the reports employ trimethylsilyl cyanide as the source of cyanide which is toxic, volatile, expensive, and incongruous for large scale synthesis. Although excellent results have been achieved by using other cyanide sources, such as acyl cyanides, there is still more work needed to find viable alternative to trimethylsilyl cyanide and other expensive cyanide sources.

In the context of enantioselective titanium-catalysed Strecker reactions of aldimines, several excellent results have been reported since 2008. Among them, a practical and highly efficient method for the synthesis of chiral α-arylamino nitriles catalysed by an *in situ* generated titanium complex of a structurally simple amino alcohol ligand at a low catalyst loading of 2.5 mol% was developed by Vilaivan *et al.*, providing remarkable yields and enantioselectivities. One can reasonably anticipate that future studies will provide new applications of chiral cyanohydrins and amino nitriles in total synthesis of natural products and biologically active compounds.

References

(1) (a) R. Noyori, in *Asymmetric Catalysts in Organic Synthesis*, Wiley, New-York, **1994**.
(b) *Transition Metals for Organic Synthesis*, M. Beller, C. Bolm, eds., Wiley–VCH,

Weinheim, **1998**, Vols I and II. (c) *Comprehensive Asymmetric Catalysis*, E. N. Jacobsen, A. Pfaltz, H. Yamamoto, eds., Springer, Berlin, **1999**. (d) *Catalytic Asymmetric Synthesis*, 2nd ed., I. Ojima, ed., Wiley–VCH, New-York, **2000**. (e) G. Poli, G. Giambastiani, A. Heumann, *Tetrahedron* **2000**, *56*, 5959–6150. (f) E. Negishi, in *Handbook of Organopalladium Chemistry for Organic Synthesis*, John Wiley & Sons, Inc., Hoboken NJ, **2002**, *2*, p. 1689. (g) A. de Meijere, P. von Zezschwitz, H. Nüske, B. Stulgies, *J. Organomet. Chem.* **2002**, *653*, 129–140. (h) *Transition Metals for Organic Synthesis*, 2nd ed., M. Beller, C. Bolm, eds., Wiley–VCH, Weinheim, **2004**. (i) L. F. Tietze, I. Hiriyakkanavar, H. P. Bell, *Chem. Rev.* **2004**, *104*, 3453–3516.

(2) (a) Y. Yu, K. Ding, G. Chen, in *Acid Catalysis in Modern Organic Synthesis*, H. Yamamoto, K. Ishihara, Wiley–VCH, Weinheim, **2008**; Chapter 14, pp. 721–823. (b) F. Chen, X. Feng, Y. Jiang, *Arkivoc* **2003**, *ii*, 21–31. (c) J. Cossy, S. Bouz, F. Pradaux, C. Willis, V. Bellosta, *Synlett* **2002**, 1595–1606. (d) K. Mikami, M. Terada, in *Lewis Acids in Organic Synthesis*, H. Yamamoto, ed., Wiley–VCH, Weinheim, **2000**, Vol. 2, Chapter 16, p. 799. (e) K. Mikami, M. Terada, in *Lewis Acid Reagents*, H. Yamamoto, ed., Wiley–VCH, Weinheim, **1999**, Vol. 1, Chapter 6, p. 93. (f) D. J. Ramon, M. Yus, *Rec. Res. Dev. Org. Chem.* **1998**, *2*, 489–523. (g) A. H. Hoveyda, J. P. Morken, *Angew. Chem., Int. Ed. Engl.* **1996**, *35*, 1262–1284. (h) B. M. Trost, in *Stereocontrolled Organic Synthesis*, B. M. Trost, ed., Blackwell, Oxford, **1994**, p. 17.

(3) (a) S. Suga, M. Kitamura, *Comprehensive Chirality* **2012**, *4*, 328–342. (b) T. Oshima, *Comprehensive Chirality* **2012**, *4*, 355–377. (c) E. M. Carreira, P. Aschwanden, *Science of Synthesis* **2011**, *2*, 517–529. (d) E. M. Carreira, D. E. Frantz, *Science of Synthesis* **2011**, *2*, 497–515. (e) M. Yus, J. C. Gonzales-Gomez, F. Foubelo, *Chem. Rev.* **2011**, *111*, 7774–7854. (f) G. Blay, A. Monleon, J. R. Pedro, *Curr. Org. Chem.* **2009**, *13*, 1498–1539. (g) L. Li, W. Rui, *Curr. Org. Chem.* **2009**, *13*, 1565–1576. (h) B. M. Trost, A. H. Weiss, *Adv. Synth. Catal.* **2009**, *351*, 963–983. (i) M. R. Luderer, W. F. Bailey, M. R. Luderer, J. D. Fair, R. J. Dancer, M. B. Sommer, *Tetrahedron: Asymmetry* **2009**, *20*, 981–1094. (j) K. Soai, T. Shibata, in *Comprehensive Asymmetric Catalysis Supplement I*, E. N. Jacobsen, A. Pfaltz, eds., Springer, Berlin, **2004**; p. 95. (k) S. E. Denmark, J. Fu, *Chem. Rev.* **2003**, *103*, 2763–2794. (l) L. Pu, H.-B. Yu, *Chem. Rev.* **2001**, *101*, 757–824. (m) K. Soai, in *Comprehensive Asymmetric Catalysis II*, E. N. Jacobsen, A. Pfaltz, H. Yamamoto, eds., Springer, Berlin, **1999**; p. 911. (n) P. Knochel, R. D. Singer, *Chem. Rev.* **1993**, *93*, 2117–2188. (o) R. Noyori, M. Kitamura, *Angew. Chem., Int. Ed. Engl.* **1991**, *30*, 49–69.

(4) (a) M. Dadashipour, Y. Asano, *ACS Catalysis* **2011**, *1*, 1121–1149. (b) T. Purkarthofer, W. Skranc, C. Schuster, H. Griengl, *Appl. Microbiol. Biotechnol.* **2007**, *76*, 309–320. (c) M. Sharma, N. N. Sharma, T. C. Bhalla, *Enzyme Microb. Technol.* **2005**, *37*, 279–294. (d) M. H. Fechter, H. Griengl, in *Enzyme Catalysis in Organic Synthesis*, 2nd ed., K. Drauz, H. Waldmann, eds., Vol. 2, Wiley–VCH, Weinheim, **2002**, p. 974.

(5) (a) M. North, *Angew. Chem., Int. Ed.* **2010**, *49*, 8079–8081. (b) N. H. Khan, R. I. Kureshi, S. H. R. Abdi, S. Agrawal, R. V. Jasra, *Coord. Chem. Rev.* **2008**, *252*, 593–623. (c) M. North, D. L. Usanov, C. Young, *Chem. Rev.* **2008**, *108*, 5146–5226.

(6) (a) S. E. Denmark, G. L. Beutner, *Angew. Chem., Int. Ed.* **2008**, *47*, 1560–1638. (b) E. A. C. Davie, S. M. Mennen, Y. Xu, S. J. Miller, *Chem. Rev.* **2007**, *107*, 5759–5812.

(7) For reviews on asymmetric synthesis and applications of cyanohydrins, see: (a) J. Wang, X. Liu, X. Feng, *Chem. Rev.* **2011**, *111*, 6947–6983. (b) W. Wang, W. Liu, L. Lin, X. Feng, *Eur. J. Org. Chem.* **2010**, 4751–4769. (c) C. Moberg, E. Wingstrand, *Synlett* **2010**, 355–367. (d) J. Gawronski, N. Wascinska, J. Gajewy, *Chem. Rev.* **2008**, *108*, 5227–5252. (e) M. North, *Science of Synthesis* **2004**, *19*, 235–284. (f) J.-M. Brunel, I. P. Holmes, *Angew. Chem., Int. Ed.* **2004**, *43*, 2752–2778. (g) M. North, *Tetrahedron: Asymmetry* **2003**, *14*, 147–176. (h) M. Shibasaki, M. Kanai, K. Funabashi, *J. Chem. Soc., Chem. Commun.* **2002**, 1989–1999. (i) I. Ojima, in *Catalytic Asymmetric Synthesis*, Wiley, New York, **2000**, p 235. (j) A. Mori, S. Inue, in *Comprehensive Asymmetric Catalysis II*; (eds., E. N. Jacobsen, A. Paltz, H. Yamamoto), Springer, Berlin, **1999**; p 983. (k) R. J. H. Gregory, *Chem. Rev.* **1999**, *99*, 3649–3682. (l) F. Effenberger, *Angew. Chem., Int. Ed. Engl.* **1994**, *33*, 1555–1564. (m) M. North, *Synlett* **1993**, 807–820.

(8) (a) F. F. Fleming, *Nat. Prod. Rep.* **1999**, *16*, 597–606. (b) C. J. Peterson, R. Tsao, J. R. Coats, *Pest. Manage. Sci.* **2000**, *56*, 615–617. (c) D.-S. Park, J. R. Coats, *J. Pest. Sci.* **2005**, *30*, 99–102.

(9) J. E. Poulton, *Plant. Physiol.* **1990**, *94*, 401–405.

(10) F. W. Winkler, *Liebigs Ann. Chem.* **1832**, *4*, 242.

(11) L. Rosenthaler, *Biochem. Z.* **1908**, *14*, 238–253.

(12) D. J. Ramon, M. Yus, *Chem. Rev.* **2006**, *106*, 2126–2208.

(13) M. T. Reetz, S.-H. Kyung, C. Bolm, T. Zierke, *Chem. Ind.* **1986**, 824.

(14) For a review on the uses of TADDOL and derivatives in asymmetric synthesis, see: H. Pellissier, *Tetrahedron* **2008**, *64*, 10279–10317.

(15) K. Narasaka, T. Yamada, H. Minamikawa, *Chem. Lett.* **1987**, *10*, 2073–2076.

(16) Y. N. Belokon, J. Hunt, M. North, *Synlett* **2008**, 2150–2154.

(17) J. F. Larrow, E. N. Jacobsen, Y. Gao, Y. Hong, X. Nie, C. M. Zepp, *J. Org. Chem.* **1994**, *59*, 1939–1942.

(18) Y. N. Belokon, J. Hunt, M. North, *Tetrahedron: Asymmetry* **2008**, *19*, 2804–2815.

(19) C. Lv, M. Wu, S. Wang, C. Xia, W. Sun, *Tetrahedron: Asymmetry* **2010**, *21*, 1869–1873.

(20) Y. Q. Wen, W. M. Ren, X. B. Lu, *Chin. Chem. Lett.* **2011**, *22*, 1285–1288.

(21) Y.-Q. Wen, W. M. Ren, X. B. Lu, *Org. Biomol. Chem.* **2011**, *9*, 6323–6330.

(22) C. Lv, Q. Cheng, D. Xu, S. Wang, C. Xia, W. Sun, *Eur. J. Org. Chem.* **2011**, 3407–3411.

(23) M. E. S. Serra, D. Murtinho, A. Goth, *Arkivoc* **2010**, *v*, 64–69.

(24) M. E. S. Serra, D. Murtinho, A. Goth, *Tetrahedron: Asymmetry* **2013**, *24*, 315–319.

(25) Y. N. Belokon, S. Caveda-Cepas, B. Green, N. S. Ikonnikov, V. N. Khrustalev, V. S. Larichev, M. A. Moscalenko, M. North, C. Orizu, V. I. Tararov, M. Tasinazzo, G. I. Timofeeva, L. V. Yashkina, *J. Am. Chem. Soc.* **1999**, *121*, 3968–3973.

(26) M. North, M. Omdes-Pujol, *Tetrahedron Lett.* **2009**, *50*, 4452–4454.

(27) Z. Zhang, Z. Wang, R. Zhang, K. Ding, *Angew. Chem., Int. Ed.* **2010**, *49*, 6746–6750.

(28) C. Lv, D. Xu, S. Wang, C.-X. Miao, C. Xia, W. Sun, *Catal. Commun.* **2011**, *12*, 1242–1245.

(29) F. Yang, S. Wei, C.-A. Chen, P. Xi, L. Yang, J. Lan, H.-M. Gau, J. You, *Chem. Eur. J.* **2008**, *14*, 2223–2231.

(30) H. Tang, Z. Zhang, *Heteroatom. Chem.* **2011**, *22*, 31–35.

(31) Y. N. Belokon, D. A. Chusov, T. V. Skrupskaya, D. A. Bor'kin, L. V. Yashkina, K. A. Lyssenko, M. M. Il'in, T. V. Strelkova, G. I. Timofeeva, A. S. Peregudov, M. North, *Russ. Chem. Bull., Int. Ed.* **2008**, *57*, 1981–1988.

(32) (a) Y. Miyamoto, T. Inoue, N. Oguni, *J. Chem. Soc., Chem. Commun.* **1991**, 1752–1753; (b) M. Hayashi, Y. Miyamoto, T. Inoue, N. Oguni, *J. Org. Chem.* **1993**, *58*, 1515–1522.

(33) Y. Li, B. He, B. Qin, X. Feng, G. Zhang, *J. Org. Chem.* **2004**, *69*, 7910–7913.

(34) M. Jaworska, E. Blocka, A. Kozakiewicz, M. Welniak, *Tetrahedron: Asymmetry* **2011**, *22*, 648–657.

(35) E. Blocka, M. J. Bosiak, M. Welniak, A. Ludwiczak, A. Wojtczak, *Tetrahedron: Asymmetry* **2014**, *25*, 554–562.

(36) S. Lundgren, H. Ihre, C. Moberg, *Arkivoc* **2008**, *vi*, 73–80.

(37) S. Sakthivel, T. Punniyamurthy, *Tetrahedron: Asymmetry* **2010**, *21*, 2834–2840.

(38) D. A. Evans, J. M. Hoffman, L. K. Truesdale, *J. Am. Chem. Soc.* **1973**, *95*, 5822–5823.

(39) W. Lidy, W. Sundermeyer, *Chem. Ber.* **1973**, *106*, 587–593.

(40) (a) C. S. Marvel, N. O. Brace, F. A. Miller, A. Roberts, *J. Am. Chem. Soc.* **1949**, *71*, 34–36. (b) R. Yoneda, K. Santo, S. Harusawa, T. Kurihara, *Synthesis* **1986**, 1054–1055. (c) M. Okimoto, T. Chiba, *Synthesis* **1996**, 1188. (d) Y. N. Belokon, A. J. Blacker, L. A. Clutterbuck, M. North, *Org. Lett.* **2003**, *5*, 4505–4507. (e) T. Watahiki, S. Ohba, T. Oriyama, *Org. Lett.* **2003**, *5*, 2679–2681. (f) W. Zhang, M. Shi, *Org. Biomol. Chem.* **2006**, *4*, 1671–1674.

(41) A. Sadhukhan, M. K. Choudhary, N.-U. H. Khan, R. I. Kureshy, S. H. R. Abdi, H. C. Bajaj, *ChemCatChem* **2013**, *5*, 1441–1448.

(42) M. North, J. Watson, *Curr. Org. Chem.* **2014**, *1*, 66–72.

(43) N. Ji, L. Yao, W. He, Y. Li, *Appl. Organometal. Chem.* **2013**, *27*, 209–213.

(44) (a) J.-A. Ma, D. Cahard, *Angew. Chem., Int. Ed.* **2004**, *43*, 4566–4583. (b) M. Kanai, N. Kato, E. Ichikawa, M. Shibasaki, *Synlett* **2005**, 1491–1508. (c) D. H. Paull, C. J. Abraham, M. T. Scerba, E. Alden-Danforth, T. Leckta, *Acc. Chem. Res.* **2008**, *41*, 655–663. (d) Z. Shao, H. Zhang, *Chem. Soc. Rev.* **2009**, *38*, 2745–2755. (e) C. Zhong, X. Shi, *Eur. J. Org. Chem.* **2010**, 2999–3025. (f) M. Rueping, R. M. Koenigs, I. Atodiresei, *Chem. Eur. J.* **2010**, *16*, 9350–9365. (g) J. Zhou, *Chem. Asian J.* **2010**, *5*, 422–434. (h) L. M. Ambrosini, T. H. Lambert, *ChemCatChem* **2010**, *2*, 1373–1380. (i) S. Piovesana, D. M. Scarpino Schietroma, M. Bella, *Angew. Chem., Int. Ed.* **2011**, *50*, 6216–6232. (j) N. T. Patil, V. S. Shinde, B. Gajula, *Org. Biomol. Chem.* **2012**, *10*, 211–224. (k) A. E. Allen, D. W. C. MacMillan, *Chem. Sci.* **2012**, *3*, 633–658. (l) Z. Du, Z. Shao, *Chem. Soc. Rev.* **2013**, *42*, 1337–1378. (m) H. Pellissier, *Tetrahedron* **2013**, *69*, 7171–7210. (n) *Recent Developments in Enantioselective Multicatalysed Tandem Reactions*, (ed., H. Pellissier), Royal Society of Chemistry, Cambridge, **2014**.

(45) J. Wang, W. Wang, W. Li, X. Hu, K. Shen, C. Tan, X. Liu, X. Feng, *Chem. Eur. J.* **2009**, *15*, 11642–11659.

(46) (a) A. Baeza, C. Najera, J. M. Sansano, J. M. Saa, *Tetrahedron: Asymmetry* **2011**, *22*, 1282–1291. (b) J. M. Saa, A. Baeza, C. Najera, J. M. Sansano, *Tetrahedron: Asymmetry* **2011**, *22*, 1292–1305.

(47) K. Shen, X. Liu, Q. Li, X. Feng, *Tetrahedron* **2008**, *64*, 147–153.

(48) A. Strecker, *Ann. Chem. Pharm.* **1850**, *75*, 27–45.

(49) C. Najera, J. M. Sansano, *Chem. Rev.* **2007**, *107*, 4584–4671.

(50) (a) X.-H. Cai, B. Xie, *Arkivoc* **2014**, *1*, 205–248. (b) J. Wang, X. Liu, X. Feng, *Chem. Rev.* **2011**, *111*, 6947–6983. (c) S. J. Connon, *Angew. Chem., Int. Ed.* **2008**, *47*, 1176–1178. (d) C. Spino, *Angew. Chem., Int. Ed.* **2004**, *43*, 1764–1766. (e) H. Gröger, *Chem. Rev.* **2003**, *103*, 2795–2828. (f) L. Yet, *Angew. Chem., Int. Ed.* **2001**, *40*, 875–877.

(51) M. S. Iyer, K. M. Gigstad, N. D. Namdev, M. Lipton, *J. Am. Chem. Soc.* **1996**, *118*, 4910–4911.

(52) S. Wünnemann, R. Fröhlich, D. Hoppe, *Eur. J. Org. Chem.* **2008**, 684–692.

(53) V. Banphavichit, W. Mansawat, W. Bhanthumnavin, T. Vilaivan, *Tetrahedron* **2009**, *65*, 5849–5854.

(54) A. M. Seayad, B. Ramalingam, K. Yoshinaga, T. Nagata, C. L. L. Chai, *Org. Lett.* **2010**, *12*, 264–267.

(55) B. Ramalingam, A. M. Seayad, L. Chuanzhao, M. Garland, K. Yoshinaga, M. Wadamoto, T. Nagata, C. L. L. Chai, *Adv. Synth. Catal.* **2010**, *352*, 2153–2158.

(56) A. M. Seayad, B. Ramalingam, C. L. L. Chai, C. Li, M. V. Garland, K. Yoshinaga, *Chem. Eur. J.* **2012**, *18*, 5693–5700.

Chapter 3

Enantioselective Titanium-Catalysed Thioether Oxidations

3.1. Introduction

This chapter covers efforts of the chemical community to develop novel enantioselective titanium-catalysed thioether oxidation reactions. It looks at developments since the beginning of 2008, when the field was reviewed by Yu and co-workers in a book chapter dealing with titanium Lewis acids.[1]

The chapter is divided into two parts. The first part deals with enantioselective titanium-catalysed sulfoxidations using alkyl hydroperoxides as oxidants, while the second collects enantioselective titanium-catalysed sulfoxidations employing hydrogen peroxide as oxidant.

Chiral sulfoxides belong to the class of chiral organosulfur compounds[2] which are the most widely used in asymmetric synthesis. Their application as chiral synthons has become a well-established and reliable strategy. This is mainly due to their ready availability, and the high asymmetric induction exerted by the chiral sulfinyl group, which has been established as one of the most efficient and versatile chiral controllers in carbon to carbon (C–C) and carbon to halogen (C–X) bond formations.[2,3] Moreover, the chiral sulfur groupings that induce optical activity can be removed from the molecule easily, under fairly mild conditions, thus presenting an additional advantage in the asymmetric synthesis of chiral compounds. The main advantage of sulfoxides over other sulfur functions, such as sulfides and sulfones, is indeed their chirality. The oxygen atom of a sulfoxide can be coordinated to a metal ion or a proton, and electronic and steric repulsions between nucleophiles and the substituents of a sulfoxide are also expected. The sulfinyl group acts as an electron-withdrawing group, activates a C–C double bond for conjugate addition, and stabilises the corresponding α-carbanion. The stable pyramidal structure of a chiral sulfoxide allows a diastereoselective reaction to occur at a nearby or distant reaction centre. Complexation of the sulfoxide group with a suitable metal ion forms a rigid diastereomeric intermediate, which can undergo subsequent reactions stereoselectively.

To date, a large number of chiral sulfoxides[4] have been investigated in a wide range of reactions. Enantiopure sulfoxides have become one of the most important classes of chiral auxiliaries as a result of their ease of preparation, remarkable synthetic versatility, and straightforward removal.[5] Their synthesis has been the subject of constant interest over the past two decades.[4cd,6] A real breakthrough occurred in their synthesis at the beginning of the 1990s, when various new methodologies appeared. Several methods are presently available to obtain optically active sulfoxides: optical resolution;[2d] asymmetric oxidation of nonsymmetric sulfides;[7] asymmetric biological oxidation; and nucleophilic addition of alkyl or aryl ligands to diastereochemically pure chiral sulfinates[8], as, for instance, in the Andersen procedure, which is still the most important and generally used method.[9] This method consists of a substitution at the sulfur atom of commercially available (*S*)-menthyl *p*-toluenesulfinate with an appropriate organometallic reagent, with the advantage that the substitution takes place with 100% inversion of configuration.[10]

This classic method has been used extensively to prepare *p*-tolyl alkyl or aryl sulfoxides, and the use of various organometallic nucleophiles has allowed the synthesis of a wide variety of enantiomerically pure sulfoxides. The usefulness of this method is mainly due to the accessibility of the sulfinylating agent, obtained as a mixture of sulfur epimers which are separated by repeated recrystallisations. The enantiomerically pure sulfinate is then displaced by an organomagnesium halide with complete inversion of stereochemistry at the sulfur, as first demonstrated by Mislow *et al.*[11] The separation of alkyl menthylsulfinate diastereomers appears to be tricky in this example, however, and Andersen's synthesis is not efficient enough to access enantiomerically pure dialkyl sulfoxides. Other authors have developed synthetic schemes that established the enantiomeric purity at sulfur prior to addition to the organometallic reagent.[12]

Importantly, chiral sulfoxides constitute vital intermediates in asymmetric synthesis, and bioactive ingredients in the pharmaceutical industry. For example, a recent industrial application of the enantioselective oxidation of sulfides using chiral metal catalysts was the development of two syntheses of sulindac, an efficient non-steroidal anti-inflammatory drug (NSAID). The first synthesis was based on an iron-catalysed sulfoxidation,[13] while the second involved an asymmetric Kagan sulfoxidation as the key step.[14] In addition, Matsugi *et al.* reported a practical synthesis of the platelet adhesion inhibitor, OPC-29030 via a catalytic asymmetric oxidation of sulfide with a titanium-mandelic acid complex.[15]

3.2. Alkyl Hydroperoxides as Oxidants

Catalytic asymmetric oxidation of prochiral sulfides has been a challenging goal for many years, reflecting the importance of the sulfinyl group as a chiral controller in a number of organic transformations, as well as the existence of natural and synthetic biologically active products that need a sulfinyl group with a defined configuration. Since the breakthrough from the groups of Kagan[16] and Modena[17] in 1984, who discovered that modified titanium(IV) isopropoxide-diethyltartrate catalyst systems were capable of asymmetric oxidising of prochiral sulfides with alkyl hydroperoxides, a range of modifications and applications of titanium based chiral catalytic systems for the asymmetric oxidation of sulfides into sulfoxides have emerged. Disadvantages of these processes are their low productivities, since superstoichiometric metal loadings were sometimes required; their complexity; the need for precise moisture control; and the use of nonpreferred alkyl hydroperoxides as oxidants. Nevertheless, the majority of practical applications have so far traditionally exploited various modifications of Kagan–Modena type systems, such as catalytic versions of the titanium tartrate systems, due to their well-studied substrate scope, tolerance to functional groups and ease of implementation. In this context, Volcho *et al.* have reported the synthesis of esomeprazole, an important proton pump inhibitor, based on a titanium-catalysed asymmetric sulfoxidation process

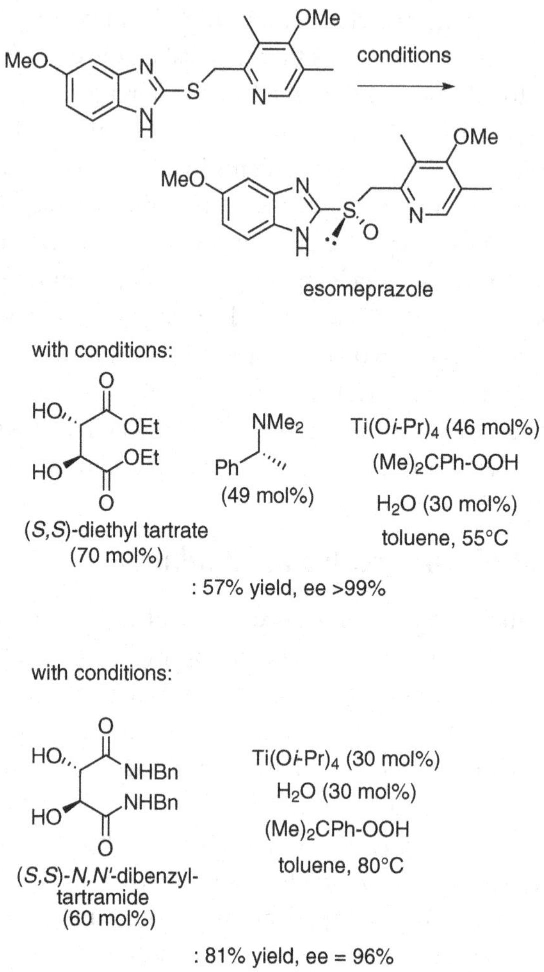
MeO
N
S
N
H
OMe
N
conditions
MeO
N
S
O
N
H
OMe
N
esomeprazole
with conditions:
HO
OEt
OEt
HO
O
O
(S,S)-diethyl tartrate
(70 mol%)
NMe2
Ph
(49 mol%)
Ti(Oi-Pr)4 (46 mol%)
(Me)2CPh-OOH
H2O (30 mol%)
toluene, 55°C
: 57% yield, ee >99%
with conditions:
HO
NHBn
NHBn
HO
O
O
(S,S)-N,N'-dibenzyl-
tartramide
(60 mol%)
Ti(Oi-Pr)4 (30 mol%)
H2O (30 mol%)
(Me)2CPh-OOH
toluene, 80°C
: 81% yield, ee = 96%

Scheme 3.1. Synthesis of esomeprazole with (*S,S*)-diethyl tartrate and (*S,S*)-*N,N′*-dibenzyl tartramide as ligands

induced by a combination of diethyl D-tartrate and (*R*)-*N,N*-dimethyl-1-phenylethanamine as chiral ligands, and employing cumene hydroperoxide as oxidant.[18] The reaction employed 46 mol% of titanium tetraisopropylate and 70 mol% of the chiral tartrate in combination with 49 mol% of (*R*)-*N,N*-dimethyl-1-phenylethanamine, affording almost enantiopure esomeprazole in 57% yield, as shown in Scheme 3.1 (first line of results). Later,

Deng *et al.* reported the use of the same alkyl hydroperoxide — cumene hydroperoxide — in the same reaction, performed in the absence of a base, albeit induced by (S,S)-N,N'-dibenzyl tartramide as chiral ligand (Scheme 3.1, second line of results).[19] In the presence of 60 mol% of this ligand, 30 mol% of titanium tetraisopropylate and 30 mol% of water, esomeprazole was achieved in both high yield and enantioselectivity of 81% and 96% ee, respectively.

Earlier, Szabo *et al.* developed another synthesis of esomeprazole, based on the use of (S,S)-diethyl tartrate as chiral ligand, employed at 60 mol% of catalyst loading in combination with 30 mol% of titanium tetraisopropylate, DIPEA as a base, cumene hydroperoxide as the oxidant, and water in toluene.[20] Under these conditions, the expected product was afforded, in 78% yield and an excellent enantioselectivity of 98% ee, as shown in Scheme 3.2. These conditions were also applied to the oxidation of a range of other imidazole-based prochiral sulfides, providing the corresponding chiral sulfoxides in good yields, and moderate to excellent enantioselectivities of up to 98% ee (Scheme 3.2). It should be noted that methylimidazole and methylbenzimidazole gave the same level of enantioselectivity (96–98% ee). Substitution of the carbon atoms of the imidazole or benzimidazole rings was found to have minor effects on the enantioselectivity of the reaction. This remarkable maintenance of the enantioselectivity was apparently due to the effects of the imidazole ring, since benzylphenylsulfide was oxidised with a low enantiomeric excess (10% ee) under the same reaction conditions.

In addition to tartrates, a range of chiral 1,2-diols have been appied as ligands to enantioselective titanium-catalysed oxidation of sulfides. For example, chiral 1,2-diarylethane-1,2-diols were studied by Rosini *et al.* in these oxidations, performed with *tert*-butyl hydroperoxide as oxidant, investigating the effects of the substitution on the aryl moiety on the asymmetric oxidation of sulfides.[21] It was shown that the substitution of the aryl ring of the diol with both electron-withdrawing and electron-donating substituents generally decreased the enantioselectivity with respect to the use of unsubstituted 1,2-diphenylethane-1,2-diol. 1,2-Di(4-*t*-butyl)phenyl-1,2-diol **1** was identified as the most efficient ligand, providing good to excellent enantioselectivities of up to >99% ee, as shown in Scheme 3.3. The best enantioselectivities ranging from 90 to> 99% ee were

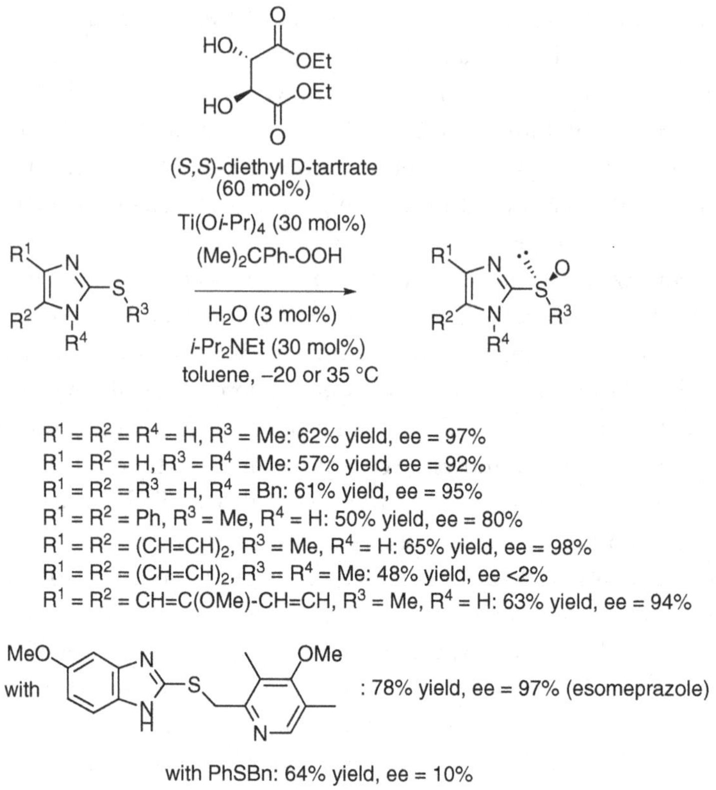

(S,S)-diethyl D-tartrate
(60 mol%)

Ti(Oi-Pr)$_4$ (30 mol%)

(Me)$_2$CPh-OOH

$\xrightarrow{\hspace{2cm}}$

H$_2$O (3 mol%)

i-Pr$_2$NEt (30 mol%)

toluene, −20 or 35 °C

R^1 = R^2 = R^4 = H, R^3 = Me: 62% yield, ee = 97%
R^1 = R^2 = H, R^3 = R^4 = Me: 57% yield, ee = 92%
R^1 = R^2 = R^3 = H, R^4 = Bn: 61% yield, ee = 95%
R^1 = R^2 = Ph, R^3 = Me, R^4 = H: 50% yield, ee = 80%
R^1 = R^2 = (CH=CH)$_2$, R^3 = Me, R^4 = H: 65% yield, ee = 98%
R^1 = R^2 = (CH=CH)$_2$, R^3 = R^4 = Me: 48% yield, ee <2%
R^1 = R^2 = CH=C(OMe)-CH=CH, R^3 = Me, R^4 = H: 63% yield, ee = 94%

with : 78% yield, ee = 97% (esomeprazole)

with PhSBn: 64% yield, ee = 10%

Scheme 3.2. Oxidation of imidazole-based sulfides with (S,S)-diethyl tartrate as ligand

reached with aryl methyl and aryl benzyl sulfides, whereas lower enantioselectivities were achieved with larger naphthyl and aryl alkyl sulfides.

With the aim of discovering novel blockbuster gastric proton pump inhibitors such as esomeprazole, Jiang *et al.* developed titanium-catalysed oxidations for a range of heteroaromatic sulfides, especially 1H-benzimidazolylpyridinylmethyl sulfides.[22] The reactions were performed with tert-butyl peroxide as oxidant and induced by novel chiral 1,2-diphenylethane-1,2-diol ligand **2** in combination with titanium tetraisopropylate, providing the corresponding chiral sulfoxides in remarkable general yields, and enantioselectivities of up to >99% ee, as shown in Scheme 3.4. The scope of the process was successfully extended to benzyl 1H-benzimidazole sulfides, which afforded the corresponding sulfoxides in significantly lower

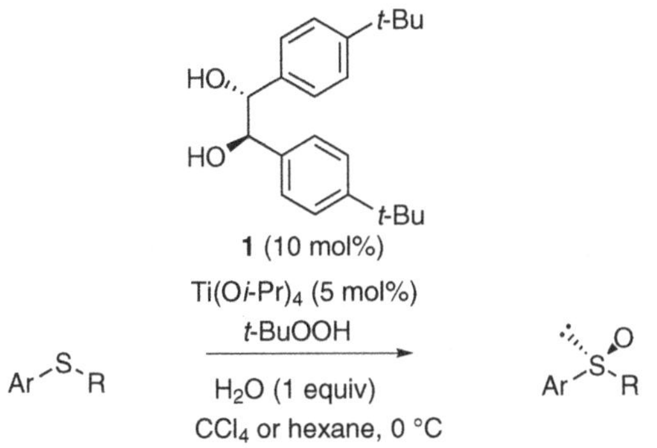

Ar = Ph, R = Me: 73% yield, ee = 82%
Ar = *p*-Tol, R = Me: 70% yield, ee = 90%
Ar = *p*-MeOC$_6$H$_4$, R = Me: 46% yield, ee = 80%
Ar = *p*-BrC$_6$H$_4$, R = Me: 52% yield, ee = 80%
Ar = 2-Naph, R = Me: 68% yield, ee = 74%
Ar = Ph, R = Bn: 65% yield, ee >99%
Ar = *p*-MeOC$_6$H$_4$, R = Bn: 72% yield, ee >99%
Ar = 2-Naph, R = Bn: 72% yield, ee = 16%

Scheme 3.3. Oxidation of aryl sulfides with chiral 1,2-diarylethane-1,2-diol **1** as ligand

enantioselectivities (≤ 73% ee), thus demonstrating that a pyridine ring in the substrate was indispensable for high enantioselectivities. Furthermore, the substituents of the benzyl, pyridinylmethyl, or benzimidazolyl moiety had an important impact on the enantioselectivity of the reactions.

Remarkable enantioselectivities were also reported by Cardellicchio *et al.* in the asymmetric oxidation of a wide variety of aryl benzyl sulfides with *tert*-butyl hydroperoxide, in the presence of a complex of titanium and chiral hydrobenzoin.[23] As shown in Scheme 3.5, independent of the nature, position, and dimensions of the substituents, aryl benzyl sulfoxides could be readily obtained in optically pure form. To explain the observed enantioselectivities, the authors had proposed a theoretical model in which interactions involving the π aryl systems played a central role. Indeed, only aryl benzyl sulfides presented the right distance to allow the required weak interactions to achieve high enantioselectivity. This was demonstrated by the low enantioselectivity (6% ee) obtained in the case of using *p*-bromophenyl phenyl sulfide as substrate (Scheme 3.5).

Besides the use of chiral 1,2-diol ligands, Zhu *et al.* have applied chiral 2,5-diisopropylcyclohexane-1,4-diol **3** to the oxidation of sulfides with

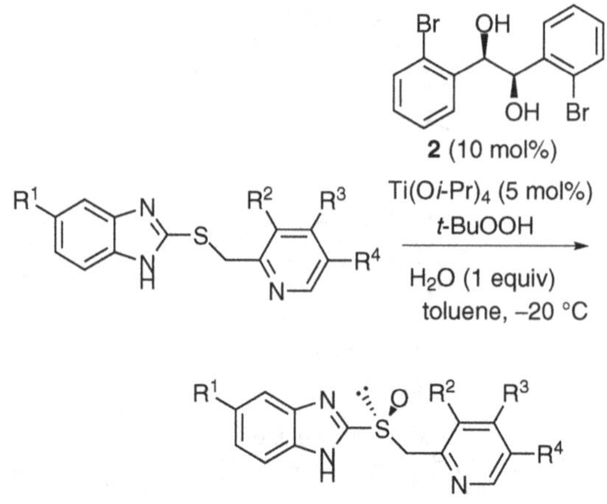

2 (10 mol%)

Ti(O*i*-Pr)$_4$ (5 mol%)

t-BuOOH

H$_2$O (1 equiv)

toluene, −20 °C

$R^1 = R^3 = OMe$, $R^2 = R^4 = Me$: 70% yield, ee >99%
$R^1 = OCF_2H$, $R^2 = R^3 = OMe$, $R^4 = H$: 90% yield, ee = 92%
$R^1 = R^4 = H$, $R^2 = Me$, $R^3 = O(CH_2)_3OMe$: 81% yield, ee = 80%
$R^1 = R^4 = H$, $R^2 = Me$, $R^3 = OCH_2CF_3$: 86% yield, ee = 90%
$R^1 = OMe$, $R^2 = Me$, $R^3 = OCH_2CF_3$, $R^4 = H$: 87% yield, ee = 92%
$R^1 = H$, $R^2 = R^4 = Me$, $R^3 = OMe$: 97% yield, ee = 98%

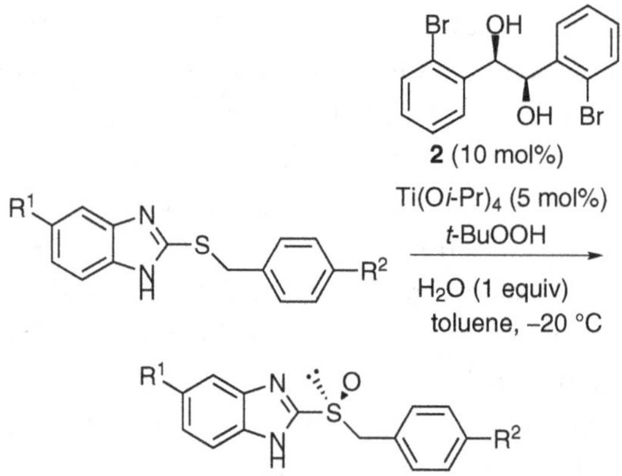

2 (10 mol%)

Ti(O*i*-Pr)$_4$ (5 mol%)

t-BuOOH

H$_2$O (1 equiv)

toluene, −20 °C

$R^1 = R^2 = H$: 37% yield, ee = 50%
$R^1 = OCF_2H$, $R^2 = H$: 63% yield, ee = 62%
$R^1 = OCF_2H$, $R^2 = Cl$: 96% yield, ee = 73%

Scheme 3.4. Oxidation of 1-*H*-benzimidazolyl pyridinylmethyl sulfides and benzyl 1-*H*-benzimidazole sulfides with chiral 1,2-diarylethane-1,2-diol **2** as ligand

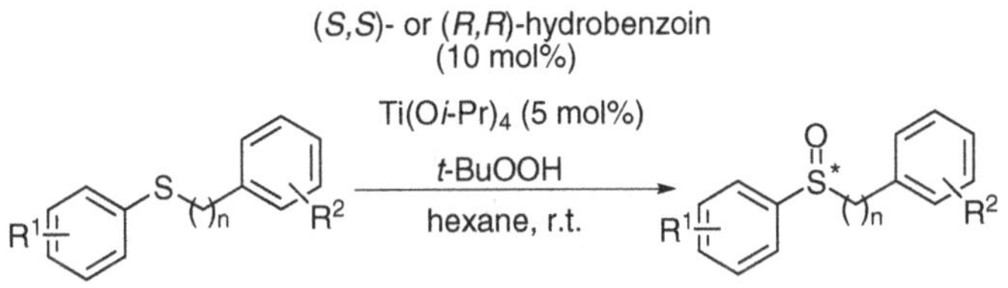

(S,S)- or (R,R)-hydrobenzoin
(10 mol%)

Ti(Oi-Pr)$_4$ (5 mol%)

t-BuOOH

hexane, r.t.

with (S,S)-hydrobenzoin as ligand:
R^1 = 2-CO$_2$Me, R^2 = H, n = 1: 77% yield, ee = 91% (R)
R^1 = 2-Cl, R^2 = H, n = 1: 92% yield, ee = 95% (R)
R^1 = 4-NO$_2$, R^2 = H, n = 1: 65% yield, ee >98% (R)
R^1 = 2-MeO, R^2 = H, n = 1: 89% yield, ee = 88% (R)
R^1 = 3-MeO, R^2 = H, n = 1: 57% yield, ee = 84% (R)
R^1 = 4-MeO, R^2 = H, n = 1: 67% yield, ee >98% (R)
R^1 = 4-Br, R^2 = 2-NO$_2$, n = 1: 72% yield, ee = 85% (R)
R^1 = 4-Br, R^2 = 4-NO$_2$, n = 1: 71% yield, ee >98% (R)
R^1 = 4-Br, R^2 = 2-MeO, n = 1: 68% yield, ee >98% (R)
R^1 = 4-Br, R^2 = 3-MeO, n = 1: 37% yield, ee = 81% (R)
R^1 = 4-Br, R^2 = 4-MeO, n = 1: 65% yield, ee >98% (R)
R^1 = 4-Br, R^2 = 3-Cl, n = 1: 66% yield, ee = 97% (R)
R^1 = 4-Br, R^2 = 2,4-Cl$_2$, n = 1: 82% yield, ee >98% (R)
R^1 = R^2 = 2,3,4,5,6-F$_5$, n = 1: 19% yield, ee = 61% (R)
R^1 = H, R^2 = 2,3,4,5,6-F$_5$, n = 1: 91% yield, ee >98% (R)
R^1 = 2-F, R^2 = 2,3,4,5,6-F$_5$, n = 1: 96% yield, ee >98% (R)
R^1 = R^2 = 4-MeO, n = 1: 72% yield, ee >98% (R)
R^1 = 4-Br, R^2 = H, n = 0: 45% yield, ee = 6% (R)
R^1 = 4-Br, R^2 = H, n = 2: 49% yield, ee = 90% (R)
R^1 = 4-Br, R^2 = H, n = 3: 46% yield, ee >98% (R)
with (R,R)-hydrobenzoin as ligand:
R^1 = 4-NO$_2$, R^2 = H, n = 1: 38% yield, ee >98% (S)
R^1 = 4-MeO, R^2 = H, n = 1: 65% yield, ee >98% (S)

Scheme 3.5. Oxidation of aryl (benzyl) sulfides with chiral hydrobenzoin as ligand

cumene hydroperoxide in the presence of MS 4Å.[24] As shown in Scheme 3.6, various chiral sulfoxides were achieved in moderate to good yields, and enantioselectivities of up to 72% and 84% ee, respectively. These mild reaction conditions were also applied to the synthesis of esomeprazole which was obtained in 72% yield and 76% ee.

Meanwhile, chiral ligands other than tartrates or diols have been investigated in enantioselective titanium-catalysed oxidations of sulfides. As an example, Bull and Davidson showed that a pseudo-C_3-symmetric titanium triflate catalyst **4** that displays propeller chirality could be used to

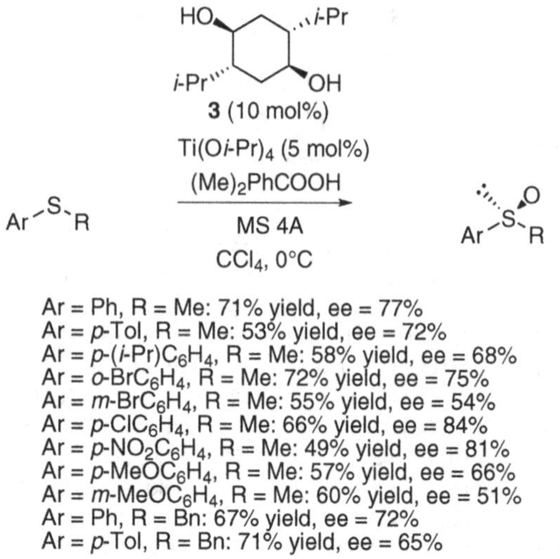

$$HO \quad i\text{-Pr}$$
$$i\text{-Pr} \quad OH$$

3 (10 mol%)
Ti(Oi-Pr)$_4$ (5 mol%)
(Me)$_2$PhCOOH
$$Ar \overset{S}{\diagdown} R \xrightarrow{\text{MS 4A} \atop \text{CCl}_4, \, 0°\text{C}} Ar \overset{O}{\overset{\parallel}{\diagdown S}} R$$

Ar = Ph, R = Me: 71% yield, ee = 77%
Ar = p-Tol, R = Me: 53% yield, ee = 72%
Ar = p-(i-Pr)C$_6$H$_4$, R = Me: 58% yield, ee = 68%
Ar = o-BrC$_6$H$_4$, R = Me: 72% yield, ee = 75%
Ar = m-BrC$_6$H$_4$, R = Me: 55% yield, ee = 54%
Ar = p-ClC$_6$H$_4$, R = Me: 66% yield, ee = 84%
Ar = p-NO$_2$C$_6$H$_4$, R = Me: 49% yield, ee = 81%
Ar = p-MeOC$_6$H$_4$, R = Me: 57% yield, ee = 66%
Ar = m-MeOC$_6$H$_4$, R = Me: 60% yield, ee = 51%
Ar = Ph, R = Bn: 67% yield, ee = 72%
Ar = p-Tol, R = Bn: 71% yield, ee = 65%

Scheme 3.6. Oxidation of aryl sulfides with chiral 2,5-diisopropylcyclohexane-1,4-diol **3** as ligand

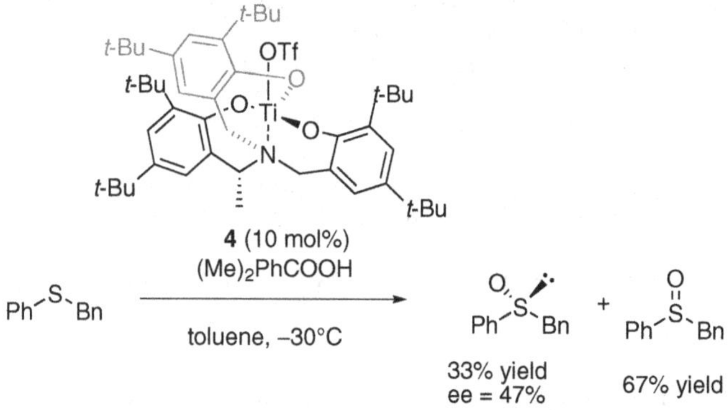

4 (10 mol%)
(Me)$_2$PhCOOH
$$Ph \overset{S}{\diagdown} Bn \xrightarrow{\text{toluene, } -30°\text{C}} Ph \overset{O}{\overset{}{\diagdown S}} Bn \quad + \quad Ph \overset{O}{\overset{\parallel}{\diagdown S}} Bn$$

33% yield   67% yield
ee = 47%

Scheme 3.7. Oxidation of benzyl phenyl sulfide with chiral pseudo-C_3-symmetric titanium triflate catalyst **4**

promote the asymmetric sulfoxidation of benzyl phenyl sulfide with moderate enantioselectivity (47% ee), using cumene hydroperoxide as oxidant (Scheme 3.7).[25]

In 2010, Zonta *et al.* applied chiral titanium trialkanolamine catalyst **5** to the asymmetric oxidation of benzyl tolyl sulfides with cumene hydroperoxide.[26] As shown in Scheme 3.8, the corresponding chiral sulfoxides

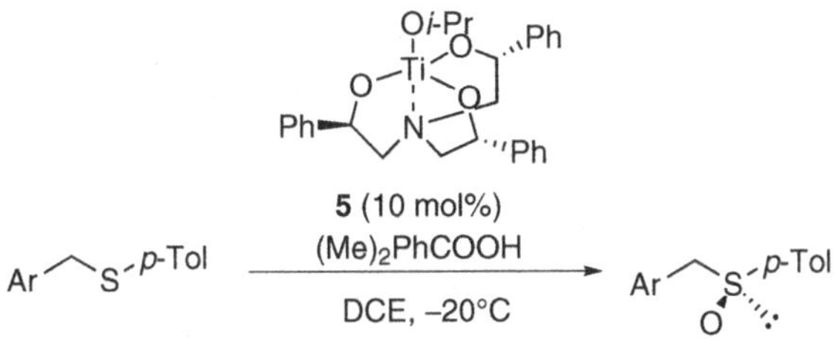

5 (10 mol%)

$Ar\diagdown S\diagup p\text{-Tol}$ $\xrightarrow[\text{DCE, }-20°C]{\text{(Me)}_2\text{PhCOOH}}$ $Ar\diagdown S\diagup p\text{-Tol}$

Ar = Ph: 98% yield, ee = 52%
Ar = *p*-Tol: 89% yield, ee = 54%
Ar = 1-Naph: 99% yield, ee = 59%
Ar = 2-Naph: 82% yield, ee = 51%
Ar = C_6F_5: 72% yield, ee = 45%
Ar = 9-anthracenyl: 99% yield, ee = 71%
Ar = *p*-$(NMe_2)C_6H_4$: 87% yield, ee = 57%
Ar = *p*-$MeOC_6H_4$: 99% yield, ee = 58%
Ar = *p*-BrC_6H_4: 99% yield, ee = 60%
Ar = *p*-$CF_3C_6H_4$: 95% yield, ee = 53%
Ar = *p*-CNC_6H_4: 92% yield, ee = 49%
Ar = 3,5-$(MeO)_2C_6H_3$: 99% yield, ee = 52%
Ar = *p*-$NO_2C_6H_4$: 73% yield, ee = 47%

Scheme 3.8. Oxidation of benzyl tolyl sulfides with chiral titanium trialkanolamine catalyst **5**

were generally obtained in very good yields, in combination with moderate to good enantioselectivities of up to 71% ee.

In addition, Laufer *et al.* have reported the asymmetric titanium-catalysed oxidation of tri- and tetra-substituted 2-thioimidazoles with *tert*-butyl hydroperoxide using chiral BINOL as ligand.[27] The process afforded the corresponding chiral sulfoxides with remarkable enantioselectivities of up to 99% ee in the cases of methyl sulfides, as shown in Scheme 3.9. Indeed, the enantioselectivities were found to be dependent on the substitution pattern of the 2-thioimidazoles. For example, if the methyl group of the sulfur atom at C2 position of the imidazole core was replaced by a bulky isopropyl group or an acetyl group, the enantioselectivities of the corresponding formed products decreased dramatically (51–62% ee). These complex polyfunctionalised chiral products constituted novel p38α mitogen-activated protein kinase inhibitors.

A chiral titanium(IV) complex of BINOL was also successfully applied to the oxidation of aryl sulfides with *tert*-butyl hydroperoxide as oxidant in toluene as the solvent,[28] as well as more recently in ionic liquids.[29] As summarised in Scheme 3.10, excellent levels of enantioselectivity,

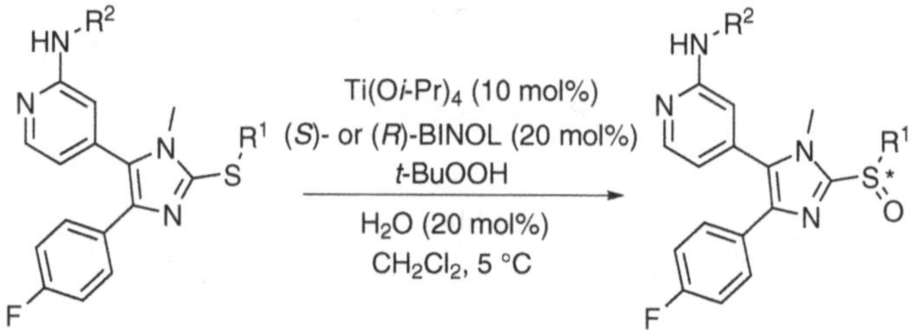

R^1 = Ac, R^2 = Me: 74% yield, ee = 51%
R^1 = Me, R^2 = *i*-Pr: 97% yield, ee = 99%
R^1 = Me, R^2 = tetrahydropyran-4-yl: 77% yield, ee = 98%
R^1 = R^2 = *i*-Pr: 87% yield, ee = 62%
R^1 = Me, R^2 = 4-fluorophenylethyl: 63% yield, ee = 95%
R^1 = Me, R^2 = (*R*)-1-phenylethyl: 92% yield, ee = 93%
R^1 = Me, R^2 = (*S*)-1-phenylethyl: 90% yield, ee = 97%
R^1 = Me, R^2 = Cy: 91% yield, ee = 97%
R^1 = Me, R^2 = (*S*)-3-methylbutan-2-yl: 90% yield, ee = 96%
R^1 = Me, R^2 = (*R*)-3-methylbutan-2-yl: 89% yield, ee = 93%
R^1 = Me, R^2 = Ac: 85% yield, ee = 89%

Scheme 3.9. Oxidation of 2-thioimidazoles with chiral BINOL as ligand

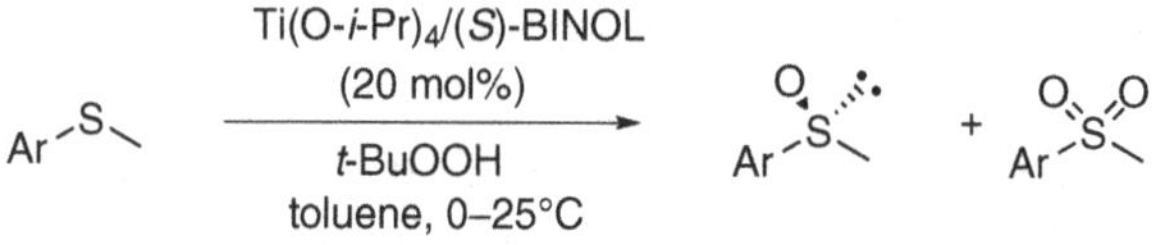

Ar = Ph: 62% ee >99%
Ar = *p*-Tol: 65% ee >99%
Ar = *p*-BrC$_6$H$_4$: 58% ee = 96%

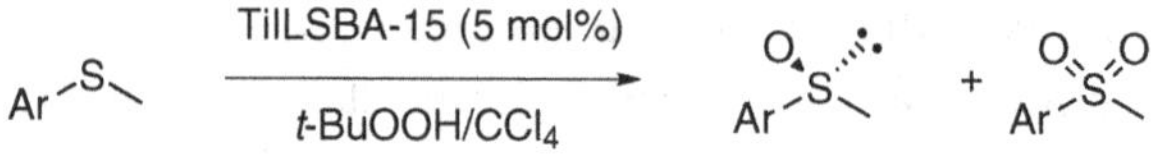

Ar = Ph: 59% ee >99%
Ar = *p*-Tol: 62% ee >99%
Ar = *p*-BrC$_6$H$_4$: 58% ee = 99%
Ar = *p*-ClC$_6$H$_4$: 61% ee = 97%
Ar = *m*-BrC$_6$H$_4$: 57% ee = 99%
Ar = *p*-MeOC$_6$H$_4$: 55% ee >99%
Ar = *p*-FC$_6$H$_4$: 59% ee = 98%

Scheme 3.10. Oxidations of aryl sulfides with chiral (immobilised) titanium BINOL complexes combined with high yields were obtained in both cases. In the case of using the chiral Ti-BINOL complex immobilised onto ionic liquid modified SBA-15 (TiILSBA-15), this catalyst could be reused in multiple catalytic runs without any loss of enantioselectivity.

3.3. Hydrogen Peroxide as Oxidant

Alkyl hydroperoxides are relatively expensive, hazardous, toxic, and produce significant amounts of organic waste in the course of catalytic oxidation. Consequently cheaper and greener oxidants are needed. In this context, environmentally benign hydrogen peroxide is a preferred oxidant, and has been widely applied in titanium-catalysed oxidation of sulfoxides, especially in the presence of chiral salen-derived ligands such as tetrahydrosalen, or salan ligands. As an example, various novel chiral binuclear titanium-salan complexes were investigated by Bryliakov *et al.* in 2008, in combination with aqueous H_2O_2 in the oxidation of benzyl phenyl sulfide.[30] In some cases of catalysts derived from chiral cyclohexane-1,2-diamine such as titanium salan complex (*S,S*)-**6**, the corresponding (*S*)-sulfoxide was obtained in good to high enantioselectivities of up to 97% ee, along with good yield (65%), as shown in Scheme 3.11 (first equation). Remarkably, these catalysts presented the advantage to be capable of performing over 500 turnovers with no loss of enantioselectivity. Later, the same authors employed enantiomer (*R,R*)-**6** under related reaction conditions to enantioselectively oxidise various bulky thioethers.[31] The reaction afforded the corresponding (*R*)-sulfoxides in moderate to good yields, and moderate to excellent enantioselectivities of up to 73% and > 98% ee, respectively (Scheme 3.11, second equation). The best results were reached in the cases of aryl benzyl sulfides as substrates. With the aim of finding a more universal titanium catalyst capable of oxidising both bulky and small sulfides, in 2013 these authors designed novel catalyst **7**, by reducing the steric bulk at the 3,3'-positions of the salicylidene rings, through the replacement of the phenyl groups with hydrogens, and by introducing mild electron acceptors at the 5,5'-positions, such as iodides.[32] As shown in Scheme 3.11 (third equation), good to high enantioselectivities of up to 93% ee were effectively achieved in the oxidation of bulky aryl benzyl sulfides as well as small alkyl phenyl sulfides by using catalyst **7**.

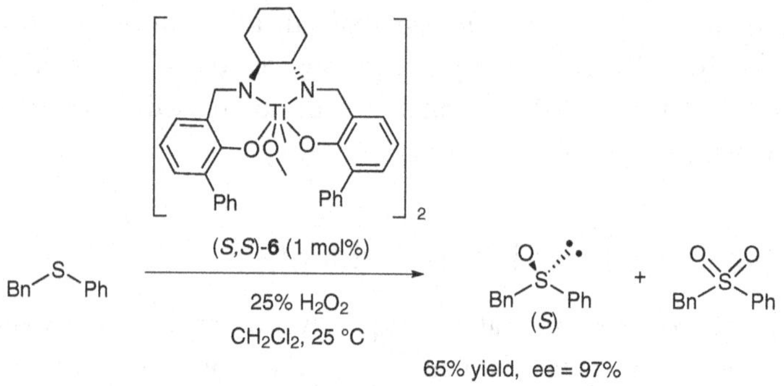
Ph
Ph
2
(S,S)-6 (1 mol%)
Bn-S-Ph
25% H2O2
CH2Cl2, 25 °C
Bn-S(O)-Ph
(S)
+
Bn-SO2-Ph
65% yield, ee = 97%

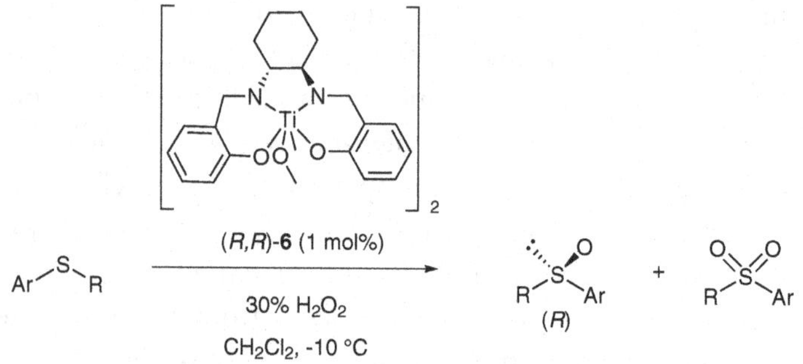
2
(R,R)-6 (1 mol%)
Ar-S-R
30% H2O2
CH2Cl2, -10 °C
R-S(O)-Ar
(R)
+
R-SO2-Ar

Ar = Ph, R = Me: 34% yield (sulfoxide), ee = 3%
Ar = Ph, R = *i*-Pr: 46% yield (sulfoxide), ee = 73%
Ar = Ph, R = Cy: 58% yield (sulfoxide), ee = 63%
Ar = Ph, R = Bn: 72% yield (sulfoxide), ee >98%
Ar = *p*-Tol, R = Bn: 73% yield (sulfoxide), ee = 96%
Ar = *p*-$NO_2C_6H_4$, R = Bn: 72% yield (sulfoxide), ee = 98%
Ar = 2-Naph, R = Bn: 67% yield (sulfoxide), ee = 97%
Ar = 2-Naph, R = Me: 48% yield (sulfoxide), ee = 15%
Ar = Ph, R = $(CH_2)_2SPh$: 26% yield (sulfoxide), ee = 53%

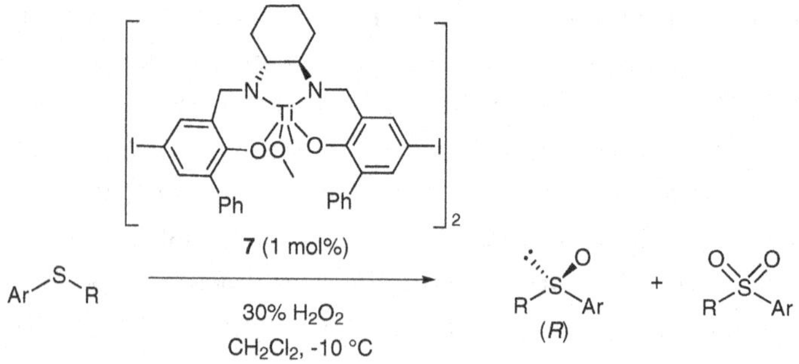
Ph
Ph
2
7 (1 mol%)
Ar-S-R
30% H2O2
CH2Cl2, -10 °C
R-S(O)-Ar
(R)
+
R-SO2-Ar

Ar = 2-Naph, R = Bn: 75% yield (sulfoxide), ee = 93%
Ar = 2-Naph, R = Me: 74% yield (sulfoxide), ee = 72%
Ar = Ph, R = *i*-Pr: 72% yield (sulfoxide), ee = 79%
Ar = Ph, R = Me: 78% yield (sulfoxide), ee = 77%

Scheme 3.11. Oxidations of aryl sulfides with chiral binuclear titanium-salan complexes **6** and **7**

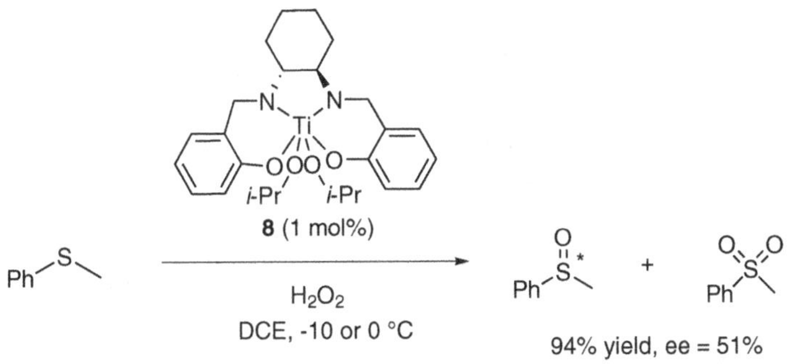

Scheme 3.12. Oxidation of thioanisole with chiral binuclear titanium-salan complex **8**

In the same area, moderate enantioselectivities of up to 51% ee in combination with excellent yields of up to 94% were reported by Pessoa and Correia, using another chiral titanium-salan complex **8**, also bearing no substituent on the aromatic groups of the ligand, in the oxidation of benzyl phenyl sulfide, as shown in Scheme 3.12.[33]

Metal-organic frameworks (MOFs) are hybrid crystalline solids composed of organic struts and inorganic nodes, and have emerged as a novel type of highly porous materials. Owing to their tunable but uniform pore sizes, and functionalisable pore walls, MOFs are attractive solid-state scaffolds for molecule transformations relevant to chemical feedstocks, energy conversions, and minimising environmental pollution. They offer advantages over other immobilised catalyst systems as crystalline structures, with high catalyst loadings, more uniformity, and isolated and accessible active sites. In 2013, Cui *et al.* reported the synthesis of a dipyridyl-functionalised chiral titanium-salan ligand **9** to make a porous MOF, containing interesting Ti_4O_6 cluster cores and showed that the framework was an efficient heterogeneous catalyst for the asymmetric oxidation of thioethers with aqueous hydrogen peroxide as oxidant.[34] As shown in Scheme 3.13, the oxidation of various aryl sulfides afforded the corresponding chiral sulfoxides in good yields and moderate to good enantioselectivities of up to 82% ee. The authors demonstrated that this heterogeneous catalyst improved the enantioselectivity of the reaction relative to the corresponding monomeric catalyst.

In the same context, Cui and Liu designed novel chiral porous metal-metallosalan frameworks on the basis of an unsymmetrical chiral pyridinecarboxylate ligand derived from titanium-salan.[35] Microporous

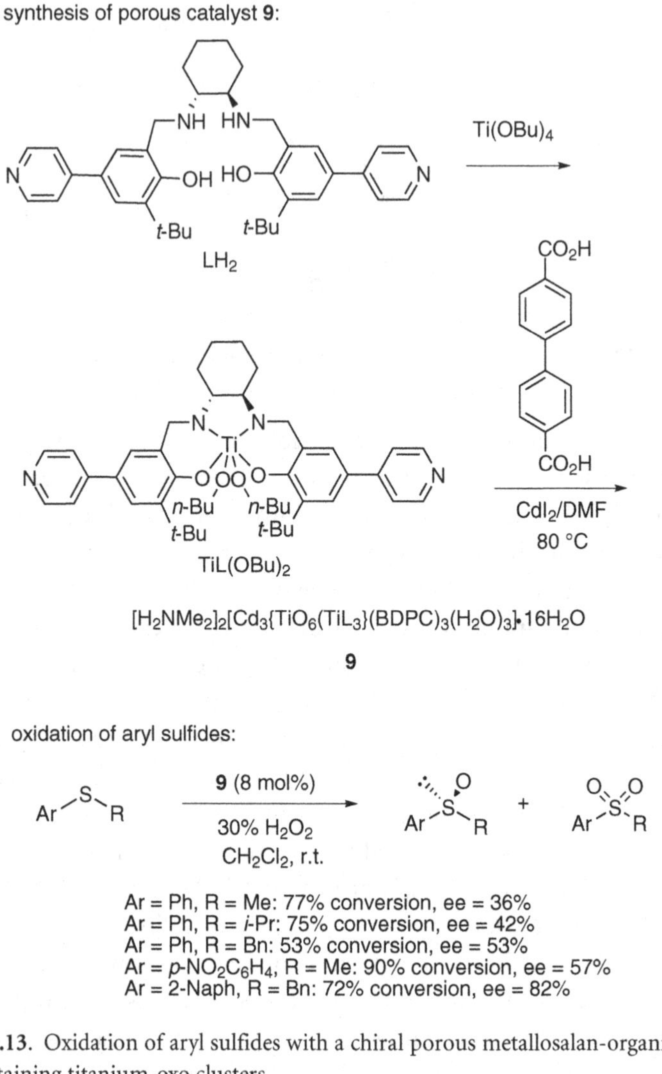
synthesis of porous catalyst 9:
NH HN
OH HO
N
N
t-Bu
t-Bu
LH2
Ti(OBu)4
CO2H
CO2H
CdI2/DMF
80 °C
Ti
n-Bu
n-Bu
TiL(OBu)2
[H2NMe2]2[Cd3{TiO6(TiL3}(BDPC)3(H2O)3]•16H2O
9
oxidation of aryl sulfides:
Ar
S
R
9 (8 mol%)
30% H2O2
CH2Cl2, r.t.
Ar = Ph, R = Me: 77% conversion, ee = 36%
Ar = Ph, R = i-Pr: 75% conversion, ee = 42%
Ar = Ph, R = Bn: 53% conversion, ee = 53%
Ar = p-NO2C6H4, R = Me: 90% conversion, ee = 57%
Ar = 2-Naph, R = Bn: 72% conversion, ee = 82%

Scheme 3.13. Oxidation of aryl sulfides with a chiral porous metallosalan-organic framework containing titanium-oxo clusters

metal-organic framework **10**, based on a pyridine- and carboxylic acid-functionalised titanium salan ligand, was demonstrated to be a heterogeneous catalyst for the asymmetric oxidation of sulfides, providing the corresponding sulfoxides in moderate enantioselectivities of up to 62% ee, as shown in Scheme 3.14. It was shown that the use of this heterogeneous catalyst allowed improved enantioselectivity to be obtained relative to the

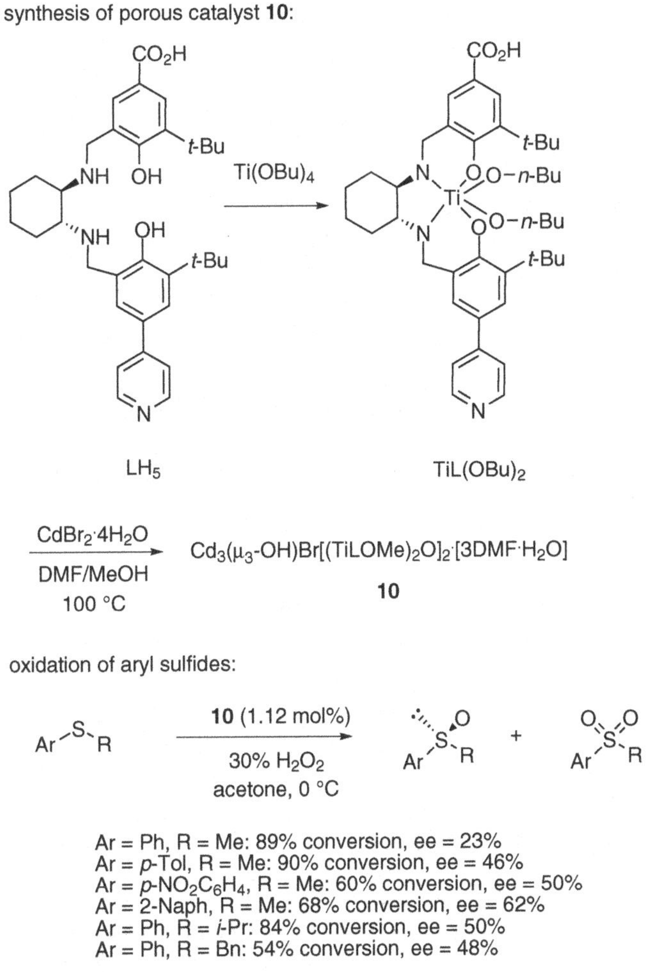
synthesis of porous catalyst 10:
CO2H
t-Bu
NH OH
NH OH
t-Bu
N
Ti(OBu)4
CO2H
t-Bu
N O O−n-Bu
Ti
N O O−n-Bu
t-Bu
N
LH5
TiL(OBu)2
CdBr2·4H2O
DMF/MeOH
100 °C
Cd3(μ3-OH)Br[(TiLOMe)2O]2·[3DMF·H2O]
10
oxidation of aryl sulfides:
Ar S R
10 (1.12 mol%)
30% H2O2
acetone, 0 °C
O
Ar S R
+
O O
Ar S R
Ar = Ph, R = Me: 89% conversion, ee = 23%
Ar = p-Tol, R = Me: 90% conversion, ee = 46%
Ar = p-NO2C6H4, R = Me: 60% conversion, ee = 50%
Ar = 2-Naph, R = Me: 68% conversion, ee = 62%
Ar = Ph, R = i-Pr: 84% conversion, ee = 50%
Ar = Ph, R = Bn: 54% conversion, ee = 48%

Scheme 3.14. Oxidation of aryl sulfides with another chiral porous metallosalan-organic framework containing titanium-oxo clusters

homogeneous catalyst. Moreover, catalyst **10** could be recovered in quantitative yield and used repeatedly without significantly degrading the catalytic performance for the following three runs.

Chiral Schiff bases have been investigated as ligands in the titanium-catalysed oxidation of aryl sulfides by Wang *et al.* in 2011.[36] Indeed, sterically hindered Schiff bases, prepared from 3,5-dicumenyl salicylaldehyde and chiral amino alcohols, were used in combination with titanium

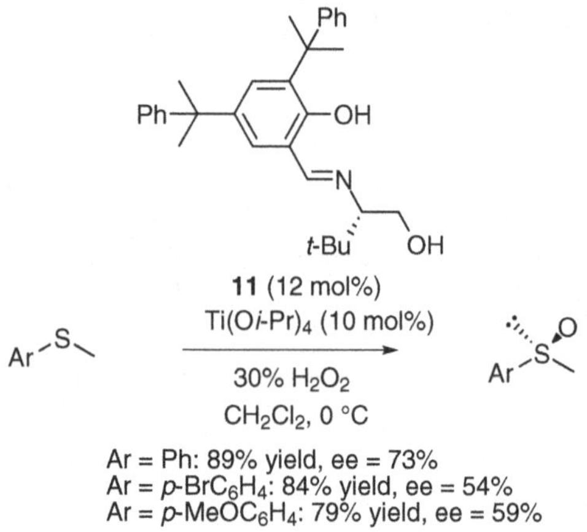

Scheme 3.15. Oxidation of aryl sulfides with Schiff base **11** as chiral ligand

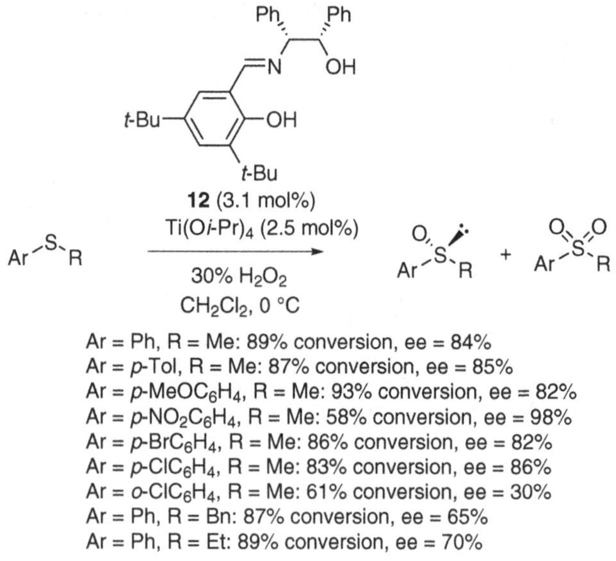

Scheme 3.16. Oxidation of aryl sulfides with Schiff base **12** as chiral ligand

tetraisopropylate with aqueous H_2O_2 as oxidant. Among these ligands, Schiff base **11** bearing a *tert*-butyl group in the chiral carbon of the amino alcohol moiety gave the best result with 89% yield and 73% ee for the sulfoxidation of thioanisole, as shown in Scheme 3.15.

In 2012, a related Schiff base ligand **12** was demonstrated by Abdfi *et al.* to give better results in comparable reactions, since enantioselectivities of up to 98% ee were achieved by using only 3.1 mol% of this ligand (Scheme 3.17).[37] On the basis of a fine tuning of the catalytic activity, the authors demonstrated a significant role of the steric influence of the substituent attached on both aryl and alkyl moiety on the enantioselectivity of the process.

3.4. Conclusions

Titanium is probably the most important metal in asymmetric sulfoxidations. In the last few years, various types of chiral titanium catalysts have been successfully developed to promote the enantioselective oxidation of sulfides into chiral sulfoxides, using alkyl hydroperoxides as well as hydrogen peroxide as the oxidant. Concerning the use of alkyl hydroperoxides — such as cumene hydroperoxide or *tert*-butyl hydroperoxide — excellent enantioselectivities of up to >99% ee were achieved by using tartrate and tartramide ligands, following modifications of the early method reported by Kagan and Modena, but also with the use of a number of chiral 1,2-diols derived from hydrobenzoin in addition to BINOL. In contrast, successful enantioselective titanium-mediated sulfoxidations using hydrogen peroxide as the oxidant had been relatively rare until the 2000s. So far, various types of titanium based catalyst systems for the enantioselective oxidation of sulfide with hydrogen peroxide have been reported, bearing either Schiff bases or salan-type ligands as chiral promoters. The best enantioselectivities of up to >98% ee were reached in the case of salan ligands.

In the past two decades, the synthesis of chiral sulfoxides has greatly expanded, significantly contributing to the production of drugs and other valuable chemical products. Despite low efficiency and other disadvantages, the modified versions from Modena and Kagan methodology remain the most applied so far, due to their well-studied substrate scope, tolerance to functional groups and ease of implementation. Time will be needed to implement more environmentally safe new developments based on the use of hydrogen peroxide. In the near future, further progress of catalyst systems of this type will have to be adapted for preparative asymmetric syntheses of biologically active sulfoxides. It is obvious that titanium/hydrogen peroxide based systems will find use in drug production, where metal contamination of the final product is critical; indeed, TiO_2 which is the product of hydrolysis of titanium complexes, is biologically inert and occurs in many pharmaceutical compositions. Chemists look forward to witnessing future studies in this direction.

References

(1) Y. Yu, K. Ding, G. Chen, in *Acid Catalysis in Modern Organic Synthesis*, H. Yamamoto, K. Ishihara, eds., Wiley-VCH: Weinheim, **2008**, Chapter 14, pp. 721–823.

(2) (a) M. Mikolajczyk, J. Drabowicz, *Top. Curr. Chem.* **1982**, *13*, 333–468. (b) M. R. Barbachyn, C. R. Johnson, *Asymmetric Synthesis* **1984**, *4*, 227–261. (c) M. C. Carreno, *Chem. Rev.* **1995**, *95*, 1717–1760. (d) M. Mikolajczk, J. Drabowicz, P. Kielbasinski in *Chiral Sulfur Reagents*, CRC Press, **1997**. (e) H. Matsuyama, *Sulfur Rep.* **1999**, *22*, 85–121. (f) S. G. Pyne, *Sulfur Rep.* **1999**, *21*, 281–334. (g) P. C. Taylor, *Sulfur Rep.* **1999**, *21*, 241–280. (h) J. C. Bayón, C. Claver, A. M. Masdeu-Bultó, *Coord. Chem. Rev.* **1999**, *193–195*, 73–145. (i) G. Solladié, *Enantiomer* **1999**, *4*, 183–193. (j) A. J. Blake, P. A. Cooke, J. D. Kendall, N. S. Simpkins, S. M. Westaway, *J. Chem. Soc., Perkin Trans. I.* **2000**, 153–163. (k) A. M. Masdeu-Bultó, M. Diéguez, E. Martin, M. Gómez, *Coord. Chem. Rev.* **2003**, *242*, 159–201. (l) H. Pellissier, ed., in *Chiral Sulfur Ligands, Asymmetric Catalysis*, Royal Society of Chemistry: Cambridge, **2009**.

(3) (a) G. Solladié, *Synthesis* **1981**, 185–196. (b) G. H. Posner, *Acc. Chem. Res.* **1987**, *20*, 72–78. (c) H. G. Posner, in *The Chemistry of Sulfones and Sulfoxides*, S. Patai, Z. Rappoport, C. J. M. Stirling, eds., Wiley and Sons: Chichester, **1988**, Chap 16, pp 823–849. (d) J. Drabowicz, P. Kiebasinki, M. Mikolajczk, in *The Chemistry of Sulfones and Sulfoxides*, S. Patai, Z. Rappoport, C. J. M. Stirling, eds., Wiley and Sons: Chichester, **1988**, Chap 16, pp. 233–378.

(4) (a) G. Kresze, *Methoden der Organischen Chemie* (Houben-Weyl), D. Klamann, ed., Georg Thieme Verlag: Stuttgart, **1985**, pp. 669–886. (b) K. K. Andersen, *The Chemistry of Sulfones and Sulfoxides*, S. Patai, Z. Rappoport, C. J. M. Stirling, eds., John Wiley and Sons: New York, **1988**, Chapter 3, pp. 56–94. (c) G. Solladié, *Comprehensive Organic Synthesis*, B. M. Trost, I. Fleming, eds., Pergamon: Oxford, **1991**, Vol. 6, Chapter 3, pp. 148–170. (d) A. J. Walker, *Tetrahedron: Asymmetry* **1992**, *3*, 961–998. (e) G. Solladié, M. C. Carreno, *Organosulfur Chemistry. Synthetic Aspects*, P. C. B. Page, ed., Academic Press: New York, **1995**, Chapter 1, pp 1–47. (f) I. Fernandez, N. Khiar, *Chem. Rev.* **2003**, *103*, 3651–3705.

(5) (a) D. J. Procter, *J. Chem. Soc., Perkin Trans. I,* **1999**, 641–668. (b) D. J. Procter, *J. Chem. Soc., Perkin Trans. I.,* **2000**, 835–871. (c) C.-C. Wang, H.-C. Huang, D. B. Reitz, *Org. Prep. Proc. Int.* **2002**, *34*, 271–319. (d) H. Pellissier, *Tetrahedron* **2006**, *62*, 5559–5601.

(6) (a) C. M. Rayner, *Contemp. Org. Synth.* **1994**, *1*, 191–203. (b) N. Khiar, I. Fernandez, A. Alcudia, F. Alcudia, in *Advances in Sulfur Chemistry 2*, C. M. Rayner, Ed., JAI Press Inc.: Stamford, **2000**, Chapter 3, p. 57. (c) D. Procter, *J. Chem. Soc., Perkin Trans. I*, **2001**, 335–354.

(7) (a) J. Legros, J. R. Dehli, C. Bolm, *Adv. Synth. Cat.* **2005**, *347*, 19–31. (b) K. P. Bryliakov, E. P. Talsi, *Curr. Org. Chem.* **2012**, *16*, 1215–1242. (c) Q. Zeng, S. Gao, A. K. Chelashaw, *Mini-Rev. Org. Chem.* **2013**, *10*, 198–206. (d) K. P. Bryliakov, *Mini-Rev. Org. Chem.* **2014**, *11*, 87–96.

(8) (a) H. B. Kagan, F. Rebiere, O. Samuel, *Phosphorus, Sulfur Silicon Relat. Elem.* **1991**, *58*, 89–110. (b) S. M. Allin, *Organosulfur Chemistry*, Academic Press, Inc., **1998**, p 41.

(9) K. K. Andersen, *Tetrahedron Lett.* **1962**, *3*, 93–95.

(10) K. K. Andersen, W. Gaffield, N. E. Papanikolau, J. W. Foley, R. I. Perkins, *J. Am. Chem. Soc.* **1964**, *86*, 5637–5646.

(11) M. Axelrod, P. Bickart, J. Jacobus, M. M. Green, K. Mislow, *J. Am. Chem. Soc.* **1969**, *90*, 4835–4842.

(12) (a) F. Wudl, T. B. K. Lee, *J. Am. Chem. Soc.* **1973**, *95*, 6349–6358. (b) F. Rebiere, H. B. Kagan, *Tetrahedron Lett.* **1989**, *30*, 3659–3662. (c) I. Fernandez, N. Khiar, J. M. Llera, F. Alcudia, *J. Org. Chem.* **1992**, *57*, 6789–6796. (d) I. Fernandez, N. Khiar, F. Alcudia, *Tetrahedron Lett.* **1994**, *35*, 5719–5722.

(13) A. Korte, J. Legros, C. Bolm, *Synlett* **2004**, *13*, 2397–2399.

(14) A. R. Maguire, S. Papot, A. Ford, S. Touhey, R. O'Connor, M. Clynes, *Synlett* **2001**, *1*, 41–44.

(15) M. Matsugi, N. Fukuda, Y. Muguruma, T. Yamaguchi, J.-I. Minamikawa, S. Otsuka, *Tetrahedron* **2001**, *57*, 2739–2744.

(16) (a) P. Pitchen, H. B. Kagan, *Tetrahedron Lett.* **1984**, *25*, 1049–1052. (b) P. Pitchen, M. Desmukh, E. Dunach, H. B. Kagan, *J. Am. Chem. Soc.* **1984**, *106*, 8188–8193.

(17) F. Di Furia, G. Modena, R. Seraglia, *Synthesis* **1984**, 325–326.

(18) T. M. Khomenko, K. P. Volcho, N. I. Komarova, N. F. Salakhutdinov, *Russ. J. Org. Chem.* **2008**, *44*, 124–127.

(19) G. Che, J. Xiang, T. Tian, Q. Huang, L. Cun, J. Liao, Q. Wang, J. Zhu, J. Deng, *Tetrahedron: Asymmetry* **2012**, *23*, 457–460.

(20) M. Seenivasaperumal, H.-J. Federsel, K. J. Szabo, *Adv. Synth. Catal.* **2009**, *351*, 903–919.

(21) S. Superchi, P. Scafato, L. Restaino, C. Rosini, *Chirality* **2008**, *20*, 592–596.

(22) B. Jiang, X.-L. Zhao, J.-J. Dong, W.-J. Wang, *Eur. J. Org. Chem.* **2009**, 987–991.

(23) (a) F. Naso, M. A. M. Capozzi, A. Bottoni, M. Calvaresi, V. Bertolasi, F. Capitelli, C. Cardellicchio, *Chem. Eur. J.* **2009**, *15*, 13417–13426. (b) C. Cardellicchio, M. A. M. Capozzi, C. Centrone, F. Naso, *Phosphorus, Sulfur Silicon Relat. Elem.* **2011**, *186*, 1193–1195. (c) M. A. M. Capozzi, C. Centrone, G. Fracchiolla, F. Naso, C. Cardellicchio, *Eur. J. Org. Chem.* **2011**, 4327–4334. (d) M. A. M. Capozzi, F. Capitelli, A. Bottoni, M. Calvaresi, C. Cardellicchio, *ChemCatChem* **2013**, *5*, 210–219.

(24) J. Sun, M. Yang, Z. Dai, C. Zhu, H. Hu, *Synthesis* **2008**, *16*, 2513–2518.

(25) P. Axe, S. D. Bull, M. G. Davidson, M. D. Jones, D. E. J. E. Robinson, W. L. Mitchell, J. E. Warren, *Dalton Trans.* **2009**, 10169–10171.

(26) G. Santoni, M. Mba, M. Bronchio, W. A. Nugent, C. Zonta, G. Licini, *Chem. Eur. J.* **2010**, *16*, 645–654.

(27) S. Bühler, M. Goettert, D. Schollmeyer, W. Albrecht, S. A. Laufer, *J. Med. Chem.* **2011**, *54*, 3283–3297.

(28) Lia, X. Li, X. Xu, L. Li, Y. Shi, Q. Au-Yeung, T. T. L. Yip, C. W. Yao, X. Chan, A. S. C. Adv. *Synth. Catal.* **2004**, *346*, 723–726.

(29) S. Sahoo, P. Kumar, F. Lefebvre, S. B. Halligudi, *J. Catal.* **2009**, *262*, 111–118.

(30) K. P. Bryliakov, E. P. Talsi, *Eur. J. Org. Chem.* **2008**, 3369–3376.

(31) K. P. Bryliakov, E. P. Talsi, *Eur. J. Org. Chem.* **2011**, 4693–4698.

(32) E. P. Talsi, K. P. Bryliakov, *Appl. Organometal. Chem.* **2013**, *27*, 239–244.

(33) P. Adao, F. Avecilla, M. Bonchio, M. Carraro, J. Costa Pessoa, I. Correia, *Eur. J. Org. Chem.* **2010**, 5568–5578.

(34) W. Xuan, C. Ye, M. Zhang, Z. Chen, Y. Cui, *Chem. Sci.* **2013**, *4*, 3154–3159.

(35) C. Zhu, X. Chen, Z. Yang, X. Du, Y. Liu, Y. Cui, *Chem. Commun.* **2013**, *49*, 7120–7122.

(36) Y. Wang, M. Wang, L. Wang, Y. Wang, X. Wang, L. Sun, *Appl. Organometal. Chem.* **2011**, *25*, 325–330.

(37) P. K. Bera, D. Ghosh, S. H. R. Abdi, N.-u. H. Khan, R. I. Kureshy, H. C. Bajaj, *J. Mol. Catal. A* **2012**, *361–362*, 36–44.

Chapter 4

Enantioselective Titanium-Catalysed Epoxidation Reactions

4.1. Introduction

This chapter covers efforts of the chemical community to develop novel enantioselective titanium-catalysed epoxidation reactions. It looks at developments since the beginning of 2008, when the field was reviewed by Yu and co-workers in a book chapter dealing with titanium Lewis acids.[1]

This chapter is divided into three parts. The first deals with enantioselective titanium-catalysed epoxidations using alkyl hydroperoxides as oxidants; the second considers enantioselective titanium-catalysed epoxidations employing hydrogen peroxide as oxidant; while the third includes enantioselective titanium-catalysed Darzens reactions.

Catalytic asymmetric epoxidations constitutes an important class of transformations in synthetic chemistry.[2] Olefins are rather inert compounds themselves, and require functionalisation including oxygenation prior to use as chemical materials, whereas enantiopure epoxides are valued as versatile and reactive — yet stable — intermediates for various small- and large-scale synthetic applications.[3] In 1980, Katsuki and Sharpless, reported the first successful catalytic enantioselective titanium-catalysed epoxidation of allylic alcohols, with *tert*-butyl hydroperoxide as oxidant using a tartrate chiral ligand, providing enantioselectivities of up to >95% ee.[4]

Since then, various chiral transition-metal-based systems including titanium complexes have been developed. The most reputable ones are the Katsuki *et al.*[5] and Jacobsen *et al.*[6] manganese-salen-based systems for the epoxidation of unfunctionalised olefins with iodosylarenes, *meta*-chloroperbenzoic acid, and hypochlorite. These have inspired the search for new catalyst systems worldwide. Meanwhile, the development of asymmetric epoxidation using cheap and environmentally friendly hydrogen peroxide as the oxidant has attracted much attention in recent studies.

4.2. Epoxidations using Hydrogen Peroxide

Surprisingly, no enantioselective epoxidation with hydrogen peroxide in the presence of titanium catalysts was reported until 2005, when Katsuki *et al.* reported that a chiral bis(μ-oxido) titanium(IV) salanen complex was capable of efficiently promoting the epoxidation of unfunctionalised olefins with 30% hydrogen peroxide, providing enantioselectivities of up to >99% ee.[7] In 2008, these authors reported the synthesis of a range of salan ligands to be investigated in enantioselective titanium-catalysed epoxidation of various olefins with aqueous hydrogen peroxide.[8] Among them, the authors selected ligand 1, bearing an ortho-methoxyphenyl group, as the most efficient one, which provided — when using at a low catalyst loading of 1.3 mol% — a series of chiral epoxides in, generally, remarkable yields, and enantioselectivities of up to 99% and 98% ee, respectively (Scheme 4.1). Notably, this chiral ligand was easily prepared in five steps from commercially available 2,2'-biphenol, rendering this epoxidation methodology truly practical.

In 2009, the same authors designed a novel series of C_1-symmetric salan ligands derived from proline that were further investigated in the enantioselective epoxidation of styrenes, with titanium tetraisopropylate and aqueous hydrogen peroxide as the oxidant.[9] Synthetically valuable chiral styrene oxides were achieved, with high enantioselectivities ranging from 96 to 98% ee by using ligand 2 at 10 mol% of catalyst loading, irrespective of the electronic nature of the substituents and substitution pattern (Scheme 4.2). In this remarkable process, the active catalyst was generated *in situ*.

The same year, the same authors extended the scope of their early chiral catalyst bis(μ-oxido) titanium (IV) salanen complex 3[6] to the epoxidation

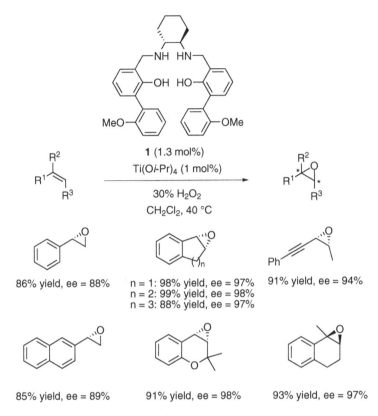

Scheme 4.1. Epoxidation of olefins with chiral salan 1 as ligand

of *cis*-alkenylsilanes.[10] Remarkably, all the formed epoxysilanes were obtained with complete enantioselectivity, in the presence of 0.5–2 mol% of the complex. As shown in Scheme 4.3, the enantiopure products were obtained in generally excellent yields, making this a powerful approach, allowing synthetically important epoxides, such as styrene oxides and geminally disubstituted epoxides, in enantiopure form, to be achieved.

In 2011, the authors demonstrated that catalyst **3** could be applied to the epoxidation of aldehyde enol esters.[11] Although (*E*)-enol esters were reluctant to proceed, (*Z*)-enol esters underwent asymmetric epoxidation to give the corresponding chiral epoxides in high yields, and enantioselectivities ranging from 63 to >99% ee in the presence of aqueous hydrogen peroxide as oxidant, as shown in Scheme 4.4. The authors showed that

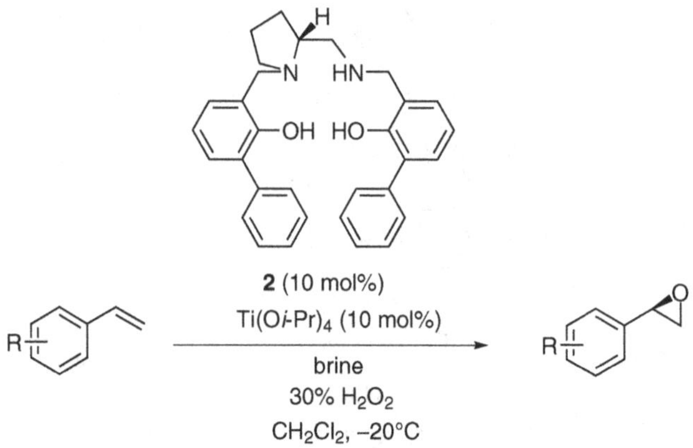

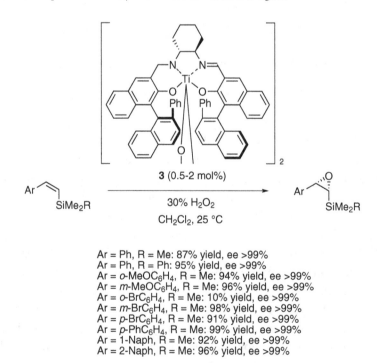

H
N
HN
OH
HO
2 (10 mol%)
Ti(Oi-Pr)4 (10 mol%)
brine
30% H2O2
CH2Cl2, –20°C
R
O

R = H: 71% yield, ee = 98%
R = *o*-Me: 70% yield, ee = 97%
R = *m*-Me: 84% yield, ee = 98%
R = *p*-Me: 80% yield, ee = 98%
R = *o*-Cl: 16% yield, ee = 96%
R = *m*-Cl: 66% yield, ee = 98%
R = *p*-Cl: 66% yield, ee = 98%

Scheme 4.2. Epoxidation of styrenes with chiral salan **2** as ligand

N
N
Ti
O
O
Ph
Ph
O
2
3 (0.5-2 mol%)
30% H2O2
CH2Cl2, 25 °C
Ar
SiMe2R

Ar = Ph, R = Me: 87% yield, ee >99%
Ar = Ph, R = Ph: 95% yield, ee >99%
Ar = *o*-MeOC$_6$H$_4$, R = Me: 94% yield, ee >99%
Ar = *m*-MeOC$_6$H$_4$, R = Me: 96% yield, ee >99%
Ar = *o*-BrC$_6$H$_4$, R = Me: 10% yield, ee >99%
Ar = *m*-BrC$_6$H$_4$, R = Me: 98% yield, ee >99%
Ar = *p*-BrC$_6$H$_4$, R = Me: 91% yield, ee >99%
Ar = *p*-PhC$_6$H$_4$, R = Me: 99% yield, ee >99%
Ar = 1-Naph, R = Me: 92% yield, ee >99%
Ar = 2-Naph, R = Me: 96% yield, ee >99%

Scheme 4.3. Epoxidation of *cis*-alkenylsilanes with chiral di-μ-oxo-titanium salalen complex **3**

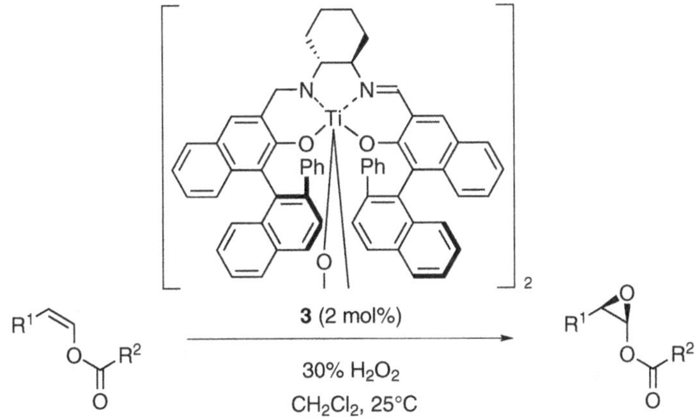

R^1 = n-Hex, R^2 = Ph: > 95% yield, ee = 92%
R^1 = n-Hex, R^2 = p-CF$_3$C$_6$H$_4$: >95% yield, ee = 87%
R^1 = n-Hex, R^2 = p-MeOC$_6$H$_4$: >95% yield, ee = 93%
R^1 = n-Hex, R^2 = o-MeOC$_6$H$_4$: >95% yield, ee = 92%
R^1 = n-Hex, R^2 = 2,4-(MeO)$_2$C$_6$H$_3$: 82% yield, ee = 93%
R^1 = n-Hex, R^2 = CH$_2$Bn: 36% yield, ee = 63%
R^1 = i-Bu, R^2 = p-MeOC$_6$H$_4$: 86% yield, ee = 98%
R^1 = Cy, R^2 = p-MeOC$_6$H$_4$: 89% yield, ee = 98%
R^1 = t-Bu, R^2 = p-MeOC$_6$H$_4$: 90% yield, ee >99%
R^1 = R^2 = Ph: 83% yield, ee = 86%
R^1 = Ph, R^2 = CH$_2$Bn: 78% yield, ee = 97%

Scheme 4.4. Epoxidation of (Z)-enol esters with chiral di-μ-oxo-titanium salalen complex **3**

ketone enol esters also underwent epoxidation with good enantioselectivity (up to 85% ee), albeit with low yield (21%). The obtained epoxides were further transformed into the corresponding 1,2-diols, with no erosion of the enantioselectivity.

In 2010, a series of chiral salan (salanen) ligands to be investigated in titanium-catalysed epoxidations of olefins with aqueous hydrogen peroxide were designed and synthesised by Sun *et al.*[12] Among these ligands derived from binaphthol, ligand **4** was found to be quite efficient in these reactions, especially for the reaction of 6-cyano-2,2-dimethyl chromene, which afforded the corresponding epoxide with an excellent enantioselectivity of 99% ee, as shown in Scheme 4.5.

In 2011, the same authors decided to bridge two salanen ligands with a biaryl linker and to synthesise their corresponding di-μ-oxotitanium complexes.[13] Among them, chiral dinuclear salalen titanium complex **5** was

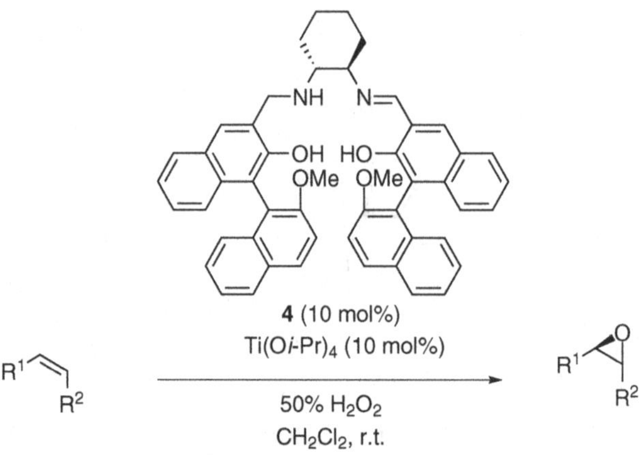

4 (10 mol%)

Ti(O*i*-Pr)$_4$ (10 mol%)

$$\xrightarrow{\qquad\text{50\% H}_2\text{O}_2\qquad}$$

CH$_2$Cl$_2$, r.t.

R^1 = Ph, R^2 = H: 61% yield, ee = 79%
R^1 = *p*-ClC$_6$H$_4$, R^2 = H: 73% yield, ee = 76%
R^1 = *m*-CNC$_6$H$_4$, R^2 = (Me)$_2$C-O: 92% yield, ee = 99%
R^1 = Ph, R^2 = Me: 95% yield, ee = 80%
R^1 = Cy, R^2 = H: 50% yield, ee = 51%

Scheme 4.5. Epoxidation of various olefins with chiral salalen ligand **4**

found to serve as an effective catalyst for the asymmetric epoxidation of a variety of terminal aromatic olefins with aqueous hydrogen peroxide. As shown in Scheme 4.6, high enantioselectivities of up to 99% ee were achieved, in combination with good yields. It must be noted that all styrene derivatives provided enantioselectivities of >90% ee, whereas epoxidation of 2-vinylnaphthalene gave the corresponding epoxide in 99% ee. The catalytic system was also successfully applied to *cis*-olefins although the reaction of *trans*-olefins proved to be very sluggish.

With the aim of discovering a novel catalyst system capable of epoxidising non-conjugated olefins, in 2013 Berkessel *et al.* disclosed a two-step approach to chiral salanen ligands derived from *cis*-1,2-diaminocyclohexane.[14] Their corresponding *in situ* generated titanium complexes were characterised by X-ray crystallography, and were investigated in the enantioselective epoxidation of a range of non-conjugated olefins with aqueous hydrogen peroxide. The readily accessible salanen ligand **6**, bearing cyclohexyl substituents *ortho* to the phenolic hydroxyl groups, proved particularly effective. As shown in Scheme 4.7, a range of chiral epoxides

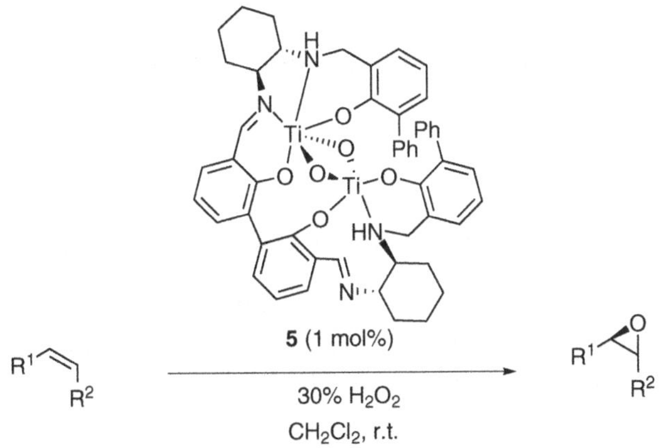

5 (1 mol%)

R^1 ⟶ R^2

30% H_2O_2
CH_2Cl_2, r.t.

⟶

R^1 epoxide R^2

R^1 = Ph, R^2 = H: 70% yield, ee = 90%
R^1 = p-ClC$_6$H$_4$, R^2 = H: 68% yield, ee = 94%
R^1 = p-BrC$_6$H$_4$, R^2 = H: 74% yield, ee = 92%
R^1 = 2-Naph, R^2 = H: 93% yield, ee = 99%
R^1 = p-CNC$_6$H$_4$, R^2 = (Me)$_2$C-O: 90% yield, ee = 99%
R^1 = Ph, R^2 = Me: 90% yield, ee = 96%
R^1 = PhC≡C, R^2 = Me: 92% yield, ee = 92%
R^1 = Cy, R^2 = H: 10% yield, ee = 78%

Scheme 4.6. Epoxidation of various olefins with chiral dinuclear salalen titanium complex **5**

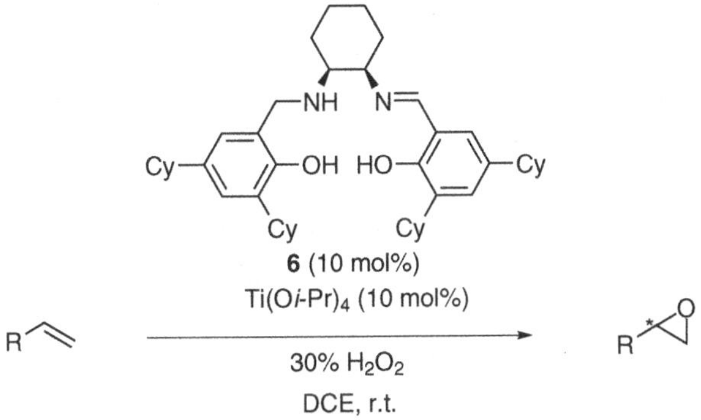

6 (10 mol%)
Ti(Oi-Pr)$_4$ (10 mol%)

R ⟶

30% H_2O_2
DCE, r.t.

⟶

R epoxide

R = n-Hex: 82% yield, ee = 95%
R = Cy: 90% yield, ee = 90%
R = n-Bu: 70% yield, ee = 89%
R = (CH$_2$)$_4$OH: 34% yield, ee = 90%
R = (CH$_2$)$_4$OBz: 94% yield, ee = 90%
R = (CH$_2$)$_4$OBn: 91% yield, ee = 89%

Scheme 4.7. Epoxidation of non-conjugated olefins with chiral salalen ligand **6**

derived from non-conjugated olefins were achieved in moderate to high yields, albeit with general high enantioselectivities of up to 95% ee. This novel process presented the advantage of being effective for a notoriously difficult substrate class.

4.3. Epoxidations using Alkyl Hydroperoxides

The asymmetric epoxidation of allylic alcohols by Sharpless has become a classic in asymmetric catalysis.[4] In this process, the chiral ligand plays a critical role in the enantioselectivity. The catalytically active species is a titanium dimer, containing two dialkyltartrate ligands, with both the oxidant and the allylic alcohol coordinated. However, there are few reports on the assemblies of chiral titanium complexes over solid supports in order to achieve the heterogenisation of these catalyst components.

In 2009, Ballesteros *et al.* described the synthesis of well-defined organotitanium compounds tethered onto capped MCM-41, capable of acting as active catalysts in the Sharpless epoxidation process.[15] As shown in Scheme 4.8, titanium-triazine-based MCM-41 hybrid material 7 was generated from triazine propyl triethoxysilane (CyPTS), as functional linker, and hexamethyldisilazane (HMDS) as capped agent, to increase the hydrophobicity of the support and mask the remaining silanol groups. Also, titanium tetraisopropylate was heterogenised by reaction with the modified MCM-41. Finally, after the immobilisation of titanium tetraisopropylate on to the organomodified support, the reaction with the chiral auxiliary (*S,S*)-diethyl tartrate was accomplished. When applied to the asymmetric epoxidation of cinnamyl alcohol with *tert*-butyl hydroperoxide as oxidant, catalyst 7 allowed the corresponding epoxide to be achieved in low yield (24%), but with good enantioselectivity of 81% ee (Scheme 4.8).

In 2012, the same authors investigated a novel family of chiral alkyl lactate mono- and disubstituted isopropoxo titanium complexes with the same process.[16] The obtained results (≤26% ee) showed that these complexes with dynamic behaviour in solution exerted a low control of the enantioselectivity of the Sharpless epoxidation of cinnamyl alcohol.

Finally, the Sharpless epoxidation was applied in 2014 by Hou and Bian in an efficient total synthesis of the Colorado potato beetle pheromone.[17]

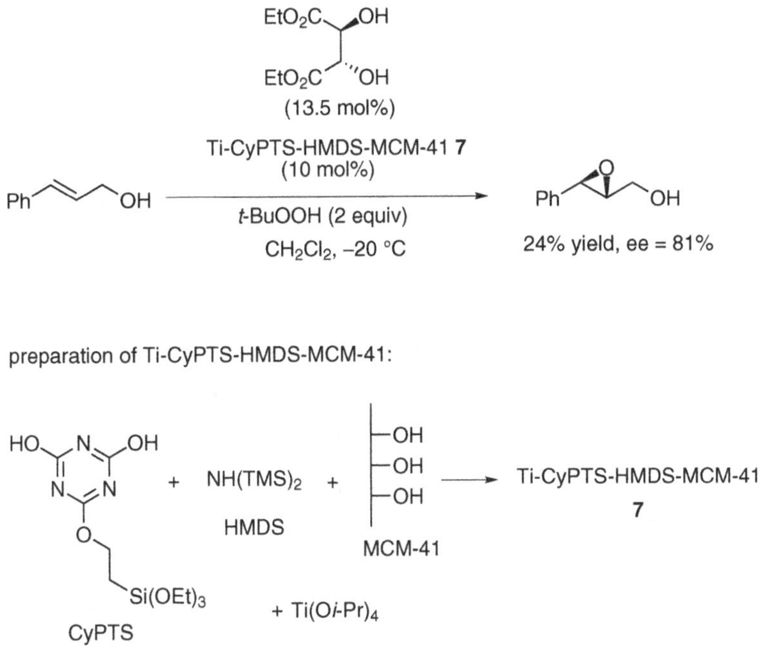

Scheme 4.8. Epoxidation of cinnamyl alcohol with an *in situ* generated chiral titanium modified MCM-41 catalyst **7**

As shown in Scheme 4.9, the key step of this synthesis was the epoxidation of geraniol, providing the almost enantiopure required (R,R)-epoxide in 95% yield, using (R,R)-diethyl tartrate at 8.4 mol% of catalyst loading. This epoxide was subsequently transformed into the expected Colorado potato beetle pheromone. Following the same method by using (S,S)-diethyl tartrate as ligand allowed the corresponding enantiopure (S,S)-geraniol epoxide in 93% yield, which then also converted into the enantiomer of the Colorado potato beetle pheromone (Scheme 4.9).

4.4. Darzens Reactions

The Darzens reaction represents a robust alternative to epoxidation for producing α,β-epoxy carbonyl and related compounds, starting from readily available aldehydes.[18] Chiral glycidic esters and amides, as important types of epoxides, have broad applications, because they can easily

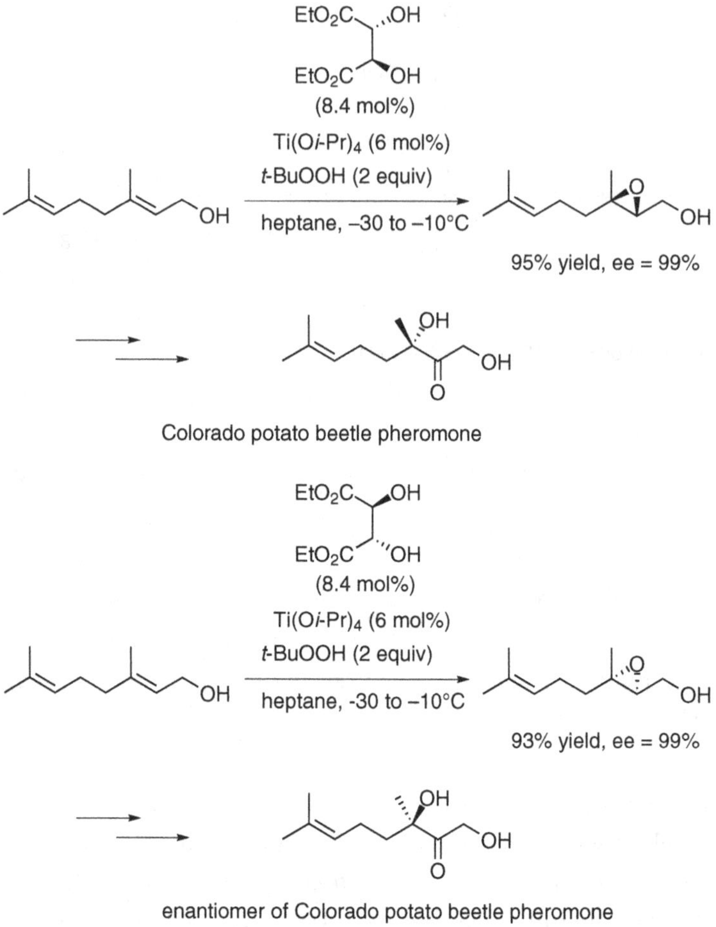

EtO2C OH
EtO2C OH
(8.4 mol%)
Ti(Oi-Pr)4 (6 mol%)
t-BuOOH (2 equiv)
OH
heptane, -30 to -10°C
O
OH
95% yield, ee = 99%
OH
OH
O
Colorado potato beetle pheromone
EtO2C OH
EtO2C OH
(8.4 mol%)
Ti(Oi-Pr)4 (6 mol%)
t-BuOOH (2 equiv)
OH
heptane, -30 to -10°C
O
OH
93% yield, ee = 99%
OH
OH
O
enantiomer of Colorado potato beetle pheromone

Scheme 4.9. Sharpless epoxidation of geraniol and synthesis of Colorado potato beetle pheromone and its enantiomer

be converted into various target molecules by simple transformation.[19] In 2009, Gong *et al.* reported a synthesis of chiral *cis*-glycidic amides based on an enantioselective titanium-catalysed Darzens reaction of diazoacetamides, such as phenyl diazoacetamide, with aldehydes.[20] This remarkable process was promoted by a catalyst generated *in situ* from (*R*)-BINOL and titanium tetraisopropylate, allowing a wide range of *cis*-glycidic amides to be achieved in generally high yields, with excellent enantioselectivities of up to > 99% ee, as shown in Scheme 4.10. Indeed,

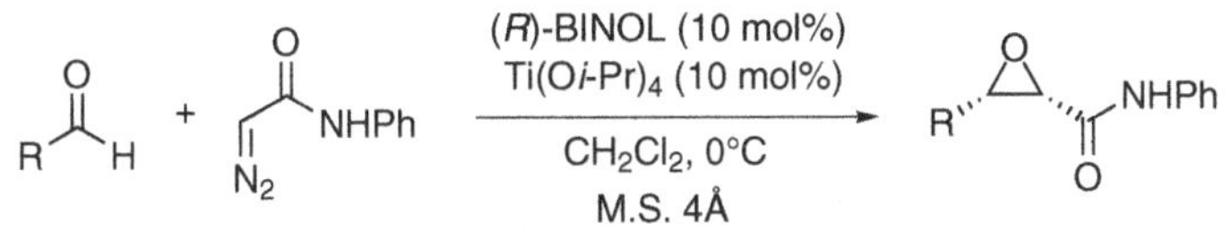

R = Ph: 88% yield, ee = 99%
R = p-NO$_2$C$_6$H$_4$: 91% yield, ee = 99%
R = p-CNC$_6$H$_4$: 89% yield, ee = 99%
R = p-CO$_2$MeC$_6$H$_4$: 80% yield, ee = 99%
R = p-BrC$_6$H$_4$: 83% yield, ee = 99%
R = p-ClC$_6$H$_4$: 86% yield, ee >99%
R = o-NO$_2$C$_6$H$_4$: 84% yield, ee = 96%
R = m-CNC$_6$H$_4$: 95% yield, ee = 99%
R = p-Tol: 81% yield, ee = 96%
R = m-AcC$_6$H$_4$: 87% yield, ee = 97%
R = 1-Naph: 93% yield, ee = 97%
R = 2-Naph: 83% yield, ee = 98%
R = CH$_2$=C(Me): 62% yield, ee = 98%
R = n-Pr: 76% yield, ee >99%
R = Et: 86% yield, ee >99%
R = CH$_2$Bn: 88% yield, ee = 97%
R = Cy: 67% yield, ee = 87%
R = CH$_2$OBn: 93% yield, ee = 94%

Scheme 4.10. Darzens reaction of aldehydes with phenyl diazoacetamide with (R)-BINOL as ligand

the protocol tolerated an impressive broad variety of structurally diverse aldehydes, including aromatic, unsaturated and aliphatic aldehydes. It must be noted that the reactions were completely diastereoselective, with no formation of the corresponding *trans*-epoxides in all substrates studied. This new method has great potential in the enantioselective synthesis of biologically active substances, as illustrated by the preparation of chiral building blocks for the side chain of taxol and (−)-bestatin.

More recently, Sun *et al.* developed comparable Darzens reactions by using a chiral titanium catalyst formed from commercially available (+)-pinanediol **8** and titanium tetraisopropylate *in situ*.[21] Once again, the reactions of a series of aldehydes with phenyl diazoacetamide led to the corresponding *cis*-glycidic amides as soles diastereomers. When aliphatic aldehydes were used, the *cis*-glycidic amides were obtained in good yields of up to 92%, with excellent enantioselectivities of up to 99% ee. For the aromatic aldehydes, the Darzens reaction products achieved good yields and with moderate to high enantioselectivities, of up to 95% ee (Scheme 4.11). Compared with Gong's previous method using (R)-BINOL

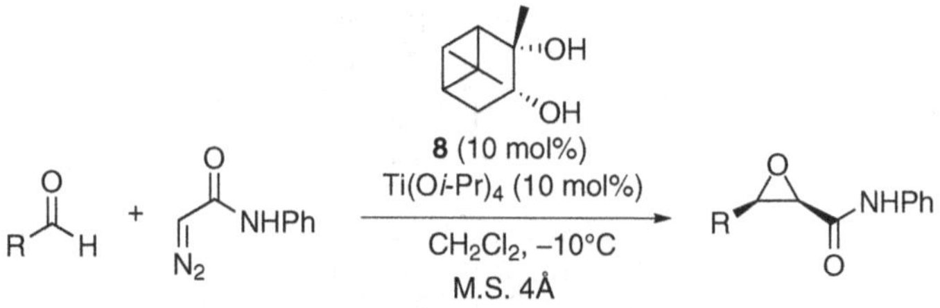

R = Et: 92% yield, ee = 98%
R = *n*-Pr: 90% yield, ee = 99%
R = *n*-Bu: 91% yield, ee = 97%
R = *n*-Pent: 88% yield, ee = 97%
R = Bn: 83% yield, ee = 95%
R = *i*-Bu: 81% yield, ee = 98%
R = *i*-Pr: 83% yield, ee = 98%
R = Cy: 85% yield, ee = 96%
R = Ph: 81% yield, ee = 85%
R = *m*-NO$_2$C$_6$H$_4$: 90% yield, ee = 88%
R = *p*-NO$_2$C$_6$H$_4$: 88% yield, ee = 95%
R = *p*-*t*-BuC$_6$H$_4$: 73% yield, ee = 82%
R = *m*-MeOC$_6$H$_4$: 83% yield, ee = 93%
R = *p*-CNC$_6$H$_4$: 84% yield, ee = 92%
R = *m*-FC$_6$H$_4$: 78% yield, ee = 89%
R = 1-Naph: 67% yield, ee = 70%
R = *p*-Tol: 76% yield, ee = 84%

Scheme 4.11. Darzens reaction of aldehydes with phenyl diazoacetamide with (+)-pinanediol as ligand

as ligand, this new method gave better yields and slightly better enantioselectivities for some aliphatic aldehydes, but lower enantioselectivities for most of the aromatic aldehydes.

4.5. Conclusions

It is quite surprising that no enantioselective epoxidation using inexpensive, safe and easy to handle hydrogen peroxide as oxidant in the presence of titanium catalysts was reported until 2005, when Katsuki *et al.* reported that a bis(μ-oxido) titanium(IV) salanen complex was capable of efficiently promoting the epoxidation of unfunctionalised olefins with enantioselectivities of up to >99% ee. Ever since, a series of highly efficient salan and salanen chiral ligands have been successfully developed by these authors and others, to be applied in the epoxidation of conjugated as well as non-conjugated olefins, such as aliphatic alkenes which are known to be poorly reactive toward electrophilic oxidants, often providing enantioselectivities of up to >99% ee. On the other hand, in the context of the use of alkyl hydroperoxides as oxidants, the Sharpless process is the most commonly used asymmetric epoxidation method. Finally, excellent results dealing with epoxidations evolving through enantioselective titanium-catalysed Darzens reactions of aldehydes with diazoacetamides have been recently reported, providing remarkable enantioselectivities of up to >99% ee achieved by using chiral BINOL or pinanediol as ligand.

References

(1) Y. Yu, K. Ding, G. Chen, in *Acid Catalysis in Modern Organic Synthesis*, H. Yamamoto, K. Ishihara, eds., Wiley–VCH, Weinheim, **2008**; Chapter 14, pp. 721–823.

(2) (a) K. P. Bryliakov, in *Comprehensive Inorganic Chemistry II, From Elements to Applications*, J. Reedijk, K. Poppelmeier, eds., **2013**, Vol. 6, pp. 625–664. (b) G. De Faveri, G. Ilyashenko, M. Watkinson, *Chem. Soc. Rev.*, **2011**, *40*, 1722–1760. (c) K. Matsumoto, T. Katsuki, in, *Asymmetric Synthesis* (2nd edition), M. Christmann, S. Brase, eds., **2008**, pp. 223–127.

(3) H. C. Kolb, M. G. Finn, K. B. Sharpless, *Angew. Chem., Int. Ed.*, **2001**, *40*, 2004–2021.

(4) T. Katsuki, K. B. Sharpless, *J. Am. Chem. Soc.*, **1980**, *102*, 5974–5976.

(5) (a) R. Irie, K. Noda, Y. Ito, N. Matsumoto, T. Katsuki, *Tetrahedron Lett.*, **1990**, *31*, 7345–7348. (b) R. Idie, K. Noda, Y. Ito, T. Katsuki, *Tetrahedron Lett.*, **1990**, *32*, 1055–1058. (c) R. Irie, K. Noda, Y. Ito, N. Matsumoto, T. Katsuki, *Tetrahedron: Asymmetry*, **1991**, *2*, 481–494. (d) R. Irie, Y. Ito, T. Katsuki, *Synlett*, **1991**, 265–266.

(6) (a) W. Zhang, J. L. Loebach, S. R. Wilson, E. N. Jacobsen, *J. Am. Chem. Soc.*, **1990**, *112*, 2801–2803. (b) W. Zhang, E. N. Jacobsen, *J. Org. Chem.*, **1991**, *56*, 2296–2298. (c) E. N. Jacobsen, W. Zhang, A. R. Muci, J. R. Ecker, L. Deng, *J. Am. Chem. Soc.*, **1991**, *113*, 7063–7064. (d) M. Palucki, P. J. Pospisil, W. Zhang, E. N. Jacobsen, *J. Am. Chem. Soc.*, **1994**, *116*, 9333–9334.

(7) K. Matsumodo, Y. Sawada, B. Saito, K. Sakai, T. Katsuki, *Angew. Chem., Int. Ed.*, **2005**, *44*, 4935–4939.

(8) (a) K. Matsumoto, Y. Sawada, T. Katsuki, *Pure Appl. Chem.*, **2008**, *80*, 1071–1077. (b) S. Kondo, K. Saruhashi, K. Seki, K. Matsubara, K. Miyaji, T. Kubo, K. Matsumoto, T. Katsuki, *Angew. Chem., Int. Ed.*, **2008**, *47*, 10195–10198.

(9) K. Matsumoto, T. Oguma, T. Katsuki, *Angew. Chem., Int. Ed.*, **2009**, *48*, 7432–7435.

(10) K. Matsumoto, T. Kubo, T. Katsuki, *Chem. Eur. J.*, **2009**, *15*, 6573–6575.

(11) K. Matsumoto, C. Feng, S. Handa, T. Oguma, T. Katsuki, *Tetrahedron*, **2011**, *67*, 6474–6478.

(12) D. Xiong, M. Wu, S. Wang, F. Li, C. Xia, W. Sun, *Tetrahedron: Asymmetry*, **2010**, *21*, 374–378.

(13) D. Xiong, X. Hu, S. Wang, C.-X. Miao, C. Xia, W. Sun, *Eur. J. Org. Chem.*, **2011**, 4289–4292.

(14) A. Berkessel, T. Günther, Q. Wang, J.-M. Neudörfl, *Angew. Chem., Int. Ed.*, **2013**, *52*, 8467–8471.

(15) R. Ballesteros, M. Fajardo, I. Sierra, I. del Hierro, *J. Mol. Catal. A*, **2009**, *310*, 83–92.

(16) Y. Pérez, I. del Hierro, M. Fajardo, *J. Organometal. Chem.*, **2012**, *717*, 172–179.

(17) S.-N. Li, L.-L. Fang, J.-C. Zhong, J.-J. Shen, H. Xu, Y.-Q. Yang, S.-C. Hou, Q.-h. Bian, *Tetrahedron: Asymmetry*, **2014**, *25*, 591–595.

(18) T. Rosen, in *Comprehensive Organic Synthesis, Vol. 2*, B. M. Trost, I. Fleming, C. H. Heathcock, eds., Pergamon, Oxford, **1991**, pp 409–439.

(19) D. Diez, M. G. Nunez, A. B. Anton, P. Garcia, R. F. Moro, N. M. Garrido, I. S. Marcos, P. Basabe, J. G. Urones, *Curr. Org. Synth.*, **2008**, *5*, 186–216.

(20) W.-J. Liu, B.-D. Lv, L.-Z. Gong, *Angew. Chem., Int. Ed.*, **2009**, *48*, 6503–6506.

(21) G. Liu, D. Zhang, J. Li, G. Xu, J. Sun, *Org. Biomol. Chem.*, **2013**, *11*, 900–904.

Chapter 5

Enantioselective Titanium-Catalysed Cycloaddition Reactions

5.1. Introduction

This chapter covers efforts of the chemical community to develop novel enantioselective titanium-catalysed cycloaddition reactions. It looks at developments since the beginning of 2008, when the field was reviewed by Yu and co-workers in a book chapter dealing with titanium Lewis acids.[1]

The chapter is divided into three parts. The first part deals with enantioselective titanium-catalysed (hetero)-Diels–Alder cycloadditions; the second collects enantioselective titanium-catalysed 1,3-dipolar cycloadditions; while the third includes enantioselective titanium-catalysed cyclopropanations.

Cycloadditions are of unique value for increasing molecular complexity and thereby achieving step brevity.[2] Reduction in the number of synthetic steps, mainly those that employ unusual conditions, minimisation of waste production and maximisation of efficiency are some of the more relevant challenges for the new age of organic synthesis. One of the best ways to address these challenges relies on the development of methods that allow a maximum increase in target-relevant molecular complexity per synthetic operation, while generating minimal amounts of by-products. Undoubtedly, cycloaddition reactions, by virtue of allowing the regio- and stereo-selective construction of new rings, by simple addition of two or more molecules, occupy a leading position among the tools available to the synthetic chemist to best meet the above requirements. These reactions allow the assembly of complex ring systems to be achieved in a convergent and often selective fashion, generally from simple, readily available building blocks. The stereocontrolled construction of chiral cyclic carbo- and hetero-cycles is a topic of paramount importance in modern organic synthesis, driven by the predominance of chiral mono- and poly-cyclic systems in natural products, and in chiral pharmaceuticals.[3]

5.2. (Hetero)-Diels–Alder Cycloadditions

Since its discovery in 1928 by Diels and Alder,[4] the Diels–Alder reaction has become one of the cornerstone reactions in organic chemistry for the construction of six-membered rings. The high regio- and stereo-selectivity typically displayed by this cycloaddition, the ease of its execution, and the feature that, during its course, up to four new stereocentres may be created simultaneously, has resulted in innumerable applications of this transfor-mation in the construction of highly complex targets. Indeed, this reaction has undergone intensive development, becoming of fundamental impor-tance for synthetic, physical and theoretical chemists. Today, this powerful reaction is one of the most examined and well appreciated reactions, hav-ing an enormous spectrum of applications in chemistry.

The preparation of numerous compounds of academic and industrial interest is widely based on cycloaddition reactions to carbonyl compounds. The asymmetric version of the Diels–Alder reaction is one of the most important organic transformations and a powerful tool for the formation of carbon to carbon (C–C) bonds in synthetic organic chemistry, allowing a facile entry into chiral cyclic and polycyclic compounds.[5] Tremendous efforts have been devoted to the development of catalytic versions of this reaction and asymmetric variants of [4+2]-catalysed cycloadditions have received unprecedented attention, presumably due to their capacity to rapidly afford complex enantioenriched carbocycles from simple substrates,[6] and provide a large number of important building blocks and intermedi-ates for the total synthesis of bioactive natural products.[6d,7] In particular, the asymmetric hetero-Diels–Alder reaction is among the most powerful available methodologies for the construction of optically active six-membered heterocycles, with extensive synthetic applications in natural or unnatural products with a wide range of biological activity.[6d,7c,8] Indeed, the asymmetric hetero-Diels–Alder reaction, involving carbonyl compounds as the heterodienophiles or the heterodienes, has allowed the preparation of numerous chiral six-membered oxygen-containing heterocycles.[6d,7c,9]

Since the discovery of the accelerating effect of Lewis acids on the Diels–Alder reaction of α,β-unsaturated carbonyl compounds, its broad and fine application under mild reaction conditions has been amply demonstrated.[6a,10] In addition to the acceleration effect on the reaction, the other important role of Lewis acid in the Diels–Alder cycloaddition is its alteration of chemo-, regio-, and diastereo-selectivity. The titanium catalysts commonly used in Diels–Alder reaction are titanium halides, alkoxides, or their mixed salts.[10b] In the last few years, various BINOL derivatives have been investigated as chiral ligands in enantioselective titanium-catalysed hetero-Diels–Alder cycloadditions. As an example, Xu *et al.* have designed a novel chiral H'_4-NOBIN Schiff base **1** to be used as chiral ligand in hetero-Diels–Alder reaction of Danishefsky's diene with aldehydes.[11] As shown in Scheme 5.1, the process provided the corresponding dihydropyranones in moderate to high yields of up to 99%, and enantioselectivities of up to 84% ee when using 2-naphthoic acid **2** as an additive. The results clearly showed that aromatic aldehydes provided much higher yields and enantiomeric excesses than aliphatic aldehydes.

In 2008, Yu *et al.* showed that the titanium(IV) catalyst derived from a 3-monosubstituted (*R*)-BINOL **3** could exhibit an enhanced catalytic activity for the hetero-Diels–Alder reaction of both aromatic and aliphatic aldehydes, with *trans*-1-methoxy-2-methyl-3-trimethylsiloxybuta-1,3-diene (Scheme 5.2).[12] Excellent yields and enantioselectivities of up to 99% ee were obtained for almost all of the substrates investigated, as shown in Scheme 5.2. When the corresponding 3,3′-bisubstituted BINOL titanium complex was used as a catalyst for similar reactions, moderate yields (up to 72%) and lower enantioselectivities (up to 80% ee) were observed by the same group.[13]

The same year, Feng *et al.* developed the enantioselective hetero-Diels–Alder reaction of Brassard's diene with a series of aliphatic aldehydes, using

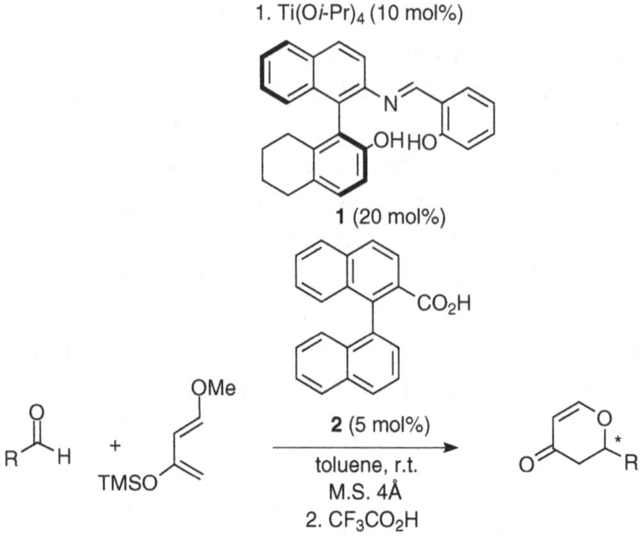

R = Ph: 98% yield, ee = 66%
R = *p*-BrC_6H_4: 99% yield, ee = 84%
R = *p*-ClC_6H_4: 99% yield, ee = 79%
R = *p*-$CF_3C_6H_4$: 99% yield, ee = 72%
R = *p*-$NO_2C_6H_4$: 96% yield, ee = 61%
R = *m*-Tol: 80% yield, ee = 45%
R = *m*-$MeOC_6H_4$: 89% yield, ee = 70%
R = (*E*)-PhCH=CH: 91% yield, ee = 62%
R = Cy: 59% yield, ee = 11%

Scheme 5.1. Hetero-Diels–Alder reaction of Danishefsky's diene with aldehydes catalysed by an *in situ* generated titanium catalyst of chiral H'_4-NOBIN Schiff base **1**

(*R*)-BINOL as chiral ligand.[14] Indeed, the catalyst was *in situ* generated from (*R*)-BINOL, Ti(O*i*-Pr)$_4$ and 4-picolyl chloride hydrochloride, allowing the corresponding cycloadducts to be formed in moderate-to-good yields, and with good enantioselectivities of up to 88% ee (Scheme 5.3). The utility of this procedure was demonstrated in one-step syntheses of two natural products, (+)-kavain and (+)-dihydrokavain (Scheme 5.3).

Chiral metal–organic frameworks (MOFs) constitute a unique class of multifunctional hybrid materials, and are seen as a versatile tool for various enantioselective applications, including asymmetric catalytic reactions. In this context, Jeong *et al.* have designed and synthesised a novel mesoporous chiral titanium organic framework, (*S*)-KUMOF-1 in which

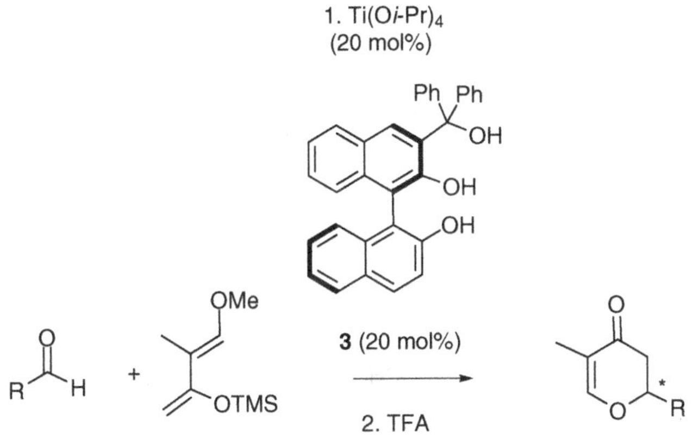

1. Ti(O*i*-Pr)$_4$
(20 mol%)

3 (20 mol%)
⟶
2. TFA

R = Ph: 98% yield, ee = 94%
R = *p*-NO$_2$C$_6$H$_4$: 91% yield, ee = 99%
R = *m*-NO$_2$C$_6$H$_4$: 86% yield, ee = 96%
R = *o*-NO$_2$C$_6$H$_4$: 94% yield, ee = 92%
R = *p*-ClC$_6$H$_4$: 99% yield, ee = 93%
R = *m*-ClC$_6$H$_4$: 99% yield, ee = 95%
R = *p*-Tol: 89% yield, ee = 93%
R = 2-Naph: 99% yield, ee = 93%
R = *p*-CNC$_6$H$_4$: 92% yield, ee = 99%
R = *p*-BrC$_6$H$_4$: 98% yield, ee = 92%
R = *n*-Pr: 91% yield, ee = 99%
R = *i*-Pr: 75% yield, ee = 92%
R = *n*-Hex: 92% yield, ee = 97%
R = *(E)*-MeCH=CH: 86% yield, ee = 91%

Scheme 5.2. Hetero-Diels–Alder reaction of *trans*-1-methoxy-2-methyl-3-trimethylsi-loxybuta-1,3-diene with aldehydes catalysed by an *in situ* generated titanium catalyst of chiral 3-substituted BINOL **3**

a non-interpenetrating NbO type framework provided a spacious pore, and was equipped with potential catalytic sites exposed into the pore.[15] This MOF turned out to be an excellent platform of heterogeneous catalysts for enantioselective hetero-Diels–Alder cycloadditions, after proper function-alisation for endowing Lewis-acidity through titanium-hydrogen exchange. Using this novel catalyst, the hetero-Diels–Alder reaction of Danishefsky's diene with benzaldehyde afforded the corresponding cycloadduct in good yield (80%) and moderate enantioselectivity of 55% ee, as shown in Scheme 5.4. The authors observed that the reaction occurred entirely inside

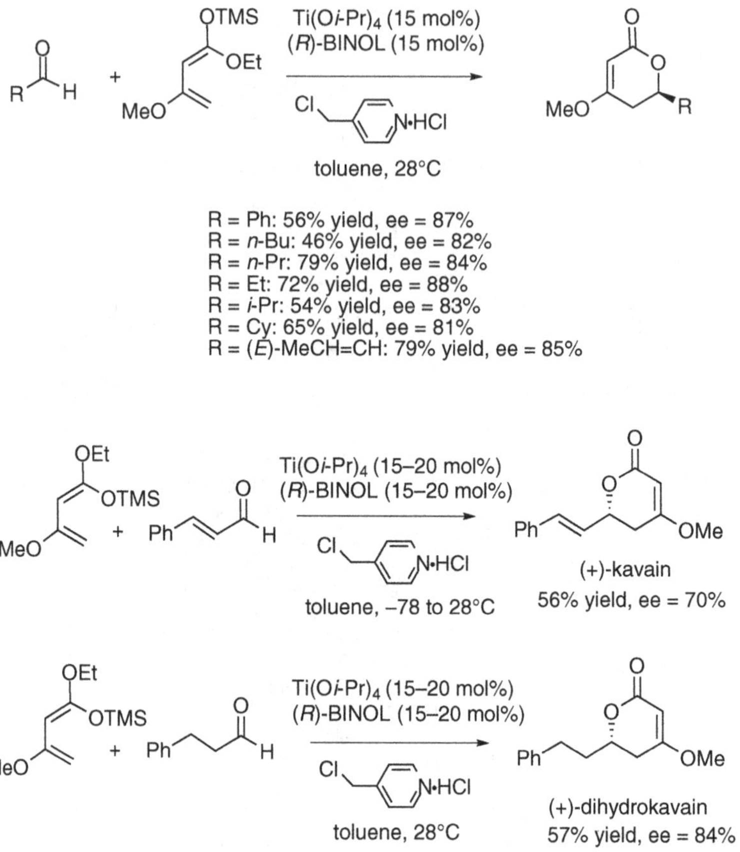

R = Ph: 56% yield, ee = 87%
R = *n*-Bu: 46% yield, ee = 82%
R = *n*-Pr: 79% yield, ee = 84%
R = Et: 72% yield, ee = 88%
R = *i*-Pr: 54% yield, ee = 83%
R = Cy: 65% yield, ee = 81%
R = (*E*)-MeCH=CH: 79% yield, ee = 85%

Scheme 5.3. Hetero-Diels–Alder reactions of Brassard's diene with aldehydes catalysed by an *in situ* generated (*R*)-BINOL-titanium complex and syntheses of (+)-kavain and (+)-dihydrokavain

the pores and the reaction rate of the heterogeneous reaction was comparable to that of its homogeneous counterpart. Also, it was observed that both the yield and enantioselectivity of the homogeneous reaction were much lower (58% yield and 18% ee) than those of the heterogeneous reaction.

In recent years, enantioselective carbon-Diels–Alder cycloaddition reactions have also been catalysed by chiral titanium complexes. As an example, Yu and co-workers have developed the Diels–Alder reaction of 2-alkenoyl pyridines with cyclopentadiene, using a catalyst *in situ* generated

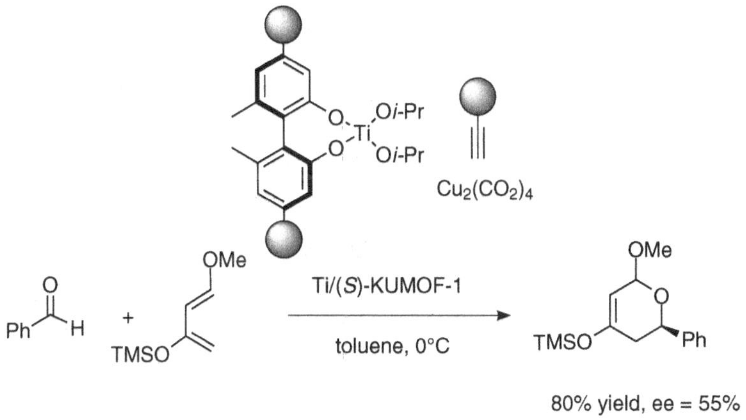
Oi-Pr
Ti
Oi-Pr
$Cu_2(CO_2)_4$
OMe
O
Ph H
+
TMSO
Ti/(S)-KUMOF-1
toluene, 0°C
OMe
O
TMSO
Ph
80% yield, ee = 55%

Scheme 5.4. Hetero-Diels–Alder reaction of Danishefsky's diene with benzaldehyde catalysed by a mesoporous chiral titanium organic framework

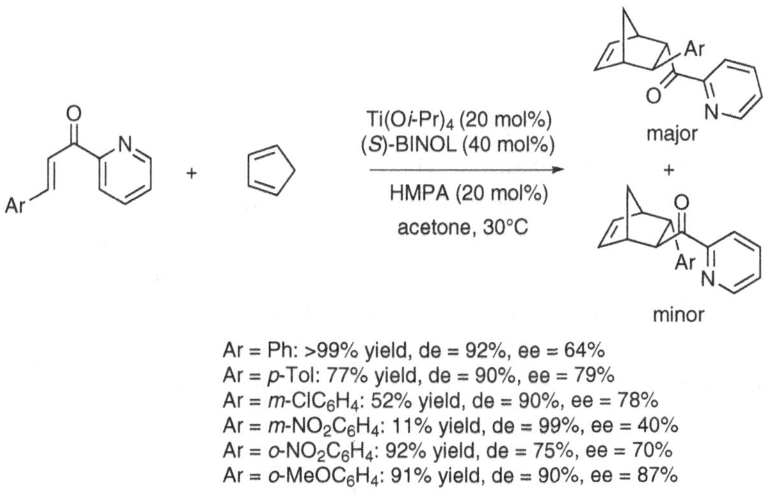
O
N
Ar
+
Ti(Oi-Pr)4 (20 mol%)
(S)-BINOL (40 mol%)
HMPA (20 mol%)
acetone, 30°C
Ar
O
N
major
+
O
Ar
N
minor
Ar = Ph: >99% yield, de = 92%, ee = 64%
Ar = p-Tol: 77% yield, de = 90%, ee = 79%
Ar = m-ClC6H4: 52% yield, de = 90%, ee = 78%
Ar = m-NO2C6H4: 11% yield, de = 99%, ee = 40%
Ar = o-NO2C6H4: 92% yield, de = 75%, ee = 70%
Ar = o-MeOC6H4: 91% yield, de = 90%, ee = 87%

Scheme 5.5. Diels–Alder reaction of 2-alkenoyl pyridine with cyclopentadiene catalysed by an *in situ* generated titanium catalyst of (*S*)-BINOL

from (*S*)-BINOL and titanium tetraisopropylate.[16] When the reaction was performed in the presence of HMPA as an additive, it afforded the corresponding cycloadducts in moderate to quantitative yields, good to complete diastereoselectivities, along with moderate to high enantioselectivities, as shown in Scheme 5.5. The 2-methoxy-substituted substrate gave the

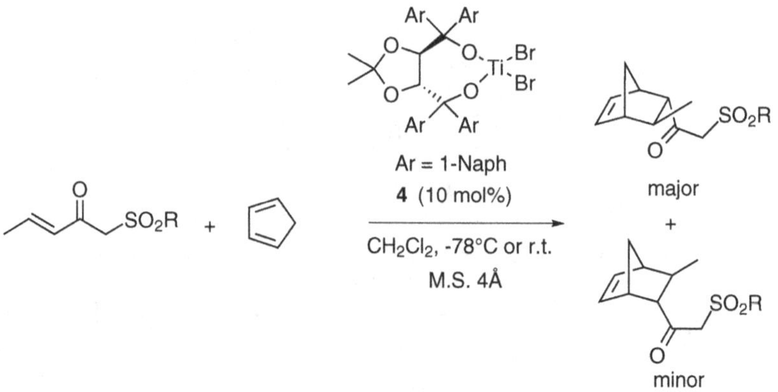

R = Ph: 92% yield, *endo/exo* = 99:1, ee >99%
R = Me: 91% yield, *endo/exo* = 99:1, ee >99%
R = *p*-Tol: 90% yield, *endo/exo* = 99:1, ee >99%

Scheme 5.6. Diels–Alder reaction of sulfonyl-functionalised α,β-unsaturated ketones with cyclopentadiene catalysed by chiral titanium complex **4**

corresponding product with 91% yield and 87% ee, which were the best results of all tested substrates.

Another recent example of enantioselective carbon-Diels–Alder cycloaddition reaction was reported by Sun *et al.* in 2008.[17] In this case, sulfonyl-functionalised α,β-unsaturated ketones, such as (*E*)-1-substituted sulfonyl-3-penten-2-ones, reacted remarkably with cyclopentadiene in the presence of a preformed chiral titanium catalyst **4** derived from tartrates, to give the corresponding cycloadducts in high yields (90–92%) and with both excellent diastereo- and enantio-selectivities of 98% de and >99% ee, respectively, as shown in Scheme 5.6.

5.3. 1,3-Dipolar Cycloadditions

The 1,3-dipolar cycloaddition, also known as the Huisgen cycloaddition,[18] is a classic reaction in organic chemistry, consisting in the reaction of a dipolarophile with a 1,3-dipolar compound, allowing the production of various five-membered heterocycles.[19] This reaction represents one of the most productive fields of modern synthetic organic chemistry.[20] Most of dipolarophiles are alkenes, alkynes and molecules, possessing related heteroatom functional groups (such as carbonyls and nitriles).

The 1,3-dipoles can be basically divided into two different types, the allyl anion type such as nitrones, azomethine ylides, nitro compounds (bearing a nitrogen atom in the middle of the dipole), carbonyl ylides or carbonyl imines (bearing an oxygen atom in the middle of the dipole), and the linear propargyl/allenyl anion type such as nitrile oxides, nitrilimines, nitrile ylides, diazoalkanes, or azides. Two π-electrons of the dipolarophile and four electrons of the dipolar compound participate in a concerted pericyclic shift. The addition is stereoconservative (suprafacial), and the reaction is therefore a $[2_S + 4_S]$ cycloaddition.

Since 2008, asymmetric 1,3-dipolar cycloadditions have become one of the most powerful tools for the construction of enantiomerically pure five-membered heterocycles.[19a,21] Up to five stereocentres can be introduced in a stereoselective manner in only one single step. In addition, a range of different substituents can be included in the dipole and the dipolarophile, resulting in a broad range of possible cycloadducts, which can serve as useful synthetic building blocks. The 1,3-dipolar or [3 + 2] cycloaddition reaction of nitrones with dipolarophiles such as alkenes has received considerable attention in asymmetric synthesis up to 2014.[21a–21b,22] One of the reasons for the success of the synthetic applications of nitrones is that, contrary to the majority of the other 1,3-dipoles, most nitrones are stable compounds that do not require an *in situ* formation.

Another synthetic utility of this reaction is the variety of attractive nitrogenated compounds which are available from the thus-formed isoxazolidines, with up to three contiguous stereocentres, such as 1,3-aminoalcohols, amino acids, azasugars and alkaloids. Chiral titanium catalysts based on various types of ligands have been used to promote 1,3-dipolar cycloadditions. As an example, Maruoka *et al.* have designed a bis-titanium chiral catalyst **5** containing two oxygen-bridged titanium centres, which was successfully applied to the asymmetric 1,3-dipolar cycloaddition of nitrones with α,β-unsaturated aldehydes.[23] It was shown that the introduction of the diphenylmethyl (DPM) group as the N substituent on the nitrones — with the aim of destabilising the nitrone–Lewis acid complex — led to the drastic enhancement of not only the reactivity but also the enantioselectivity. Therefore, by employing this approach, the 1,3-dipolar cycloaddition of *N*-diphenylmethyl nitrones with acrolein gave the corresponding functionalised cycloadducts with a unique regioselectivity and excellent enantioselectivities of up to 98% ee (Scheme 5.7).

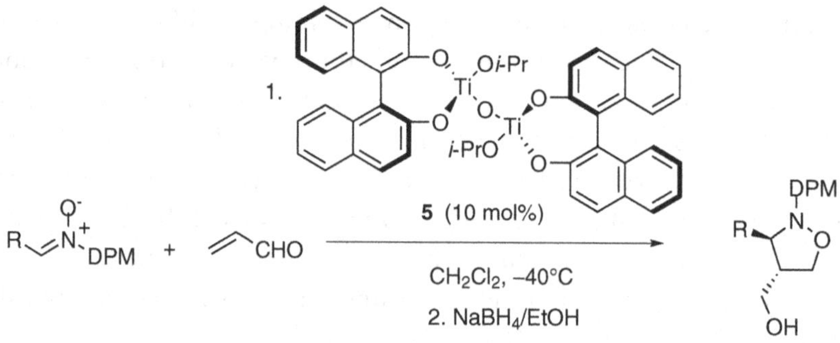

R = Ph: 91% yield, ee = 97%
R = *p*-Tol: 93% yield, ee = 98%
R = *p*-$MeOC_6H_4$: 99% yield, ee = 98%
R = Cy: 96% yield, ee = 97%
R = *i*-Pr: 83% yield, ee = 97%

Scheme 5.7. 1,3-Dipolar cycloaddition of *N*-diphenylmethyl nitrones with acrolein catalysed by a chiral bis-titanium complex **5**

The scope of the methodology could also be extended to the reaction of the rather unreactive methacrolein which provided the corresponding cycloadducts bearing one all-carbon quaternary centre in comparable enantioselectivities of up to 99% ee albeit lower yields (46–80%).

By the accommodation of modified BINOLs as chiral ligands, the enantioselectivities and yields in the bis-titanium chiral complex-catalysed 1,3-dipolar cycloaddition of *N*-diphenylmethyl nitrones with methacrolein could be slightly improved by the same authors.[24] Indeed, using the closely related chiral bis-titanium catalyst **6**, the reactions afforded a range of oxazolidines as single regio- and diastereoisomers in moderate to good yields (up to 84%) and generally high enantioselectivities (up to 98% ee as shown in Scheme 5.8).

The use of azomethine ylides has received much attention. Azomethine ylides are planar 1,3-dipoles, composed of a central nitrogen atom and two terminal sp^2 carbon atoms. Their cycloaddition to olefinic dipolarophiles provides a direct and general method for the synthesis of pyrrolidine derivatives. Normally the azomethine ylides are generated *in situ*, and trapped by a multiple C–C or carbon to halogen (C–X) bond. In 2010, Maruoka *et al.* reported the exploitation of yet unexplored C,N-cyclic azomethine imines in highly enantioselective 1,3-dipolar cycloadditions,

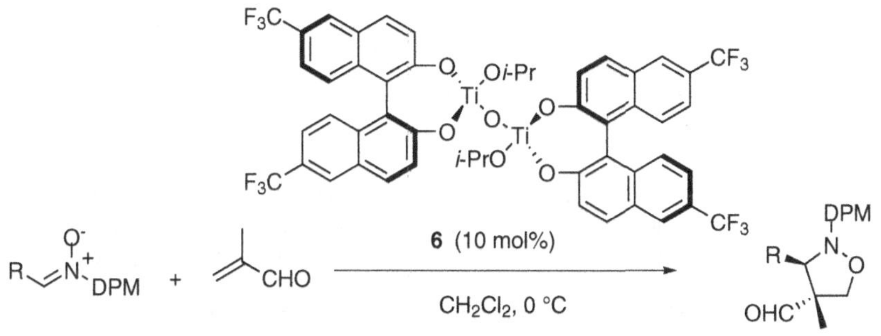

$R = $ Ph: 84% yield, ee = 97%
$R = m$-Tol: 75% yield, ee = 97%
$R = p$-Tol: 76% yield, ee = 94%
$R = p$-ClC$_6$H$_4$: 42% yield, ee = 91%
$R = $ (Me)C=CHPh: 70% yield, ee = 83%

Scheme 5.8. 1,3-Dipolar cycloaddition of *N*-diphenylmethyl nitrones with methacrolein catalysed by a chiral bis-titanium complex **6**

catalysed by an *in situ* generated BINOL-based titanium catalyst.[25] As shown in Scheme 5.9, the reaction of these C,N-cyclic azomethine imines with a range of α,β-unsaturated aldehydes provided the corresponding chiral tetrahydroisoquinolines bearing a 1,3-diamine unit along with three contiguous stereogenic centres. These important polyfunctionalised complex products were obtained in general high yields and enantioselectivities of up to 99% ee. Only in the case of α,β-unsaturated aldehydes lacking a β-substituent, almost equal amounts of two diastereomers were obtained moreover in moderate enantioselectivities (62% ee), suggesting the importance of the steric factor for the exoselectivity.

The scope of the precedent methodology could be extended to C,N-cyclic azomethine imines which were not fused to the aromatic ring. However, attempted isolation of the free azomethine imines failed in this case. To address this issue, the authors implemented their *in situ* generation from the corresponding hydrobromic acid salts, using DTBMP as a base. Under these conditions, the reaction afforded the corresponding piperidines having a 1,3-diamine unit in moderate to excellent yields (36–99%) as mixtures of *exo-* and *endo*-cycloadducts in ratios ranging from 7:93 to 88:12 (for *exo:endo*), and with good to excellent enantioselectivities (75–98% ee), as shown in Scheme 5.10.[24]

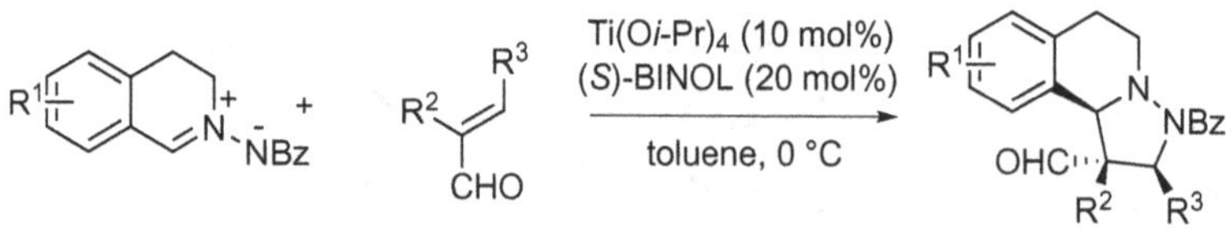

R¹ = 5-Me, R² = H, R³ = Me: 85% yield, *exo/endo* >95:5, ee = 89%
R¹ = 6-Me, R² = H, R³ = Me: 99% yield, *exo/endo* >95:5, ee = 92%
R¹ = 6-OMe, R² = H, R³ = Me: 96% yield, *exo/endo* >95:5, ee = 82%
R¹ = 6-Br, R² = H, R³ = Me: 93% yield, *exo/endo* >95:5, ee = 95%
R¹ = 7-Br, R² = H, R³ = Me: 92% yield, *exo/endo* >95:5, ee = 93%
R¹ = R² = H, R³ = Ph: 94% yield, *exo/endo* >95:5, ee = 99%
R¹ = H, R² = R³ = Me: 85% yield, *exo/endo* >95:5, ee = 88%
R¹ = H, R²,R³ = (CH₂)₃: 93% yield, *exo/endo* = 92:8, ee = 89%
R¹ = R² = R³ = H: 97% yield, *exo/endo* = 61:39, ee = 62%

Scheme 5.9. 1,3-Dipolar cycloaddition of C,N-cyclic azomethine imines with α,β-unsaturated aldehydes catalysed by an *in situ* generated (*S*)-BINOL-titanium complex

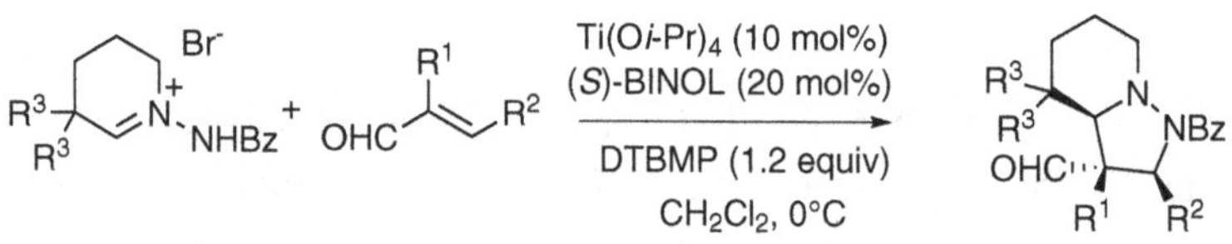

R¹ = H, R² = Me, R³,R³ = (CH₂)₄: 74% yield, *exo/endo* : 81:19, ee (*exo*) = 90%
R¹ = H, R² = Me, R³,R³ = (CH₂)₃: 64% yield, *exo/endo* : 77:23, ee (*exo*) = 86%
R¹ = H, R² = R³ = Me: 65% yield, *exo/endo* : 88:12, ee (*exo*) = 76%
R¹ = H, R² = *n*-Pr, R³,R³ = (CH₂)₄: 36% yield, *exo/endo* : 67:33, ee (*exo*) = 75%
R¹ = Me, R² = H, R³,R³ = (CH₂)₄: 98% yield, *exo/endo* : 22:78, ee (*exo*) = 96%,
ee (*endo*) = 97%
R¹ = R³ = Me, R² = H: 99% yield, *exo/endo* : 11:89, ee (*exo*) = ee (*endo*) = 98%
R¹ = R² = H, R³,R³ = (CH₂)₄: 98% yield, *exo/endo* : 15:85, ee = 83%, ee (*endo*) = 96%
R¹ = R² = H, R³,R³ = (CH₂)₃: 97% yield, *exo/endo* : 13:87, ee = 89%, ee (*endo*) = 96%
R¹ = R² = H, R³ = Me: 98% yield, *exo/endo* : 7:93, ee = 75%, ee (*endo*) = 97%

Scheme 5.10. 1,3-Dipolar cycloaddition of *in situ* generated C,N-cyclic azomethine imines with α,β-unsaturated aldehydes catalysed by an *in situ* generated (*S*)-BINOL-titanium complex

5.4. Cyclopropanations

Organic chemists have always been fascinated by the cyclopropane sub-unit[26] which has played and continues to play a prominent role in organic chemistry.[27] While the cyclopropane ring is a highly strained entity, it is nonetheless found in a wide variety of naturally occurring compounds including terpenes, pheromones, fatty acid metabolites and unusual amino acids.[28] Indeed, the prevalence of cyclopropane-containing compounds with biological activity, whether isolated from natural sources or rationally designed pharmaceutical agents, has inspired chemists to find novel and diverse approaches to their synthesis.[7b,29] In 2010, de Meijere *et al.* reported a synthesis of cyclopropylamines based on reactions of Grignard reagents with *N,N*-dialkylcarboxamides.[30] An asymmetric version of this process was developed with the transformation of *N,N*-dibenzylformamide with *n*-hexylmagnesium bromide as a model system. After studying various chiral titanium ligands, such as (*S*)-BINOL and various chiral diamines, a chiral TADDOL derivative was selected as most efficient ligand to promote the reaction. As shown in Scheme 5.11, when catalysed by TADDOL-derived catalyst **7**, the reaction afforded chiral 2-butyl-*N,N*-dibenzylcyclopropylamine in moderate yield (47%) as a 3:1 mixture of *trans-* and *cis*-diastereomers, each obtained in good enantioselectivities of 77% and 84% ee, respectively. It must be noted that the process employed a stoichiometric amount of catalyst and efforts to decrease this quantity led to disappointingly low yields.

Finally, in 2013 Kananovich and Kulinkovich reported the use of chiral bis(TADDOL)ate-titanium complex **8** to promote enantioselective cyclopropanation between carboxylic esters and alkyl magnesium bromides.[31] As shown in Scheme 5.12, the reaction of isopropyl 4-chlorobutyrate with *n*-BuMgBr in the presence of one equivalent of catalyst **8** led to

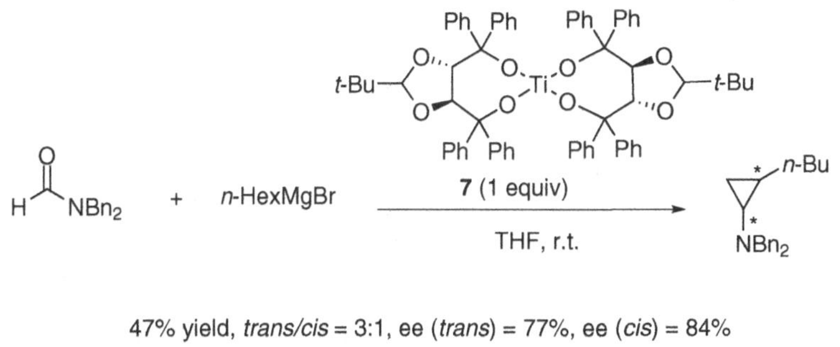

47% yield, *trans/cis* = 3:1, ee (*trans*) = 77%, ee (*cis*) = 84%

Scheme 5.11. Cyclopropanation of *N,N*-dibenzylformamide with hexyl magnesium bromide catalysed by chiral *bis*(TADDOL)ate-titanium complex **7**

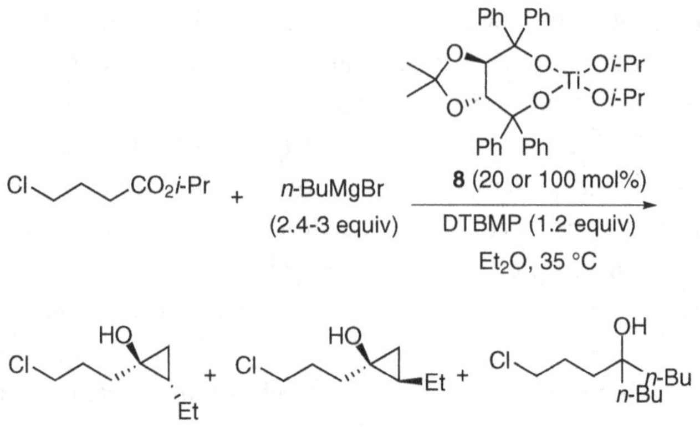

cis-**9** *trans*-**9** **10**

with 20 mol% of **8**:
55% yield of *cis*-**9**, 9% yield of *trans*-**9** (*cis*-**9**:*trans*-**9**) = 6:1,
10% yield in **10**, ee (*cis*) = 65%
with 100 mol% of **8**:
55% yield of *cis*-**9**, 7% yield of *trans*-**9** (*cis*-**9**:*trans*-**9**) = 8:1,
5% yield in **10**, ee (*cis*) = 65%

Scheme 5.12. Cyclopropanation of a carboxylic ester with *n*-butyl magnesium bromide catalysed by chiral TADDOLate-titanium complex **8**

the formation of the corresponding cyclopropanol *cis*-**9** as major product in 55% yield and enantiomeric excess of 65% ee, along with minor diastereomer *trans*-9 and tertiary alcohol **10.** No significant losses in yield and enantioselectivity were observed when the reaction was performed with only a catalytic amount of **8** (20 mol%), but the *cis/trans* ratio decreased from 8:1 to 6:1.

5.5. Conclusions

In the last few years, the use of chiral titanium catalysts in enantioselective cycloadditions has allowed progress to be achieved. At first, a universal catalyst derived from BINOL applicable to different class of aldehydes in enantioselective hetero-Diels–Alder reactions with Danishefsky-type dienes was described by Yu, allowing an easy access to chiral dihydropyranones with high yields and enantioselectivities of up to 99% ee to be achieved. A second work was reported by Sun, dealing with carbon-Diels–Alder cycloaddition reactions of sulfonyl-functionalised α,β-unsaturated ketones, with cyclopentadiene catalysed by a chiral titanium catalyst derived from tartrates, providing the corresponding cycloadducts in high yields and with excellent diastereo- and enantio-selectivities of 98% de and >99% ee, respectively.

Important results have also been described by Maruoka in enantioselective titanium-catalysed 1,3-dipolar cycloadditions. Therefore, a bis-titanium chiral catalyst containing two oxygen-bridged titanium centres was successfully applied to the asymmetric 1,3-dipolar cycloaddition of *N*-diphenylmethyl nitrones with α,β-unsaturated aldehydes with a unique regioselectivity and excellent enantioselectivities of up to 99% ee. More importantly, the same author reported the first exploitation of C,N-cyclic azomethine imines in highly enantioselective 1,3-dipolar cycloadditions, with α,β-unsaturated aldehydes catalysed by an *in situ* generated BINOL-based titanium catalyst, providing important chiral tetrahydroisoquinolines with a 1,3-diamine unit and bearing three contiguous stereogenic centres in general high yields and enantioselectivities of up to 99% ee.

References

(1) Y. Yu, K. Ding, G. Chen, in *Acid Catalysis in Modern Organic Synthesis*, H. Yamamoto, K. Ishihara, eds., Wiley–VCH, Weinheim, **2008**, Chapter 14, pp. 721–823.

(2) (a) W. Carruthers, in *Cycloaddition Reactions in Organic Synthesis*, Pergamon, Oxford, **1990**. (b) L. Ghosez, in *Stereocontrolled Organic Synthesis*, B. M. Trost, ed., Blackwell Science, Oxford, **1994**, pp 193–233. (c) C. P. Dell, *Contemp. Org. Synth.*, **1997**, *4*, 87–117. (d) C. P. Dell, *J. Chem. Soc. Perkin Trans. 1*, **1998**, 3873–3905. (e) *Advances in Cycloaddition*, JAI Press, **1988–1999**, Vols. 1–6. (f) *Cycloaddition Reactions in Organic Synthesis*, S. Kobayashi, K. A. Jorgensen, eds., Wiley–VCH: Weinheim, **2002**. (g) A. Moyano, R. Rios, *Chem. Rev.*, **2011**, *111*, 4703–4832.

(3) T. Hudlicky, J. W. Reed, *The Way of Synthesis*, Wiley–VCH, Weinheim, **2007**.

(4) O. Diels, K. Alder, *Liebigs Ann. Chem.*, **1928**, *460*, 98–122.

(5) (a) W. Oppolzer, in *Comprehensive Organic Synthesis*, B. M. Trost, I. Fleming, eds., Vol. 5, Pergamon: Oxford, **1991**, p 315. (b) U. Pindur, G. Lutz, C. Otto, *Chem. Rev.*, **1993**, *93*, 741–761.

(6) (a) W. Oppolzer, in *Comprehensive Organic Synthesis*, B. M. Trost, I. Fleming, eds., Pergamon Press: New York, **1991**, Vol. 5, Chapter 4.1. (b) L. C. Dias, *J. Braz. Chem. Soc.*, **1997**, *8*, 289–332. (c) D. A. Evans, J. S. Johnson, in *Comprehensive Asymmetric Catalysis*, E. N. Jacobsen, A. Pfaltz, H. Yamamoto, eds., Springer, New York, **1999**, Vol. 3, Chapter 33.1. (d) E. J. Corey, *Ang. Chem., Int. Ed.*, **2002**, *41*, 1650–1667.

(7) (a) H. B. Kagan, O. Riant, *Chem. Rev.,* **1992**, *92,* 1007–1019. (b) M. Lautens, W. Klute, W. Tam, *Chem. Rev.,* **1996**, *96,* 49–92. (c) K. C. Nicolaou, S. A. Snyder, T. Montagnon, G. Vassilikogiannakis, *Ang. Chem., Int. Ed.,* **2002**, *41,* 1668–1698. (d) K.-I. Takao, R. Munakata, K.-i. Tadano, *Chem. Rev.,* **2005**, *105,* 4779–4807.

(8) (a) L. F. Tietze, G. Kettschau, *Top. Curr. Chem.,* **1997**, *189,* 1–120. (b) D. L. Boger, S. M. Weinreb, *Hetero-Diels–Alder Methodology in Organic Synthesis,* H. H. Wasserman, ed., Academic Press, San Diego, **1987**.

(9) (a) S. Danishefsky, *Acc. Chem. Res.,* **1981**, *14,* 400–406. (b) S. J. Danishefsky, M. P. DeNinno, *Angew. Chem., Int. Ed. Engl.,* **1987**, *26,* 15–23. (c) S. Danishefsky, *Chemtracts,* **1989**, 273–297. (d) D. L. Boger, in *Comprehensive Organic Synthesis,* Vol. 5, B. M. Trost, I. Fleming, eds., Pergamon, New York, **1991**, p 451. (e) R. Noyori, *Asymmetric Catalysis in Organic Synthesis,* Wiley, New York, **1994**. (f) L. F. Tietze, G. Kettschau, in *Stereoselective Heterocyclic Synthesis 1,* Vol. 189, P. Metz, ed., Springer, Berlin, **1997**, p 1. (g) K. A. Jorgensen, M. Johannsen, S. Yao, H. Audrain, J. Thorhauge, *Acc. Chem. Res.,* **1999**, *32,* 605–613. (h) D. Carmona, M. P. Lamata, L. A. Oro, *Coord. Chem. Rev.,* **2000**, 717–772. (i) K. A. Jorgensen, *Ang. Chem., Int. Ed.,* **2000**, *39,* 3558–3588. (j) L. L. Lin, X. H. Liu, X. M. Feng, *Synlett,* **2007**, 2147–2157. (k) H. Pellissier, *Tetrahedron,* **2009**, *65,* 2839–2877.

(10) (a) Y. Hayashi, in *Cycloaddition Reactions in Organic Synthesis,* S. Kobayashi, K. A. Jorgensen, eds., Wiley–VCH: Weinheim, **2001**, p 5. (b) D. A. Evans, J. S. Johnson, in *Comprehensive Asymmetric Catalysis,* E. N. Jacobsen, A. Pfaltz, H. Yamamoto, eds., Springer: New York, **1999**, Vol. 3, pp. 1177–1235. (c) S. Reymond, J. Cossy, *Chem. Rev.,* **2008**, *108,* 5359–5406. (d) M. P. Sibi, L. M. Stanley, C. P. Jasperse, *J. Am. Chem. Soc.,* **2005**, *127,* 8276–8277. (e) S. Shirikawa, P. J. Lombardi, J. L. Leighton, *J. Am. Chem. Soc.,* **2005**, *127,* 16394–16395. (f) T. Kano, T. Hashimoto, K. Maruoka, *J. Am. Chem. Soc.,* **2006**, *128,* 2174–2175. (g) K. Maruoka, in *Catalytic Asymmetric Synthesis,* I. Ojima, ed., Wiley: New York, **2000**, 2nd ed., pp. 467–491.

(11) L. X. Sheng, X. Meng, H. Su, X. Wu, D. Xu, *Synlett,* **2008**, *6,* 857–860.

(12) X.-B Yang, J. Feng, J. Zhang, N. Wang, L. Wang, J.-L. Liu, X.-Q. Yu, *Org. Lett.,* **2008**, *10,* 1299–1302.

(13) H. Yu, J. Zhang, Y.-C. Zhao, N. Wang, Q. Wang, X.-B. Yang, X.-Q. Yu, *Chem. Papers,* **2008**, *62,* 187–193.

(14) L. Lin, Z. Chen, X. Yang, X. Liu, X. Feng, *Org. Lett.,* **2008**, *10,* 1311–1314.

(15) K. S. Jeong, Y. B. Go, S. M. Shin, S. J. Lee, J. Kim, O. M. Yaghi, N. Jeong, *Chem. Sci.,* **2011**, *2,* 877–882.

(16) L. Wang, J. Zhang, N. Wang, X.-B. Yang, Q. Wang, X.-Q. Yu, *Lett. Org. Chem.,* **2009**, *6,* 392–396.

(17) W. Pei, Y.-G. Wang, Y.-J. Wang, L. Sun, *Synthesis,* **2008**, *21,* 3383–3388.

(18) R. Huisgen, *Ang. Chem., Int. Ed. Engl.,* **1963**, *10,* 565–598.

(19) (a) C. Najera, J. M. Sansano, *Ang. Chem., Int. Ed.,* **2005**, *44,* 6272–6276. (b) M. Pichon, B. Figadere, *Tetrahedron: Asymmetry,* **1996**, *7,* 927–964.

(20) (a) G. Broggini, G. Molteni, A. Terraneo, G. Zecchi, *Heterocycles*, **2003**, *59*, 823–858. (b) I. N. N. Namboothiri, A. Hassner, *Top. Curr. Chem.*, **2001**, *216*, 1–49.

(21) (a) K. V. Gothelf, K. A. Jorgensen, *Chem. Rev.*, **1998**, *98*, 863–909. (b) K. V. Gothelf, K. A. Jorgensen, *Chem. Commun.*, **2000**, 1449–1458. (c) S. Karlsson, H.-E. Högberg, *Org. Prep. Proc. Int.*, **2001**, *33*, 103–172. (d) S. Kanemasa, *Synlett*, **2002**, *9*, 1371–1387. (e) K. V. Gothelf, *Synthesis*, **2002**, 211–247. (f) C. Najera, J. M. Sansano, *Curr. Org. Chem.*, **2003**, *7*, 1105–1150. (g) Y. Ukajima, K. Inomata, *Synlett*, **2003**, 1075–1087. (h) *Synthetic Applications of 1,3-Dipolar Cycloaddition Chemistry toward Heterocycles and Natural Products*, A. Padwa, W. H. Pearson, eds., Wiley,: New York, **2003**. (i) H. Pellissier, *Tetrahedron*, **2007**, *63*, 3235–3285. (j) C. Najera, J. M. Sansano, M. Yus, *J. Braz. Chem. Soc.*, **2010**, *21*, 377–412.

(22) M. Frederickson, *Tetrahedron*, **1997**, *53*, 403–425.

(23) T. Hashimoto, M. Omote, Y. Hato, T. Kano, K. Maruoka, *Chem. Asian J.*, **2008**, *3*, 407–412.

(24) T. Hashimoto, M. Omote, K. Maruoka, *Org. Biomol. Chem.*, **2008**, *6*, 2263–2265.

(25) T. Hashimoto, Y. Maeda, M. Omote, H. Nakatsu, K. Maruoka, *J. Am. Chem. Soc.*, **2010**, *132*, 4076–4077.

(26) (a) S. Patai, Z. Rappoport, *The Chemistry of the Cyclopropyl Group*, Wiley and Sons: New York, **1987**. (b) *Small Ring Compounds in Organic Synthesis VI*, A. de Meijere, ed., Spinger, Berlin, **2000**, Vol. 207.

(27) M. Rubin, M. Rubina, V. Gevorgyan, *Chem. Rev.*, **2007**, *107*, 3117–3179.

(28) (a) J. Salaün, *Top. Curr. Chem.*, **2000**, *207*, 1–67. (b) R. Faust, *Ang. Chem., Int. Ed.*, **2001**, *40*, 2251–2253. (c) F. Gnad, O. Reiser, *Chem. Rev.*, **2003**, *103*, 1603–1623.

(29) (a) J. Salaün, *Chem. Rev.*, **1989**, *89*, 1247–1270. (b) A. H. Hoveyda, D. A. Evans, G. C. Fu, *Chem. Rev.*, **1993**, *93*, 1307–1370. (c) H.-U. Reissig, *Ang. Chem., Int. Ed.*, **1996**, *35*, 971–973. (d) R. C. Hartley, S. T. Caldwell, *J. Chem. Soc., Perkin Trans. I*, **2000**, 477–502. (e) A. de Meijere, S. I. Kozhushkov, A. F. Khlebnivow, *Top. Curr. Chem.*, **2000**, *207*, 89–147. (f) A. de Meijere, S. I. Kozhushkov, L. P. Hadjiarapoglou, *Top. Curr. Chem.*, **2000**, *207*, 149–227. (g) W. A. Donaldson, *Tetrahedron*, **2001**, *57*, 8589–8627. (h) H. Lebel, J.-F. Marcoux, C. Molinaro, A. B. Charette, *Chem. Rev.*, **2003**, *103*, 977–1050. (i) L. A. Wessjohan, W. Brandt, T. Thiemann, *Chem. Rev.*, **2003**, *103*, 1625–1647. (j) I. L. Lysenko, H. G. Lee, J. K. Cha, *Org. Lett.*, **2006**, *8*, 2671–2673. (k) H. Pellissier, *Tetrahedron*, **2008**, *64*, 7041–7095.

(30) A. de Meijere, V. Chaplinski, H. Winsel, M. Kordes, B. Stecker, V. Gazizova, A. I. Savchenko, R. Boese, F. Schill, *Chem. Eur. J.*, **2010**, *16*, 13862–13875.

(31) Y. A. Konik, D. G. Kananovich, O. G. Kulinkovich, *Tetrahedron*, **2013**, *69*, 6673–6678.

Chapter 6

Enantioselective Titanium-Catalysed Aldol-Type Reactions

This chapter covers efforts of the chemical community to develop novel enantioselective titanium-catalysed aldol-type reactions. It looks at developments since the beginning of 2008, when the field was reviewed by Yu and co-workers in a book chapter dealing with titanium Lewis acids.[1]

The aldol condensation of an enolate — derived from a carbonyl compound with another carbonyl compound derivative, to afford the corresponding β-hydroxy carbonyl product — is a field of constant interest in organic chemistry.[2] As a powerful method for carbon to carbon (C–C) bond formation, with the potential of simultaneous construction of several stereogenic centres, the catalytic asymmetric version of the aldol reaction has been extensively investigated.[3] The first titanium-catalysed enantioselective aldol reaction was reported in 1986 by Reetz et al., using a BINOL titanium complex, and affording the aldol products of aliphatic aldehydes in low enantioselectivity (8% ee).[4] Since then, a wide number of enantioselective titanium-catalysed aldol-type reactions have been successfully developed, often employing BINOL derivatives as chiral ligands. For example, (R)-BINOL was applied as ligand by Scheidt et al. to induce the titanium-catalysed aldol recation between a dienoxy silane and a protected saturated aldehyde, to afford the corresponding secondary alcohol in 63% yield and 88% ee, as shown in Scheme 6.1.[5] This process

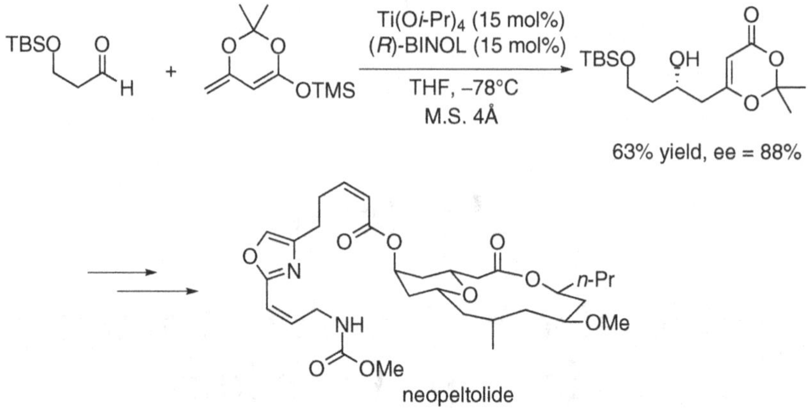

Scheme 6.1. Aldol reaction of a dienoxy silane with a functionalised aldehyde with (R)-BINOL as ligand, and synthesis of neopeltolide

constituted the key step of a total synthesis of the marine macrolide neopeltolide.

Chiral polymer-supported catalysts have been intensively investigated in asymmetric synthesis.[6] In particular, the advantages of soluble polymers, such as poly(ethylene glycol) (PEG) and its mono methyl ether (MPEG) have been explored as supports for catalyst immobilisation, and have led to a steadily increasing number of applications. These polymers can be equipped with a variety of spacers or reactive groups, and are inexpensive, thermally stable, easy to recover, and soluble in many organic solvents. Surprisingly, only a few contributions have been reported dealing with soluble PEG-supported catalysts for asymmetric aldol reactions. A rare example was described by Zimmer *et al.* who synthesised poly(ethylene glycol)-supported (R)-BINOL derivatives, such as **1**, to promote the aldol reaction between 2-styryl-oxazole-4-carbaldehyde and a ketene silyl acetal in the presence of titanium tetraisopropylate in dichloromethane as solvent.[7] As shown in Scheme 6.2, the process afforded the corresponding aldol product in 63% yield, with a high enantioselectivity of 94% ee when using chiral proline derivative **2** as an additive.

In 2010, Yang *et al.* investigated the role of achiral additives in the enantioselective aldol reaction of aldehydes, with Chan's diene catalysed by an *in situ* generated titanium complex of (S)-BINOL.[8] The authors

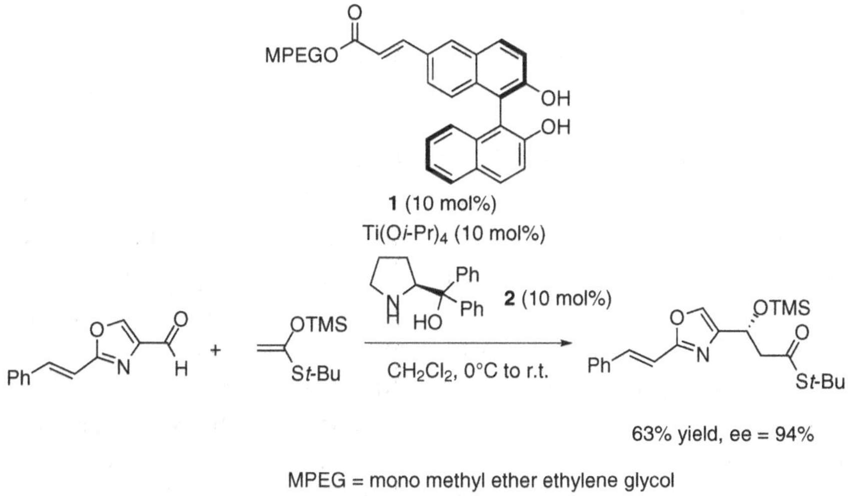
MPEGO
OH
OH
1 (10 mol%)
Ti(O*i*-Pr)4 (10 mol%)
Ph
Ph
HO
2 (10 mol%)
OTMS
S*t*-Bu
CH2Cl2, 0°C to r.t.
Ph
OTMS
S*t*-Bu
63% yield, ee = 94%
MPEG = mono methyl ether ethylene glycol

Scheme 6.2. Aldol reaction of 2-styryl-oxazole-4-carbaldehyde with a ketene silyl acetal using a poly(ethylene glycol)-supported (*R*)-BINOL-derived titanium complex

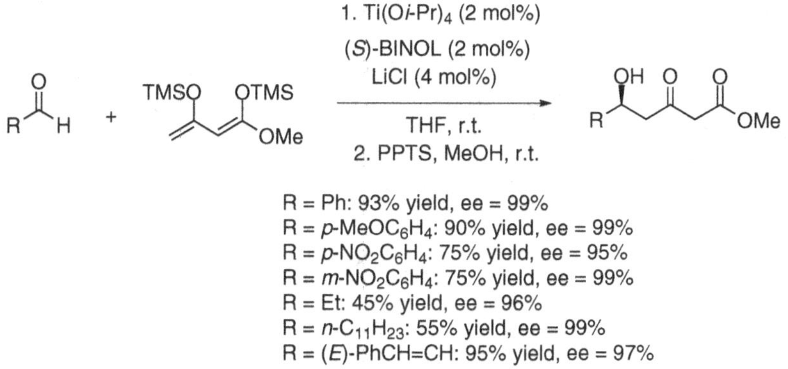
1. Ti(O*i*-Pr)4 (2 mol%)
(*S*)-BINOL (2 mol%)
LiCl (4 mol%)
THF, r.t.
2. PPTS, MeOH, r.t.
TMSO OTMS
OMe
OH O O
OMe
R = Ph: 93% yield, ee = 99%
R = *p*-MeOC6H4: 90% yield, ee = 99%
R = *p*-NO2C6H4: 75% yield, ee = 95%
R = *m*-NO2C6H4: 75% yield, ee = 99%
R = Et: 45% yield, ee = 96%
R = *n*-C11H23: 55% yield, ee = 99%
R = (*E*)-PhCH=CH: 95% yield, ee = 97%

Scheme 6.3. Aldol reaction of Chan's diene with aldehydes with (*S*)-BINOL as ligand

demonstrated that the use of achiral additives could dramatically enhance the enantioselectivities of these reactions. The best results were obtained by using LiCl as the additive, allowing the corresponding δ-hydroxy-β-ketoesters to be achieved as almost single enantiomers, in good to high yields of up to 95% after subsequent hydrolysis with PPTS (Scheme 6.3).

It must be noted that the catalyst system was highly effective since, since the catalyst loadings could be reduced from 2 mol% to 0.1 mol%, always providing excellent enantioselectivities (for example, 95% ee, instead of 97% ee in the case of cinnamaldehyde as substrate).

A related methodology using (*R*)-BINOL as chiral ligand was applied by Paterson and Paquet to develop the fisrt total synthesis of the rare marine macrolide (+)-phorbaside A.[9] Indeed, the key step of this synthesis was the enantioselective aldol reaction between Chan's diene and an α,β-unsaturated aldehyde, promoted by an *in situ* generated titanium catalyst from (*R*)-BINOL and titanium tetraisopropylate. It afforded the corresponding aldol product in 87% yield and high enantioselectivity of >95% ee (Scheme 6.4). The latter was further converted into the expected natural product, which was finally obtained in 8% overall yield over 23 steps.

The design and use of chiral homo- and hetero-bimetallic complexes, especially those in which the two metals work synergistically in a catalytic process, constitute a promising approach in asymmetric catalyst design. In this context, Qu *et al.* have developed novel BINOL-titanium catalyst

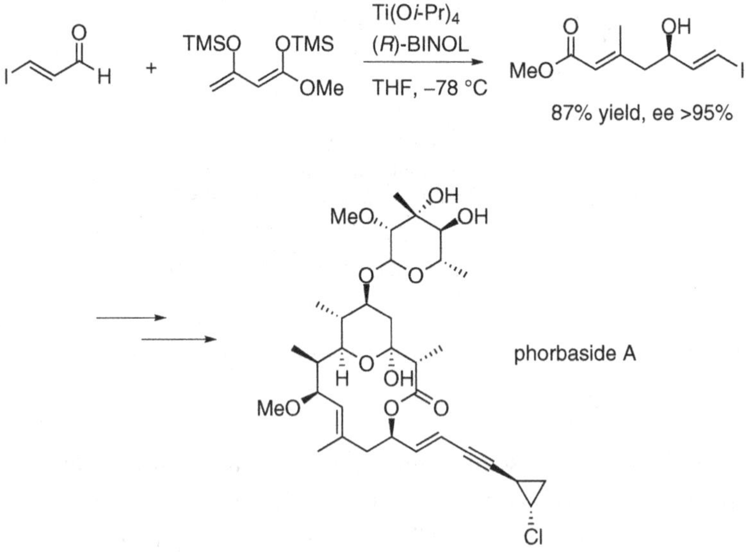

Scheme 6.4. Aldol reaction of Chan's diene with an α,β-unsaturated aldehyde with (*R*)-BINOL as ligand and synthesis of phorbaside A

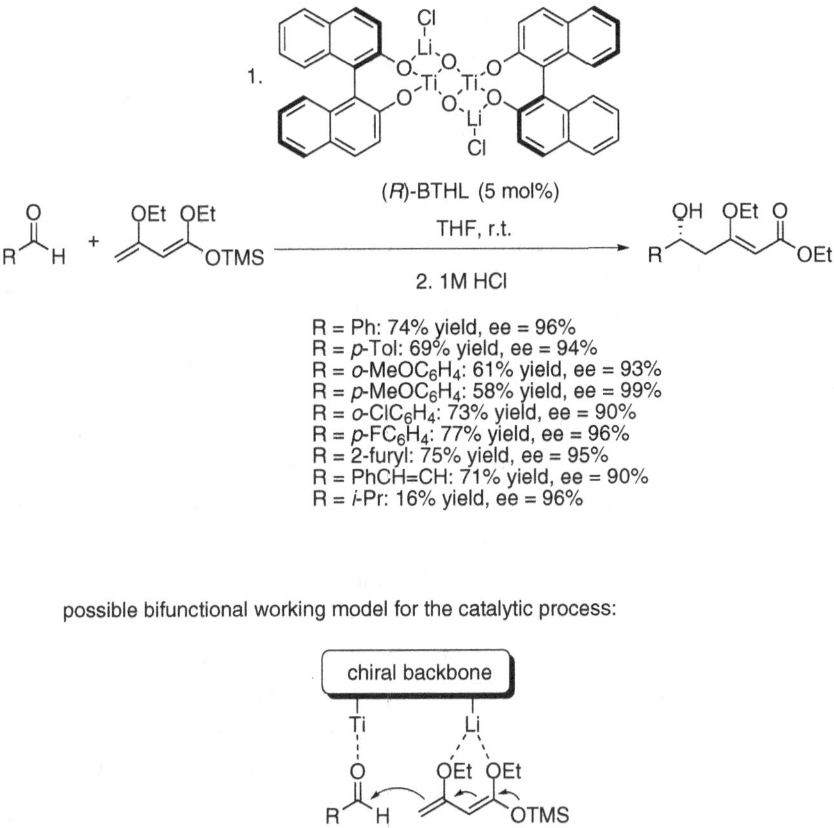
1.
Cl
Li
O
O
Ti
Ti
O
O
O
O
Li
Cl
(R)-BTHL (5 mol%)
THF, r.t.
2. 1M HCl
O
R
H
+
OEt OEt
OTMS
OH OEt O
R
OEt
R = Ph: 74% yield, ee = 96%
R = p-Tol: 69% yield, ee = 94%
R = o-MeOC6H4: 61% yield, ee = 93%
R = p-MeOC6H4: 58% yield, ee = 99%
R = o-ClC6H4: 73% yield, ee = 90%
R = p-FC6H4: 77% yield, ee = 96%
R = 2-furyl: 75% yield, ee = 95%
R = PhCH=CH: 71% yield, ee = 90%
R = i-Pr: 16% yield, ee = 96%
possible bifunctional working model for the catalytic process:
chiral backbone
Ti
Li
O
OEt OEt
R
H
OTMS

Scheme 6.5. Aldol reaction of Brassard's diene with aldehydes catalysed with (*R*)-BTHL

(*R*)-BTHL assisted by a weak Lewis acid LiCl. It worked as a Lewis acid to Lewis acid bifunctional reagent, promoting the enantioselective vinylogous aldol reaction of Brassard's diene with aromatic aldehydes.[10] This catalyst was *in situ* prepared from (*R*)-BINOL and titanium tetraisopropylate, followed by hydrolysis, to give the corresponding μ-oxo bistitanium precatalyst to which two equivalents of LiCl were added, to afford (*R*)-BTHL. Remarkable enantioselectivities of up to 99% ee for the aldol products were achieved, using only 5 mol% of catalyst loading of (*R*)-BTHL (Scheme 6.5).

The authors have proposed a Lewis acid assisted Lewis acid model for this catalytic system, based on the control experiments. When LiCl was added to the μ-oxo bistitanium precatalyst, the lithium coordinated to oxygen, owing to its strong oxygen affinity, thereby enhancing the acidity of the titanium centre through associative interaction. A possible Lewis acid to Lewis acid bifunctional working model is depicted in Scheme 6.5. The strong Lewis acid titanium activated the carbonyl group, and discriminated the prochiral face. Meanwhile, the diene coordinated to the weak Lewis acid lithium centre through a chelating manner, and was positioned close to the aldehyde. Therefore, the two metal centres functioned differently and both were essential for high selectivity.

Among all the O-silyl dienolates that have been developed, the synthetic equivalents of acetoacetate represent a group of promising reagents, as their addition to aldehydes provides an easy access to chiral δ-hydroxy-β-ketoesters — versatile key intermediates in the synthesis of biologically active natural products and commercial drugs. Meanwhile, scarce attention has been devoted to the homologous propionyl acetate derived dienolate, and no effective catalytic methodology has been exploited for the asymmetric vinylogous Mukaiyama aldol reaction with propionyl acetate derivatives in terms of enantio- and diastereoselectivities, until the work reported by Qu *et al.*, in 2012.[11] Indeed, this study proposed the first highly enantio- and diastereo-selective vinylogous Mukaiyama aldol reaction between propionyl acetate-derived Brassard's diene and aldehydes which was accomplished by titanium–lithium combined Lewis acid, using (R)-BTHL as catalyst. After subsequent hydrolysis with aqueous HCl, the corresponding δ-hydroxy-γ-methyl-β-methoxy acrylates were achieved in good to high yields, and remarkable general enantio- and *syn*-diastereo-selectivities of·up to 99% ee and 98% de, respectively (Scheme 6.6). The utility of this nice methodology, presenting wide aldehydes tolerance under mild conditions, was demonstrated in a convenient and formal synthesis of cystothiazole A and melithiazole C.

More recently, Pietruszka and Bielitza reported a total synthesis of 8-desmethoxy psymberin, constituting a biosynthetic intermediate towards the marine polyketide psymberin.[12] This synthesis was based on a highly enantioselective aldol reaction between a silylenol ether and 2-(benzyloxy) acetaldehyde, as shown in Scheme 6.7. When using 20 mol% of an *in situ* generated titanium catalyst from titanium tetraisopropylate and (S)-BINOL as promoter, the reaction afforded the corresponding key aldol product in

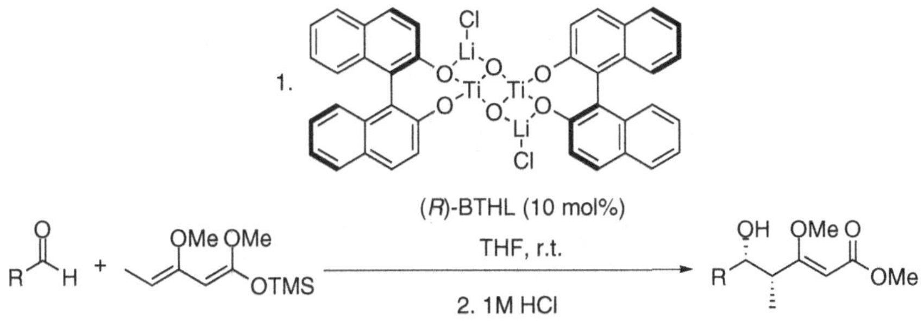

(*R*)-BTHL (10 mol%)

THF, r.t.

2. 1M HCl

R = Ph: 94% yield, *syn/anti* = 99:1, ee = 99%
R = *p*-Tol: 92% yield, *syn/anti* = 99:1, ee = 99%
R = *o*-MeOC$_6$H$_4$: 92% yield, *syn/anti* = 99:1, ee = 99%
R = *p*-MeOC$_6$H$_4$: 88% yield, *syn/anti* = 99:1, ee = 99%
R = 3,4,5-(MeO)$_3$C$_6$H$_3$: 93% yield, *syn/anti* = 99:1, ee = 99%
R = *p*-ClC$_6$H$_4$: 94% yield, *syn/anti* = 98:2, ee = 98%
R = *o*-FC$_6$H$_4$: 93% yield, *syn/anti* = 99:1, ee = 99%
R = *p*-NO$_2$C$_6$H$_4$: 82% yield, *syn/anti* = 98:2, ee = 98%
R = 2-furyl: 95% yield, *syn/anti* = 99:1, ee = 99%
R = PhCH=CH: 96% yield, *syn/anti* = 99:1, ee = 99%
R = *n*-Pr: 75% yield, *syn/anti* = 99:1, ee = 99%
R = *n*-Bu: 71% yield, *syn/anti* = 99:1, ee = 99%
R = MeCH=CH: 48% yield, *syn/anti* = 99:1, ee = 99%
R = BnCH$_2$: 86% yield, *syn/anti* = 99:1, ee = 99%

Scheme 6.6. Aldol reaction of propionyl acetate-derived Brassard's diene with aldehydes catalysed with (*R*)-BTHL

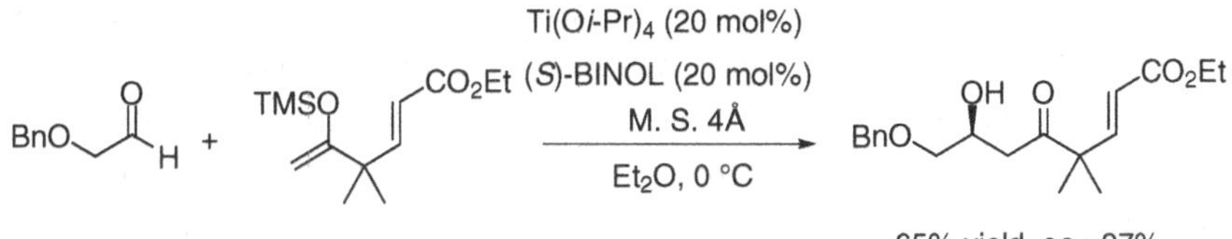

Ti(O*i*-Pr)$_4$ (20 mol%)

(*S*)-BINOL (20 mol%)

M. S. 4Å

Et$_2$O, 0 °C

65% yield, ee >97%

Scheme 6.7. Aldol reaction of a silylenol ether with 2-(benzyloxy) acetaldehyde catalysed with an *in situ* generated catalyst of (*S*)-BINOL

65% yield and excellent enantioselectivity of >97% ee. It was further converted into expected 8-desmethoxy psymberin, which was finally afforded in 25 steps.

Intramolecular version of enantioselective aldol reactions have also been successfully catalysed by chiral titanium complexes. For example, in 2012 Wang *et al.* described a general and efficient method for the synthesis

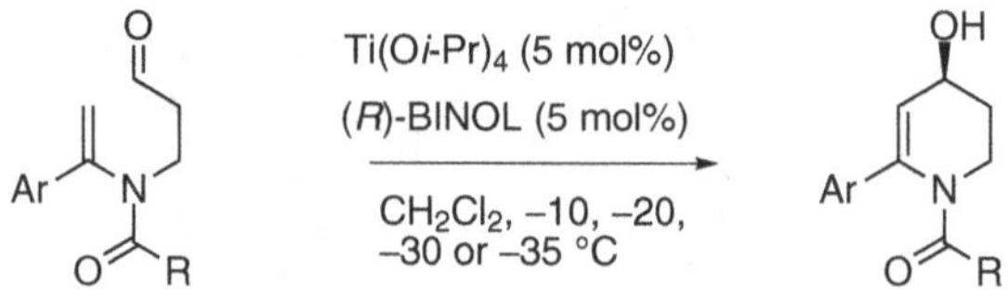

R = Ar = Ph: 91% yield, ee = 96%
R = Ph, Ar = p-ClC$_6$H$_4$: 92% yield, ee = 98%
R = Ph, Ar = p-FC$_6$H$_4$: 97% yield, ee = 93%
R = Ph, Ar = p-BrC$_6$H$_4$: 99% yield, ee = 98%
R = Ph, Ar = o-ClC$_6$H$_4$: 96% yield, ee >99%
R = Ph, Ar = 2,4-Cl$_2$C$_6$H$_3$: 97% yield, ee >99%
R = Ph, Ar = p-Tol: 77% yield, ee >99%
R = Me, Ar = Ph: 82% yield, ee = 89%
R = BnO, Ar = Ph: 69% yield, ee = 93%

Scheme 6.8. Intramolecular aldol reaction of tertiary enamides with aldehydes catalysed with an *in situ* generated catalyst of (R)-BINOL

of almost enantiopure 4-hydroxy-1,2,3,4-tetrahydropyridine derivatives on the basis of the intramolecular nucleophilic addition of enamides to aldehydes, catalysed by an *in situ* generated chiral titanium complex of (R)-BINOL.[13] As shown in Scheme 6.8, the intramolecular aldol reaction of phenyl-substituted tertiary enamides and their aromatic analogues led to the corresponding enantiopure 4-hydroxy-1,2,3,4-tetrahydropyridines in high yields of up to 99%, and remarkable enantioselectivities of up to >99% ee. N-Acetyl substituted enamides and enecarbamates were also tolerated by the catalytic system, providing slightly lower enantioselectivities of up to 93% ee. It must be highlighted that this process only employed 5 mol% of catalyst loading, and constituted a novel easy entry to the enantiopure 4-hydroxypiperidine ring core present in a number of natural products and biologically active compounds.

In 2009, Feng *et al.* used a combination of (S)-BINOL, cinchonine, and titanium tertaisopropylate, each employed at 5 mol% of catalyst loading, as a catalyst system to promote the enantioselective aldol reaction of ethyl diazoacetate with aldehydes.[14] The process, performed in the presence of 15 mol% of water, afforded the corresponding α-diazo-β-hydroxy esters in moderate to good yields of up to 74%, in combination with good to high enantioselectivities of up to 94% ee, as shown in Scheme 6.9. A wide range of aromatic, heteroaromatic and aliphatic aldehydes were found to be suitable substrates.

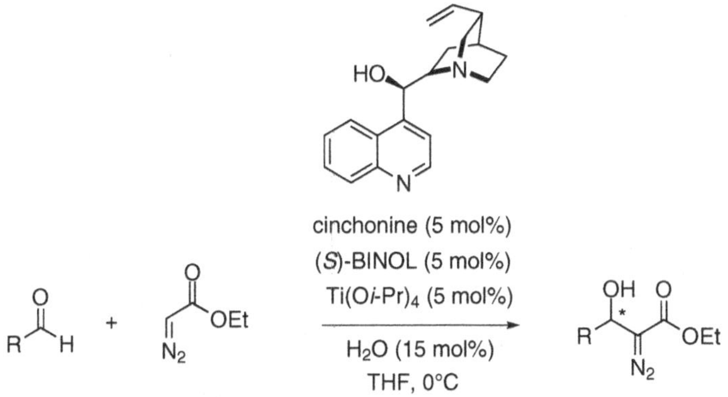

cinchonine (5 mol%)

(S)-BINOL (5 mol%)

$$R \overset{O}{\underset{H}{\bigwedge}} + \quad \overset{O}{\underset{N_2}{\bigvee}} OEt \quad \xrightarrow[\substack{\text{H}_2\text{O (15 mol%)}\\ \text{THF, 0°C}}]{\text{Ti(O}i\text{-Pr})_4 \text{ (5 mol%)}} \quad R \overset{OH}{\underset{N_2}{\bigwedge}} \overset{O}{\underset{}{\bigvee}} OEt$$

R = Ph: 62% yield, ee = 91%
R = *p*-FC$_6$H$_4$: 70% yield, ee = 90%
R = *o*-ClC$_6$H$_4$: 54% yield, ee = 94%
R = *m*-ClC$_6$H$_4$: 67% yield, ee = 89%
R = *p*-ClC$_6$H$_4$: 78% yield, ee = 87%
R = *m*-BrC$_6$H$_4$: 41% yield, ee = 86%
R = *p*-Tol: 48% yield, ee = 93%
R = 3-pyridyl: 53% yield, ee = 86%
R = 2-furyl: 40% yield, ee = 81%
R = *n*-Pent: 74% yield, ee = 64%

Scheme 6.9. Aldol reaction of ethyl diazoacetate with aldehydes catalysed with a combination of (S)-BINOL and cinchonine

In addition to BINOL-derivatives, other chiral ligands have been used in enantioselective titanium-catalysed aldol reactions. As a recent example, a chiral bifunctional cinchona alkaloid-derived thiourea **3** was applied by Shibata *et al.* as the optimal chiral ligand of titanium tetraisopropylate in enantioselective monofluoromethylation of aldehydes with 2-fluoro-1,3-benzodithiole-1,1,3,3-tetraoxide (FBDT).[15] The corresponding FBDT-products were obtained in good to high yields, and moderate to high enantioselectivities of up to 96% ee (Scheme 6.10). In the case of aliphatic aldehydes, the reactions proceeded smoothly at room temperature, however, the enantioselectivities were low compared to those achieved with aromatic aldehydes. The FBDT-products could be nicely converted into the corresponding monofluoromethyl carbinols in good yields without any loss of the enantiopurity by treatment with SmI$_2$. This result represented the first enantioselective catalytic monofluoromethylation of aldehydes with FBDT performed in the presence of a thiourea-based bifunctional cinchona alkaloid and a titanium complex.

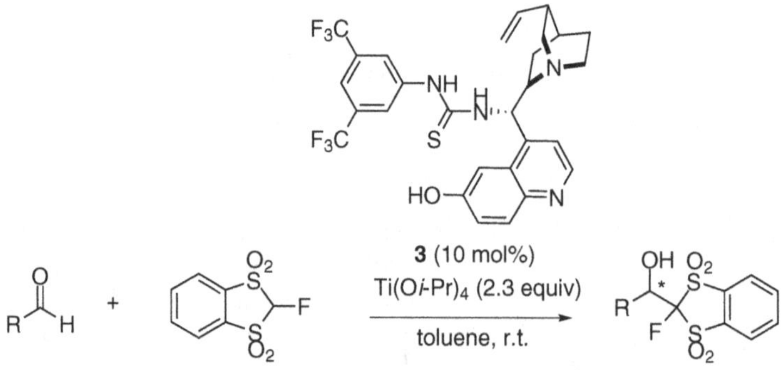

R = Ph: 84% yield, ee = 91%
R = *p*-Tol: 83% yield, ee = 60%
R = *p*-BrC$_6$H$_4$: 81% yield, ee = 56%
R = *p*-NO$_2$C$_6$H$_4$: 91% yield, ee = 60%
R = *m*-NO$_2$C$_6$H$_4$: 89% yield, ee = 96%
R = *p*-CNC$_6$H$_4$: 87% yield, ee = 67%
R = *m*-CNC$_6$H$_4$: 88% yield, ee = 80%
R = 2-Naph: 82% yield, ee = 42%
R = 3-pyridyl: 78% yield, ee = 61%
R = 2-furyl: 40% yield, ee = 81%
R = Cy: 79% yield, ee = 37%
R = *i*-Pr: 73% yield, ee = 32%

Scheme 6.10. Aldol reaction of FBDT with aldehydes catalysed with a cinchona alkaloid-derived thiourea ligand

Conclusions

The last seven years have seen several excellent works reported in the field of enantioselective titanium-catalysed aldol-type reactions, most of them employing BINOL derivatives as chiral ligands. At first, a rare example of aldol reaction performed with soluble PEG-supported catalysts was described by Zimmer, providing the aldol product arisen from 2-styryl-oxazole-4-carbaldehyde, and a ketene silyl acetal in enantioselectivity of 94% ee by using a poly(ethylene glycol)-supported (R)-BINOL-derived titanium complex. In addition, a novel BINOL-titanium catalyst (R)-BTHL, assisted by a weak Lewis acid LiCl, was introduced by Qu to promote the enantioselective vinylogous aldol reaction of Brassard's diene with aromatic aldehydes, producing excellent enantioselectivities of up to 99% ee. This method was extended to aldol reactions of propionyl acetate-derived Brassard's diene with a wide variety of aldehydes, providing almost diastereo- and enantiopure δ-hydroxy-γ-methyl-β-methoxy acrylates. Furthermore, a nice intramolecular aldol reaction of tertiary enamides was reported by Wang, allowing enantiopure 4-hydroxy-1,2,3,4-tetrahydropyridines in high yields and remarkable enantioselectivities of up to >99% ee to be achieved using (R)-BINOL as ligand. Finally, Shibata described the first enantioselective catalytic monofluoromethylation of aldehydes with FBDT performed in the presence of a thiourea-based bifunctional cinchona alkaloid and a titanium complex.

References

(1) Y. Yu, K. Ding, G. Chen, in *Acid Catalysis in Modern Organic Synthesis*, H. Yamamoto, K. Ishihara, eds., Wiley–VCH, Weinheim, **2008**, Chapter 14, pp. 721–823.

(2) (a) J.-i. Matsuo, M. Murakami, *Angew. Chem., Int. Ed.*, **2013**, *52*, 9109–9118. (b) L. C. Diaz, E. C. de Lucca, M. A. B. Ferreira, C. E. Polo, *J. Braz. Chem. Soc.*, **2012**, *23*, 2137–2158. (c) S. M. Yliniemela-Sipari, P. M. Pihko, *Science of Synthesis*, **2011**, *2*, 621–676. (d) G. Casiraghi, L. Battistini, C. Curti, G. Rassu, F. Zanardi, *Chem. Rev.*, **2011**, *111*, 3076–3154. (e) *Modern Aldol Reactions*, Vols. 1 and 2, R. Mahrwald, ed., Wiley–VCH, Weinheim, **2004**. (f) C. Gennari in *Comprehensive Organic Synthesis*, B. M. Trost, I. Fleming, eds., Pergamon Press, Oxford, Vol. 2, pp. 639–660, **1991**. (g) T. Mukaiyama in *Organic Reactions*, W. G. Dauben, ed., Wiley, New York, Vol. 28, pp. 203–331, **1982**.

(3) (a) J. Mlynarski, S. Bas, *Chem. Soc. Rev.*, **2014**, *43*, 577–587. (b) D Gryko, D. Walaszek, in *Stereoselective Organocatalysis*, R. T. Rios, ed., Wiley, Weinheim, pp 81–127, **2013**. (c) L. Liu, C.-J. Li, *Science of Synthesis*, **2011**, *2*, 585–620. (d) B. M. Trost, C. S. Brindle, *Chem. Soc. Rev.*, **2010**, *39*, 1600–1632. (e) G. Guillena, C. Najera, D. J. Ramon, *Tetrahedron: Asymmetry*, **2007**, *18*, 2249–2293. (f) C. Palomo, M. Oiarbide, J. M. Garcia, *Chem. Soc. Chem.*, **2004**, *33*, 65–72. (g) A. Soriente, M. De Rosa, R. Villano, A. Scettri, *Curr. Org. Chem.*, **2004**, *8*, 993–1007. (h) A. Abiko, *Acc. Chem. Res.*, **2004**, *37*, 387–395. (i) P. Merino, T. Tejero, *Angew. Chem., Int. Ed.* **2004**, *43*, 2995–2997. (j) C. Palomo, M. Oiarbide, J. M. Garcia, *Chem. Eur. J.*, **2002**, *8*, 36–44. (k) B. Alcaide, P. Almendros, *Eur. J. Org. Chem.*, **2002**, 1595–1601.

(4) M. T. Reetz, S.-H. Kyung, C. Bolm, T. Zierke, *Chem. Ind. (London)*, **1986**, 824.

(5) (a) D. W. Custar, T. P. Zabawa, K. A. Scheidt, *J. Am. Chem. Soc.*, **2008**, *130*, 804–805. (b) D. W. Custar, T. P. Zabawa, J. Hines, C. M. Crews, K. A. Scheidt, *J. Am. Chem. Soc.*, **2009**, *131*, 12406–12414.

(6) N. E. Leadbeater, M. Marco, *Chem. Rev.*, **2002**, *102*, 3217–3273.

(7) R. Zimmer, V. Dekaris, M. Knauer, L. Schefzig, H.-U. Reissig, *Synth. Commun.*, **2009**, *39*, 1012–1026.

(8) Q. Xu, J. Yu, F. han, J. Hu, W. Chen, L. Yang, *Tetrahedron: Asymmetry*, **2010**, *21*, 156–158.

(9)   I. Paterson, T. Paquet, *Org. Lett.*, **2010**, *12*, 2158–2161.

(10) G. Wang, J. Zhao, Y. Zhou, B. Wang, J. Qu, *J. Org. Chem.*, **2010**, *75*, 5326–5329.

(11) G. Wang, B. Wang, S. Qi, J. Zhao, Y. Zhou, J. Qu, *Org. Lett.*, **2012**, *14*, 2734–2737.

(12) M. Bielitza, J. Pietruszka, *Chem. Eur. J.*, **2013**, *19*, 8300–8308.

(13) S. Tong, D.-X. Wang, L. Zhao, J. Zhu, M.-X. Wang, *Angew. Chem., Int. Ed.* **2012**, *51*, 4417–4420.

(14) W. Wang, K. Shen, X. Hu, J. Wang, X. Liu, X. Feng, *Synlett*, **2009**, *10*, 1655–1658.

(15) H. Ma, K. Matsuzaki, Y.-D. Yang, E. Tokunaga, D. Nakane, T. Ozawa, H. Masuda, N. Shibata, *Chem. Commun.*, **2013**, *49*, 11206–11208.

Chapter 7

Enantioselective Titanium-Catalysed Reduction Reactions

7.1. Introduction

This chapter covers efforts of the chemical community to develop novel enantioselective titanium-catalysed reduction reactions. It looks at developments since the beginning of 2008, when the field was reviewed by Yu and co-workers in a book chapter dealing with titanium Lewis acids.[1]

The chapter has been divided into three parts. The first part deals with enantioselective titanium-catalysed hydroaminations; the second considers enantioselective titanium-catalysed hydrophosphonylations; while the third includes enantioselective titanium-catalysed miscellaneous reactions.

The enantioselective reduction is a fundamental process of organic synthesis,[2] in which the use of titanium complexes was very limited until the independent introduction of chiral titanocenes as chiral precatalysts for the reduction of ketones, reported by Halterman and Buchwald groups in 1994, allowing enantioselectivities of up to 90% ee to be achieved.[3] Since then, a number of other chiral titanium catalysts, including ones based on combinations of two ligands, have been developed, to be successfully applied to enantioselective titanium-catalysed reduction reactions.

7.2. Hydroaminations

Carbon–nitrogen bond formation is an important field in organic synthesis,[4] with hydroamination an atom-efficient process for the generation of amines and imines from olefins, allenes, and alkynes.[5] In particular, titanium-mediated hydroamination is among the most useful protocols developed for this aim. By using unsymmetrical olefins and alkynes, the addition of secondary amines can lead to two isomeric products, where the isomeric ratio is usually dependent on the type of the titanium catalyst used. In particular, aminoalkenes can undergo cyclisation via intramolecular hydroamination in the presence of titanium complexes, yielding predominantly pyrrolidine and piperidine derivatives. The development of chiral catalysts for intramolecular asymmetric alkene hydroamination has received significant attention over the past two decades, because hydroamination, the formal addition of an amine nitrogen to hydrogen (N–H) bond to an unsaturated carbon to carbon (C–C) bond, is a highly economical process for the formation of chiral nitrogen heterocycles, which are important building blocks for fine chemicals, and biologically active and pharmacologically active compounds.[6]

Chiral group 4 metal catalysts have been shown to be promising for this transformation, yet reports of successful catalysts affording significant enantioselectivities (> 90% ee) are still scarce. In this context, Zi *et al.* have reported the synthesis of novel chiral bis-ligated group 4 metal complexes from the reactions between the corresponding chiral C_1-symmetric ligands and $M(NMe_2)_4$ (M = Ti, Zr).[7] Among these complexes, bis-ligated amidate titanium complex **1** was demonstrated as a more effective chiral catalyst for asymmetric intramolecular hydroamination of aminoalkenes than catalysts ligated by Schiff bases ligands. As shown in Scheme 7.1, the corresponding heterocyclic products were produced with excellent yields, and low to good enantioselectivities of up to 68% ee. The authors have demonstrated that the titanium amidate complex **1** provided higher enantioselectivities than the corresponding zirconium amidate complex in all cases of substrate studied.

A better enantioselectivity was reported by the same authors using another bis-ligated chiral titanium amide catalyst **2**.[8] Indeed, the chiral cyclic amine depicted in Scheme 7.2 was formed in low yield (20%) and high enantioselectivity of 91% ee. Again, the efficiency of titanium amidate

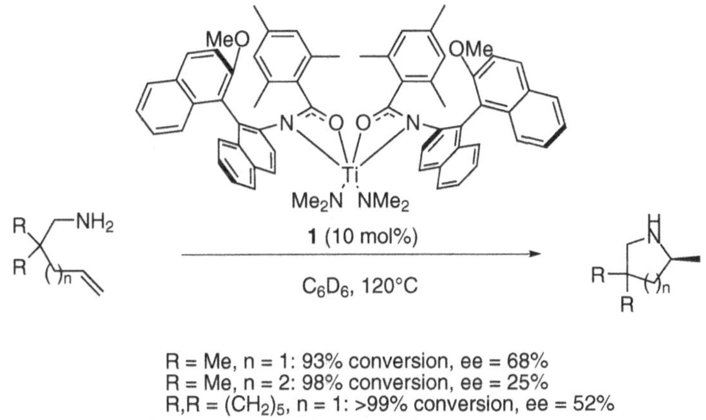
MeO
OMe
N
O
O
N
Ti
Me2N NMe2
R
NH2
R
( )n
1 (10 mol%)
C6D6, 120°C
H
N
R
( )n
R
R = Me, n = 1: 93% conversion, ee = 68%
R = Me, n = 2: 98% conversion, ee = 25%
R,R = (CH2)5, n = 1: >99% conversion, ee = 52%

Scheme 7.1. Intramolecular hydroamination of aminoalkenes with a *bis*-ligated chiral titanium amide catalyst **1**

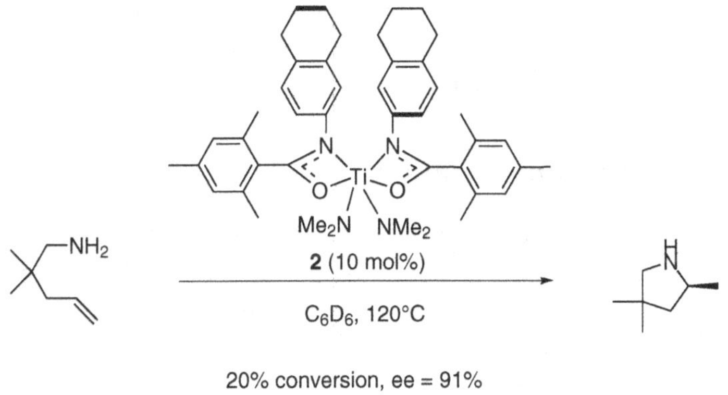
N
N
O
Ti
O
Me2N NMe2
NH2
2 (10 mol%)
C6D6, 120°C
H
N
20% conversion, ee = 91%

Scheme 7.2. Intramolecular hydroamination of an aminoalkene with another bis-ligated chiral titanium amide catalyst **2**

complex **2** was compared to that of the corresponding zirconium complex, and it was shown that **2** provided slightly lower enantioselectivities, and much lower yields than the zirconium catalyst, which gave a quantitative yield combined with an enantioselectivity of 93% ee.

The hydroamination of allenes is thermodynamically more favourable than that of alkenes, and the intramolecular reaction of 4,5-dienylamines can give nitrogen-containing heterocycles with a pendant vinyl group for further substitution. In this context, Johnson *et al.* have investigated the

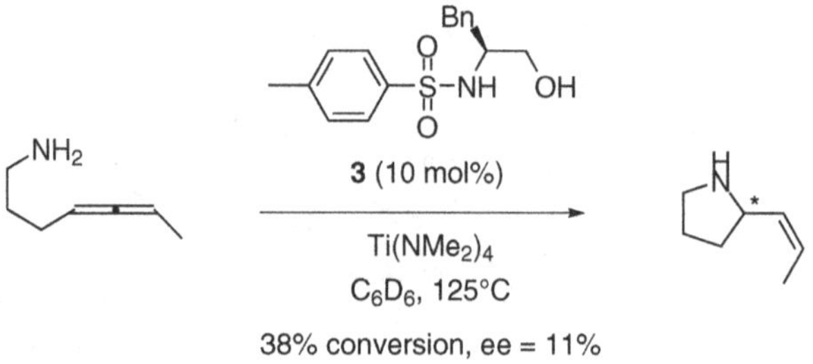

Scheme 7.3. Intramolecular hydroamination of an aminoallene with chiral sulfonamide alcohol ligand **3**

intramolecular hydroamination reaction of aminoallenes in the presence of *in situ* generated titanium catalysts from Ti(NMe$_2$)$_4$, and a series of chiral bidentate sulfonamide alcohol ligands with varying steric and electronic properties.[9] Among these ligands, sulfonamide alcohol **3** was selected as the most efficient, allowing the cyclic amine derived from 6-methyl-hepta-4,5-dienylamine to be achieved in moderate yield (38%) with low enantioselectivity of 11% ee (Scheme 7.3).

7.3. Hydrophosphonylations

Hydrophosphonylation is an economical process in which a dialkyl phosphite is added to carbonyl or imine compounds (Pudovik reaction), leading to the formation of α-hydroxy or α-amino phosphonates. Chiral α-hydroxy-functionalised phosphonates and phosphonic acids have been widely employed to synthesise pharmaceutically and biologically active compounds.[10] Catalytic asymmetric hydrophosphonylation, by addition of an appropriate phosphorus nucleophile to the carbonyl bond can provide a very convenient access to the corresponding chiral α-hydroxy phosphonates, which is probably the most general and widely applied approach.[11] Since the pioneering work of Shibuya *et al.* in 1993,[12] many chiral catalysts, based on various metals, have been extensively studied to promote enantioselective hydrophosphonylation reactions. Among these, the chiral titanium catalysts have been identified as very promising candidates for this transformation, but even within this class, successful catalysts affording significant stereoselectivity are still rare.[13] Indeed, the development of new chiral titanium catalysts for enantioselective hydrophosphonylation is still a challenge.

In this context, You *et al.* have developed a new type of bifunctional catalyst, generated from the metal–organic self-assembly of substituted chiral BINOLS and cinchona alkaloids, in combination with titanium tetraisopropylate, for the highly efficient asymmetric hydrophosphonylation of aldehydes with dimethyl phosphite.[14] The protocols are capable of tolerating a wide range of substrates, even in the presence of 2.5 mol% of catalyst loading. In particular, their modular nature enables easy tuning of the steric and electronic properties of each moiety. As shown in Scheme 7.4, employing a mixture of BINOL derivatives **4** or **5** with cinchonidine **6**, allows a range of chiral α-hydroxy phosphonates to be achieved from the

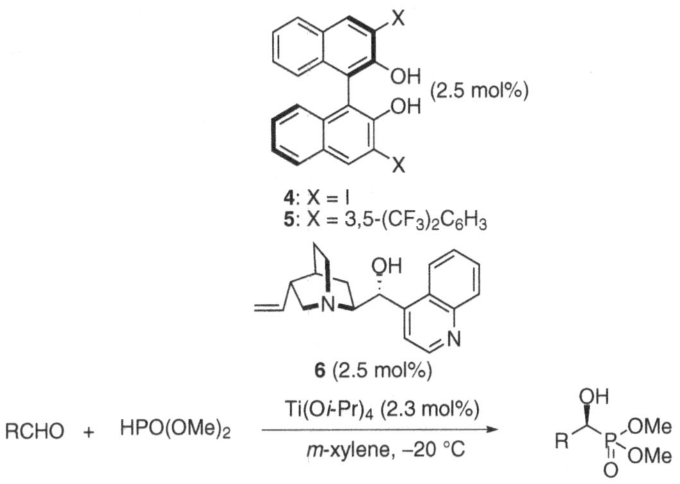

$$\text{4: X = I}$$
$$\text{5: X = 3,5-(CF}_3)_2\text{C}_6\text{H}_3$$

6 (2.5 mol%)

$$\text{RCHO} + \text{HPO(OMe)}_2 \xrightarrow[\text{m-xylene, }-20\,°\text{C}]{\text{Ti(O}i\text{-Pr)}_4 \text{ (2.3 mol%)}}$$

with ligand **4**:
R = Ph: 92% yield, ee = 94%
R = *m*-Tol: 90% yield, ee = 95%
R = *p*-Tol: 92% yield, ee = 94%
R = *o*-MeOC$_6$H$_4$: 95% yield, ee = 96%
R = *o*-ClC$_6$H$_4$: 99% yield, ee = 90%
R = *p*-ClC$_6$H$_4$: 87% yield, ee = 90%
R = *p*-NO$_2$C$_6$H$_4$: 99% yield, ee = 90%
R = 1-Naph: 97% yield, ee >99%
R = 2-Naph: 97% yield, ee >99%
with ligand **5**:
R = BnCH$_2$: 96% yield, ee = 92%
R = Cy: 93% yield, ee = 92%
R = *n*-Oct: 98% yield, ee = 94%
R = *i*-Pr: 90% yield, ee = 94%

Scheme 7.4. Hydrophosphonylation of aldehydes with a combination of cinchonidine and a (*R*)-BINOL derivative as ligands corresponding aldehydes in both remarkable yields and enantioselectivities of up to 99% and >99% ee, respectively. Importantly, this powerful process presented the advantage of employing commercially available ligands.

In 2013, Zi *et al.* reported the synthesis of a series of chiral titanium complexes prepared from the reaction between titanium tetraisopropylate and chiral Schiff base ligands, which were then further investigated in enantioselective hydrophosphonylation of aldehydes with dimethyl phosphite.[15] Among a range of C_1- and C_2-symmetric mono-ligated titanium complexes and *bis*-ligated titanium complexes, catalyst 7 was selected as the most efficient to promote the hydrophosphonylation reaction of aromatic aldehydes, providing moderate enantioselectivities of up to 66% ee, as shown in Scheme 7.5. The steric hindrance of the substrate was shown to have a noticeable effect on the enantioselectivity of the reaction, and the best enantioselectivity was reached with 2-chlorobenzaldehyde, whereas the electronic effect on the enantioselectivity was not noticed.

Due to the low reactivity of ketones in comparison with aldehydes, effective methods for their enantioselective hydrophosphonylation are

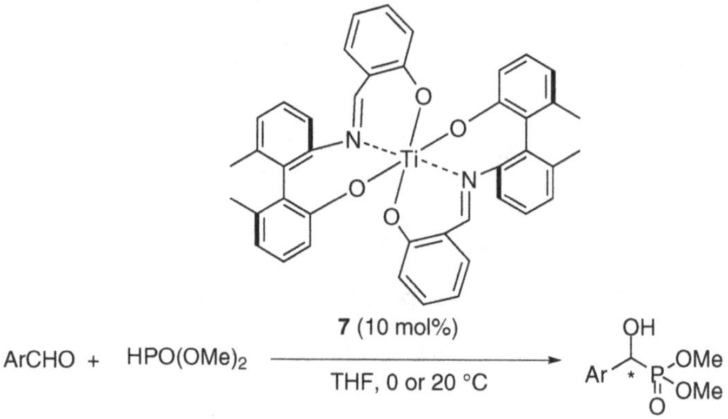

7 (10 mol%)

ArCHO + HPO(OMe)$_2$ $\longrightarrow$ THF, 0 or 20 °C

Ar = Ph: 42% yield, ee = 49%
Ar = *p*-Tol: 84% yield, ee = 44%
Ar = *o*-MeOC$_6$H$_4$: 88% yield, ee = 56%
Ar = *o*-ClC$_6$H$_4$: 87% yield, ee = 66%
Ar = *m*-ClC$_6$H$_4$: 89% yield, ee = 61%
Ar = *p*-ClC$_6$H$_4$: 95% yield, ee = 40%
Ar = *p*-FC$_6$H$_4$: 90% yield, ee = 45%

Scheme 7.5. Hydrophosphonylation of aromatic aldehydes with chiral Schiff base bis-ligated titanium catalyst 7

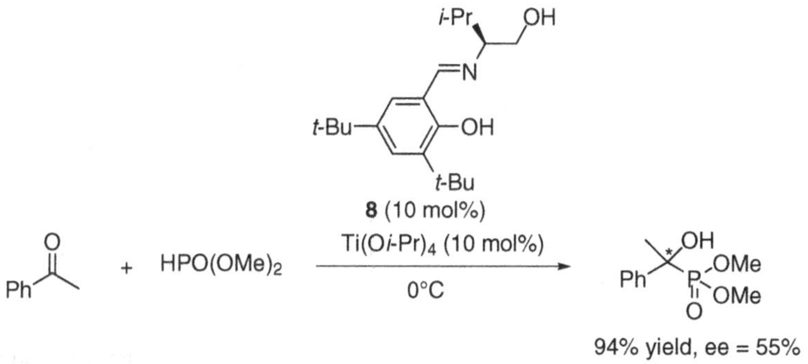

Scheme 7.6. Hydrophosphonylation of acetophenone with chiral Schiff base ligand **8**

even more scarcely developed, and finding a catalyst that can achieve high reactivity and wide applicability, especially for functionalised ketones, is still challenging. In this context, Feng *et al.* have reported on the hydrophosphonylation of acetophenone with dimethyl phosphite, performed in the presence of an *in situ* generated titanium catalyst, derived from titanium tetraisopropylate and chiral tridentate Schiff base **8**.[16] Under neat conditions, the process smoothly afforded the corresponding quaternary α-hydroxy phosphonate in high yield (94%) and moderate enantioselectivity of 55% ee (Scheme 7.6). Although the enantioselectivity of this process was moderate, this result constituted the first example of catalytic asymmetric hydrophosphonylation of an unactivated ketone.

7.4. Miscellaneous Reductions

Chiral amines are important building blocks for the synthesis of a wide variety of biologically active products. Thus, considerable efforts have been devoted to the development of efficient and selective methods for the catalytic asymmetric reduction of imines. In addition to catalysts based on late-transition metals, many titanocenes for enantioselective hydrosilylations of imines have been developed. However, these catalysts are expensive, and they are usually prepared under aggressive conditions (e.g. hydrofluoric acid complexes to prepare difluoride titanium catalysts). In 2009, Khinast *et al.* reported an enantioselective hydrosilylation reaction of imines using an organolithium reagent (*n*-BuLi) and a silane (PhSiH₃),

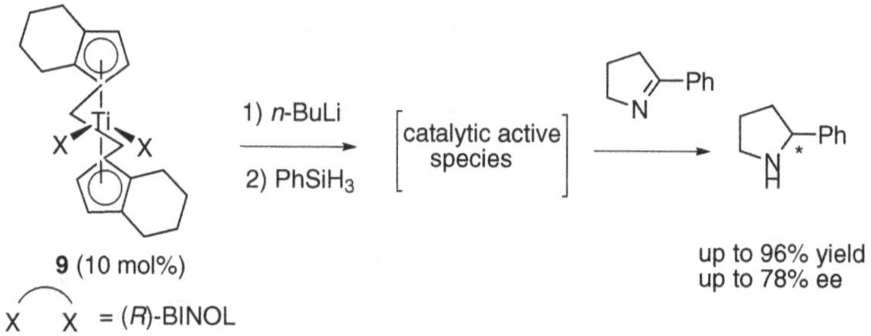

Scheme 7.7. Hydrosilylation of 2-phenylpyrroline with chiral ethylenebis(tetrahydroindenyl) titanocene catalyst **9**

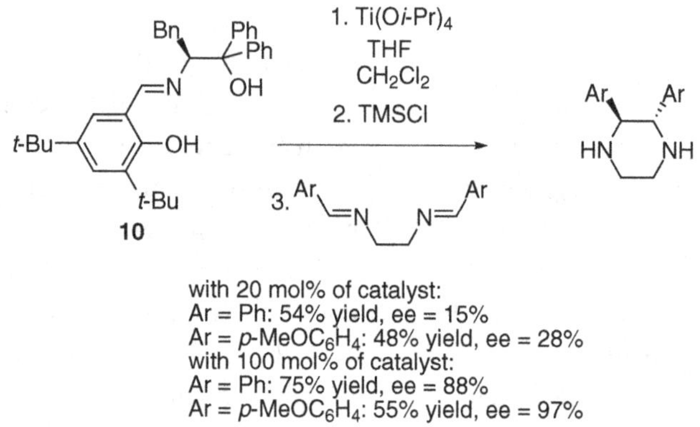

Scheme 7.8. Reductive cyclisation of diimines with chiral hemisalen ligand **10**

giving an active species that catalysed the hydrosilylation reaction.[17] As shown in Scheme 7.7, the use of (R,R)-ethylene-1,2-bis(μ5-4,5,6,7-tetrahydro-1-indenyl)titanium-1,1'-binaphth-2-olate **9** as catalyst precursor allowed 2-phenylpyrroline to be reduced into the corresponding chiral pyrrolidine in excellent yield of 96% with good enantioselectivity of 78% ee.

In 2008, Periasamy and Vairaprakash developed new titanium catalysts, based on chiral hemisalen ligands derived from chiral β-amino alcohols for the enantioselective intramolecular reductive cyclisations of diimines, yielding chiral *trans*-2,3-diarylpiperazines.[18] As shown in Scheme 7.8, the reactions of diimines derived from ethylenediamine afforded, in the presence of ligand **10** and titanium tetraisopropylate, the corresponding chiral *trans*-2,3-diarylpiperazines in low enantioselectivities of up to 28% ee, when a catalyst loading of 20 mol% of the catalytic system was used; while employing a stoichiometric amount of catalyst allowed excellent enantiose-lectivities of up to 97% ee to be achieved in combination with good yields. This novel methodology opened a new route to chiral biologically impor-tant 2,3-diarylpiperazines.

The reductive cyclisation of ω-ketonitriles is of particular interest because the resulting α-hydroxyketone fragment is present in over 1500 known natural products. In this context, Streuff *et al.* have developed a titanium(III)-catalysed asymmetric reductive coupling of ketones with nitriles.[19] The cyclic α-hydroxyketone products were obtained in good yields and high enantioselectivities of up to 94% ee, when the reaction was catalysed by chiral titanocene catalyst **11** in the presence of triethylamine hydrochloride as an additive (Scheme 7.9). A number of aromatic and ali-phatic substitution patterns, heterocyclic groups, and annulated rings were tolerated in the process. It was shown that other titanium catalysts such as TADDOL- or salen-based complexes did not mediate the cyclisation.

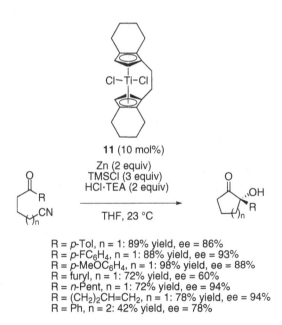

11 (10 mol%)
Zn (2 equiv)
TMSCl (3 equiv)
HCl·TEA (2 equiv)
THF, 23 °C

R = *p*-Tol, n = 1: 89% yield, ee = 86%
R = *p*-FC$_6$H$_4$, n = 1: 88% yield, ee = 93%
R = *p*-MeOC$_6$H$_4$, n = 1: 98% yield, ee = 88%
R = furyl, n = 1: 72% yield, ee = 60%
R = *n*-Pent, n = 1: 72% yield, ee = 94%
R = (CH$_2$)$_2$CH=CH$_2$, n = 1: 78% yield, ee = 94%
R = Ph, n = 2: 42% yield, ee = 78%

Scheme 7.9. Reductive cyclisation of ketonitriles with chiral titanocene catalyst **11**

7.5. Conclusions

In the last few years, several excellent results have been reported in the context of enantioselective titanium-catalysed reduction reactions. While the development of new chiral titanium catalysts for the enantioselective hydrophosphonylation is still a challenge, You and co-workers have developed a new type of bifunctional catalyst, generated from the metal–organic self-assembly of substituted chiral BINOLs and cinchona alkaloids, in combination with titanium tetraisopropylate, for the highly efficient asymmetric hydrophosphonylation of aldehydes, allowing a range of chiral α-hydroxy phosphonates to be achieved in both remarkable yields and enantioselectivities of up to 99% and > 99% ee, respectively. Another excellent contribution in enantioselective titanium-catalysed reduction reactions was reported by Streuff *et al.* with a titanium(III)-catalysed asymmetric reductive coupling of ketones with nitriles, which afforded chiral cyclic α-hydroxyketones in good yields and high enantioselectivities of up to 94% ee, when the reaction was catalysed by a chiral titanocene catalyst.

References

(1) Y. Yu, K. Ding, G. Chen, in *Acid Catalysis in Modern Organic Synthesis*, H. Yamamoto, K. Ishihara, eds., Wiley–VCH, Weinheim, **2008**, Chapter 14, pp. 721–823.

(2) (a) *Comprehensive Asymmetric Catalysis 1*, E. N. Jacobsen, A. Pfaltz, H. Yamamoto, eds., Springer, Berlin, **2004**. (b) *Comprehensive Asymmetric Catalysis Supplement 1*, E. N. Jacobsen, A. Pfaltz, H. Yamamoto, eds., Springer, Berlin, **2004**, Chapters 6.1–6.3. (c) *Comprehensive Asymmetric Catalysis Supplement 2*, E. N. Jacobsen, A. Pfaltz, H. Yamamoto, eds., Springer, Berlin, **2004**, Chapters 5.2 and 6.4. (d) P. Daverio, M. Zanda, *Tetrahedron: Asymmetry*, **2001**, *12*, 2225–2259. (e) J. M. Brunel, *Recent Res. Dev. Org. Chem.*, **2003**, *7*, 155–190. (f) O. Riant, J. C. Mostefaï, N. J. Coumarcel, *Synthesis*, **2004**, 2943–2958.

(3) (a) R. L. Halterman, T. M. Ramsey, Z. Chen, *J. Org. Chem.*, **1994**, *59*, 2642–2644. (b) M. B. Carter, B. Schiott, A. Gutiérrez, S. L. Buchwald, *J. Am. Chem. Soc.*, **1994**, *116*, 11667–11670.

(4) (a) J. Seayad, A. Tillack, M. Beller, *Angew. Chem., Int. Ed.*, **2004**, *43*, 3368–3398. (b) D. J. Ramon, M. Yus, *Chem. Rev.*, **2006**, *106*, 2126–2208.

(5) (a) T. E. Müller, M. Beller, *Chem. Rev.*, **1998**, *98*, 675–704. (b) M. Nobis, B. Driessen-Hölscher, *Angew. Chem., Int. Ed.*, **2001**, *40*, 3983–3985. (c) J. Seayad, A. Tillack, C. G. Hartung, M. Beller, *Adv. Synth. Catal.*, **2002**, *344*, 795–813. (d) F. Pohlki, S. Doye,

Chem. Soc. Rev., **2003**, *32*, 104–114. (e) I. Bytschkov, S. Doye, *Eur. J. Org. Chem.,* **2003**, 935–946. (f) S. Doye, *Synlett,* **2004**, 1653–1672. (g) A. L. Odom, *Dalton Trans.,* **2005**, 225–233. (h) T. E. Müller, K. C. Hultzsch, M. Yus, F. Foulebo, M. Tada, *Chem. Rev.,* **2008**, *108*, 3795–3892.

(6) (a) P. W. Roesky, T. E. Müller, *Angew. Chem., Int. Ed.,* **2003**, *42*, 2708–2710. (b) S. Hong, T. J. Marks, *Acc. Chem. Res.,* **2004**, *37*, 673–686. (c) K. C. Hultzsch, *Org. Biomol. Chem.,* **2005**, *3*, 1819–1824. (d) K. C. Hultzsch, D. V. Gribkov, F. Hampel, *J. Organomet. Chem.,* **2005**, *690*, 4441–4452. (e) K. C. Hultzsch, *Adv. Synth. Catal.,* **2005**, *347*, 367–391. (f) K. K. Hii, *Pure Appl. Chem.,* **2006**, *78*, 341–349. (g) I. Aillaud, J. Collin, J. Hannedouche, E. Schulz, *Dalton Trans.,* **2007**, 5105–5118. (h) R. A. Widenhoefer, *Chem. Eur. J.,* **2008**, *14*, 5382–5391. (i) S. R. Chemler, *Org. Biomol. Chem.,* **2009**, *7*, 3009–3019. (j) G. Zi, *Dalton Trans.,* **2009**, 9101–9109.

(7) Q. Wang, H. Song, G. Zi, *J. Organomet. Chem.,* **2010**, *695*, 1583–1591.

(8) G. Zi, F. Zhang, L. Xiang, Y. Chen, W. Fang, H. Song, *Dalton Trans.,* **2010**, *39*, 4048–4061.

(9) K. E. Near, B. M. Chapin, D. C. McAnnally-Linz, A. R. Johnson, *J. Organomet. Chem.,* **2011**, *696*, 81–86.

(10) C. Queffélec, M. Petit, P. Janvier, D. A. Knight, B. Bujoli, *Chem. Rev.,* **2012**, *112*, 3777–3807.

(11) H. Gröger, B. Hammer, *Chem. Eur.,* **2000**, *6*, 943–948.

(12) (a) T. Yokomatsu, T. Yamagishi, S. Shibuya, *Tetrahedron: Asymmetry,* **1993**, *4*, 1779–1782. (b) T. Yokomatsu, T. Yamagishi, S. Shibuya, *Tetrahedron: Asymmetry,* **1993**, *4*, 1783–1784.

(13) F. Yang, D. Zhao, J. Lan, P. Xi, L. Yang, S. Xiang, J. You, *Angew. Chem., Int. Ed.,* **2008**, *47*, 5646–5649.

(14) F. Yang, D. Zhao, J. Lan, P. Xi, L. Yang, S. Xiang, J. You, *Angew. Chem., Int. Ed.,* **2008**, *47*, 5646–5649.

(15) L. Chen, N. Zhao, Q. Wang, G. Hou, H. Song, G. Zi, *Inorg. Chim. Acta,* **2013**, *402*, 140–155.

(16) X. Zhou, Y. Liu, L. Chang, J. Zhao, D. Shang, X. Liu, L. Lin, X. Feng, *Adv. Synth. Catal.,* **2009**, *351*, 2567–2572.

(17) H. Gruber-Woelfler, J. G. Khinast, *Organometallics,* **2009**, *28*, 2546–2553.

(18) P. Vairaprakash, M. Periasamy, *Tetrahedron Lett.,* **2008**, *49*, 1233–1236.

(19) J. Streuff, M. Feurer, P. Bichovski, G. Frey, U. Gellrich, *Angew. Chem., Int. Ed.,* **2012**, *51*, 8661–8664.

Chapter 8

Enantioselective Titanium-Catalysed Ring-Opening Reactions of Epoxides and Aziridines

8.1. Introduction

This chapter covers efforts of the chemical community to develop novel enantioselective titanium-catalysed ring-opening reactions of epoxides and aziridines. It looks at developments since the beginning of 2008, when the field was reviewed by Yu and co-workers in a book chapter dealing with titanium Lewis acids.[1]

The chapter has been divided into two parts. The first part deals with enantioselective titanium-catalysed ring-opening reactions of epoxides; while the second collects enantioselective titanium-catalysed ring-opening reactions of aziridines.

8.2. Ring-Opening Reactions of Epoxides

Epoxides are versatile building blocks that have been extensively used in the synthesis of complex organic compounds. Their utility as valuable intermediates has been further expanded upon with the advent of asymmetric catalytic methods for their synthesis. The asymmetric ring-opening reaction of racemic or *meso* epoxides has been achieved with a variety of nucleophiles in the presence of chiral Lewis acids, to afford enantioenriched alcohols with various functionalities, among which chiral titanium complexes represent one type of promising catalysts for enantiocontrol of the processes.[2] In addition to the regio- and stereo-selective ring-opening of epoxides,[3] the desymmetrisation of *meso* epoxides through asymmetric ring-opening with nucleophiles has been extensively investigated since 2008, especially by using chiral titanium catalysts.[4] In the presence of a chiral titanium catalyst, the metal centre of the catalyst coordinates to the oxygen atom of the epoxide. This is followed by enantioselective ring-opening of the epoxide with an achiral nucleophile through nucleophilic substitution. One of the earliest examples of this type of reaction was reported by Oguni, dealing with the asymmetric ring-opening of cyclohexene oxide with trimethylsilyl azide as the nucleophilic agent performed in the presence of a chiral titanium catalyst derived from tartrates, which afforded 2-azidocyclohexanol in 90% yield and 63% ee.[5]

The catalytic asymmetric ring-opening reaction of an epoxide with an amine is of prime interest as it allows the formation of chiral β-amino alcohol units to be achieved. These constitute valuable building blocks for many biologically active compounds and chiral auxiliaries in asymmetric synthesis.[6] Various efficient catalytic methods have been reported for the asymmetric ring-opening of *meso*-epoxides with alkyl and aryl amines, using catalysts based on various chiral ligands with different metal ions, including titanium, to provide chiral β-amino alcohols in excellent yield and enantioselectivity.[7] Among recent works, Kureshy *et al.* have reported the highly enantioselective ring-opening of *meso*-stilbene oxide with anilines catalysed by titanium complexes of novel chiral Schiff base ligands.[8] These catalysts were *in situ* generated through the interaction of titanium tetraisopropylate, with novel chiral Schiff bases prepared by condensation of 3,3'-di-*tert*-butyl-5,5'-methylenebis(salicylaldehyde) with 3,3'-dimethyl-5,5'-methylenebis(salicylaldehyde) with (1*R*,2*S*)-(−)-2-amino-1,2-diphenylethanol. High yields of chiral β-amino alcohols were achieved with excellent enantioselectivities of up to >99% ee when using chiral Schiff base **1** as ligand in the presence of chiral imine (*S*)-**2** as an additive (Scheme 8.1). This chiral catalyst had the advantage of being recoverable and recyclable several times while retaining its performance. Moreover,

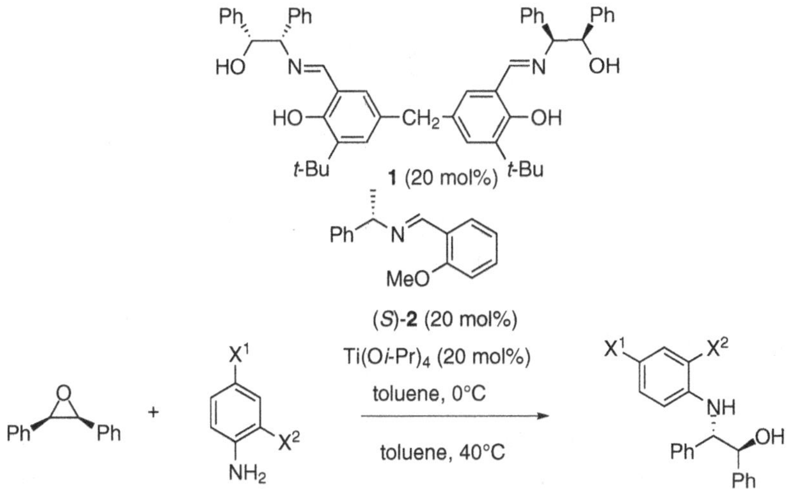

$X^1 = X^2 = H$: 98% yield, ee >99%
$X^1 = H$, $X^2 = OMe$: 95% yield, ee = 85%
$X^1 = OMe$, $X^2 = H$: 96% yield, ee = 98%
$X^1 = Me$, $X^2 = H$: 75% yield, ee >99%
$X^1 = Cl$, $X^2 = H$: 65% yield, ee >99%

Scheme 8.1. Aminolysis of *meso*-stilbene oxide with a chiral Schiff base ligand

the scope of the process was extended to other *meso*-epoxides, such as cyclohexene oxide, cyclooctene oxide, and *cis*-butene oxide, which afforded the corresponding β-amino alcohols in both good to high yields (95, 59 and 88%, respectively) and enantioselectivities (83, 78, and >99% ee, respectively).

The same year, Wang and Ding also reported excellent enantioselectivities of up to 99% ee in the enantioselective aminolysis of 4,4-dimethyl-3,5,8-trioxabicyclo[5.1.0]octane.[9] In this case, the reaction was performed in the presence of an *in situ* generated titanium catalyst of (*R*)-BINOL at a low catalyst loading of only 1 mol%, as shown in Scheme 8.2. The presence of 10 mol% of water was demonstrated to have a beneficial effect on both the reactivity, and enantioselectivity of the reaction, probably by facilitating the formation of oxo- and/or hydroxo-bridged aggregates/oligomers through partial hydrolysis of the titanium alkoxide complexes. A range of aliphatic amines could be used, providing the corresponding chiral β-amino alcohols in moderate to high yields, and

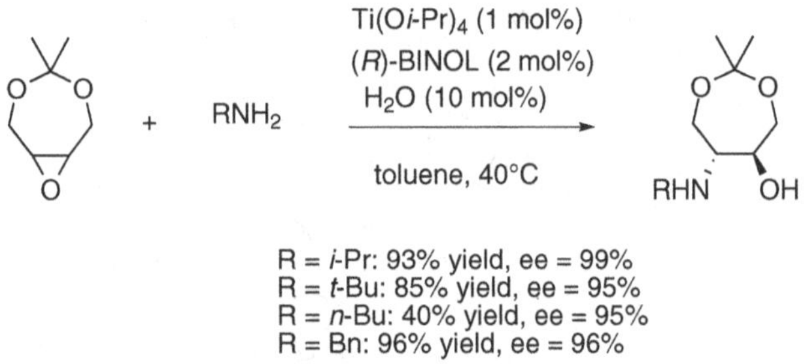

Scheme 8.2. Aminolysis of 4,4-dimethyl-3,5,8-trioxabicyclo[5.1.0]octane with (*R*)-BINOL as ligand

excellent enantioselectivities of up to 99% ee, irrespective of the steric bulkiness of the alkyl group near the reactive centre (NH_2). The authors could not explain why no reaction occurred between 4,4-dimethyl-3,5,8-trioxabicyclo[5.1.0]octane and aniline, whereas the similar reaction with benzylamine proceeded efficiently (96% yield, ee = 96%) under the same reaction conditions.

Later, Kureshy *et al.* described the first report on asymmetric ring-opening of *meso*-epoxides using a polymeric titanium (IV) salen complex as catalyst, which showed steady performance in reuse experiments (six times).[10] This active polymeric catalyst **3** was actually *in situ* generated by reaction of poly-[(*R,R*)-*N,N'*-bis-{3-(1,1-dimethylethyl)-5-methylene sal-icylidene} cyclohexane-1,2-diamine] with titanium tetraisopropylate. When applied to the asymmetric ring-opening of *meso*-stilbene oxide with aniline, it provided better results in the presence of (*R*)-**2** as an additive. Remarkably, the corresponding enantiopure β-amino alcohol was yielded in almost quantitative yield, as shown in Scheme 8.3. However, the same protocol was less effective for the asymmetric ring-opening of other cyclic epoxides, with the best results obtained for cyclohexene oxide as substrate (90% yield, 67% ee). The authors have compared the efficiency of catalyst **3** with that of its corresponding monomeric catalyst, and found that they worked in a similar manner except that the polymeric catalyst was more active and recycled several times with retention of enantioselectivity when compared with the monomeric catalyst, which was nonrecyclable.

More recently, the same authors investigated the same reaction in the presence of novel Schiff base ligands bearing multiple stereogenic centres,

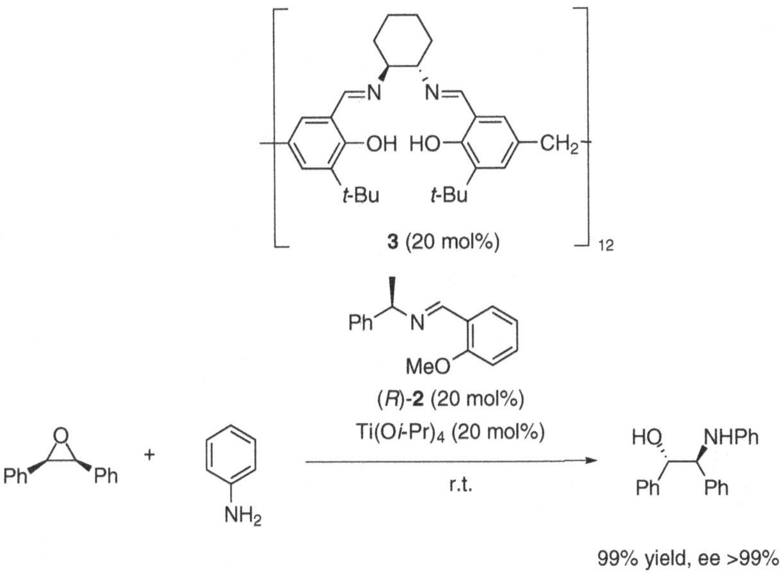

Scheme 8.3. Aminolysis of *meso*-stilbene oxide with a polymeric chiral titanium salen catalyst

arising from various diastereomeric combinations of aminoalcohol functionality with (R)- and (S)-BINOL.[11] Catalytically active titanium complexes were generated *in situ* in the presence of water for the enantioselective ring-opening reaction of *meso*-stilbene oxide with different anilines. Significantly, the catalyst derived from ligand **4** produced at room temperature the corresponding *syn*-β-amino alcohol in both remarkable yields and enantioselectivities of up to 98% and 99% ee, respectively (Scheme 8.4). Furthermore, this catalyst was successfully subjected to catalyst recovery and recyclability experiments over six cycles, with retention of performance. The scope of the methodology was extended to other *meso*-epoxides, such as cyclohexene oxide, cyclopentene oxide, and *cis*-butene oxide with various anilines, providing the corresponding *syn*-β-amino alcohols in high yields (92–98%) and with good to high enantioselectivities of up to 85% ee, 84% ee, and 93% ee, respectively.

The desymmetrisation of *meso* epoxides by ring-opening reaction using a thiol as nucleophile, catalysed by chiral salen based titanium catalysts, was first reported by Hou *et al.* in 1998, providing enantioselectivities

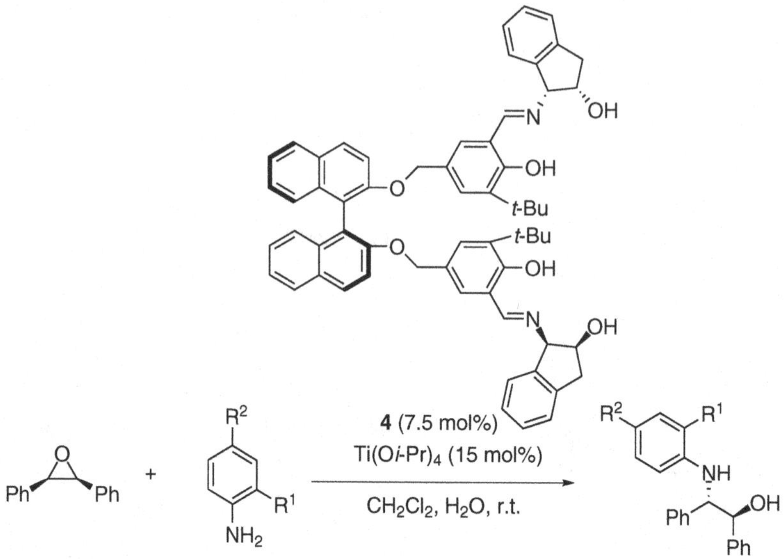

$R^1 = R^2 = H$: 98% yield, ee = 99%
$R^1 = OMe$, $R^2 = H$: 94% yield, ee = 85%
$R^1 = H$, $R^2 = OMe$: 87% yield, ee = 97%
$R^1 = H$, $R^2 = Me$: 94% yield, ee = 99%
$R^1 = Cl$, $R^2 = H$: 81% yield, ee = 95%

Scheme 8.4. Aminolysis of *meso*-stilbene oxide with a (*S*)-BINOL-derived Schiff base ligand

of up to 63% ee.[12] More recently, Zhu *et al.* described moderate to high enantioselectivities of up to 92% ee in the ring-opening of *meso*-epoxides, with aryl thiols promoted by a chiral heterobimetallic Ti–Ga-salen catalyst **5**.[13] This catalyst was *in situ* generated from the corresponding salen ligand, $GaMe_3$, and $Ti(Oi\text{-}Pr)_4$. Applied to the reaction of a range of *meso*-epoxides with aryl thiols, it allowed chiral β-hydroxy sulfides to be synthesised in good to high yields and moderate to high enantioselectivities, as shown in Scheme 8.5. It was found that cyclic *meso*-epoxides displayed better asymmetric induction than acyclic ones under the same conditions. Moreover, with more sterically hindered epoxides, lower enantioselectivities were observed. The authors demonstrated a strong synergistic cooperation between the two metals, finding that both were essential to achieve high enantioselectivity.

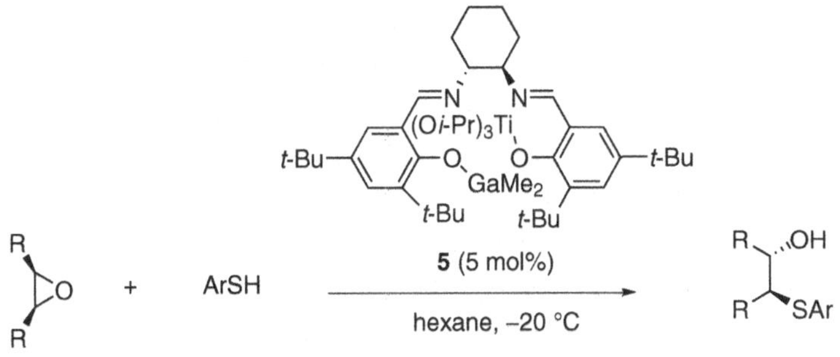

R,R = (CH$_2$)$_4$, Ar = Ph: 95% yield, ee = 84%
R,R = (CH$_2$)$_4$, Ar = p-Tol: 95% yield, ee = 87%
R,R = (CH$_2$)$_4$, Ar = p-ClC$_6$H$_4$: 97% yield, ee = 92%
R,R = (CH$_2$)$_3$, Ar = Ph: 96% yield, ee = 71%
R,R = (CH$_2$)$_3$, Ar = p-Tol: 95% yield, ee = 82%
R = Ar = Ph: 91% yield, ee = 74%
R = Ph, Ar = p-Tol: 95% yield, ee = 84%
R = Ph, Ar = p-ClC$_6$H$_4$: 90% yield, ee = 85%
R,R = (CH$_2$)$_6$, Ar = Ph: 50% yield, ee = 80%

Scheme 8.5. Ring-opening of *meso*-epoxides with aryl thiols in the presence of a chiral heterobimetallic Ti-Ga-salen catalyst

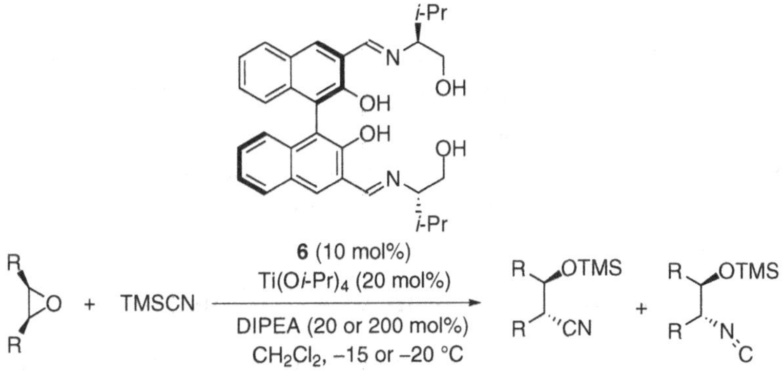

R,R = (CH$_2$)$_3$: 98% conversion, nitrile/isonitrile = 4:1
ee (nitrile) = 94%, ee (isonitrile) = 93%
R,R = (CH$_2$)$_5$: >99% conversion, nitrile/isonitrile = 5:1
ee (nitrile) = 96%
R,R = (CH$_2$)$_2$CH=CH(CH$_2$)$_2$: 10% conversion, nitrile/isonitrile = 1:0,
ee (nitrile) >99%
R,R = CH$_2$CH=CHCH$_2$: 81% conversion, nitrile/isonitrile = 4:1, ee (nitrile) = 90%

Scheme 8.6. Ring-opening of *meso*-epoxides with TMSCN in the presence of a chiral hexadentate ligand

In 2009, Belokon *et al.* developed enantioselective titanium-catalysed ring-opening of *meso*-epoxides with trimethylsilyl cyanide (TMSCN) in the presence of a chiral hexadentate ligand **6**.[14] In the presence of a base such as DIPEA, the reaction afforded a mixture of the corresponding nitrile and isonitrile products in ratios of 4:1 to 1:0, both of them in high enantioselectivities of up to >99% ee and 93% ee, respectively (Scheme 8.6). Moreover, the authors performed the reaction of cyclohexene oxide with TMSCN in the presence of various chiral catalysts derived from the same chiral hexadentate ligand and aluminum, zinc or titanium ions, and demonstrated that it could be controlled to give, predominantly, either the nitrile with up to 99% ee or the isonitrile product with up to 94% ee. The metal ion, ligand stereochemistry, and base concentration were shown to all play a role in determining the product ratio.

It is also possible to enantioselectively open epoxides with water in the presence of a chiral titanium catalyst. For example, Neumann and Milo have employed a porous, homochiral titanium-phosphonate material based on a tripodal peptide scaffold as a heterogeneous reaction medium for the enantioselective hydration of styrene oxide.[15] This titanium phosphonate catalyst, TIP-leu, which was shown to contain confined chiral spaces, was prepared by polymerisation of L-leucine onto a tris(2-aminoethyl)amine initiator, followed by capping with phosphonate groups and completed by non-aqueous condensation with titanium isopropylate (Scheme 8.7). Circular dichroism confirmed that the peptide tethers yielded a secondary structure. As shown in Scheme 8.7, the reaction afforded enantiopure (*S*)-styrene glycol in low yield (20%).

8.3. Ring-Opening Reactions of Aziridines

The asymmetric ring-opening of meso-aziridines has attracted attention of the chemical community only in the 2000s, offering an elegant and direct route to chiral α-functionalised amines.[3c,16] Since 2008, a number of remarkable contributions in this area have been achieved, largely dealing with silylated nucleophiles, such as TMSCN or TMSN$_3$, which afforded 1,2-cyano amines and 1,2-azido amines, respectively, in both excellent yields and enantioselectivities.[16a] However, the use of weaker nucleophiles, such as anilines, has been less studied. In 2009, Kobayashi

synthesis of catalyst TIP-leu:

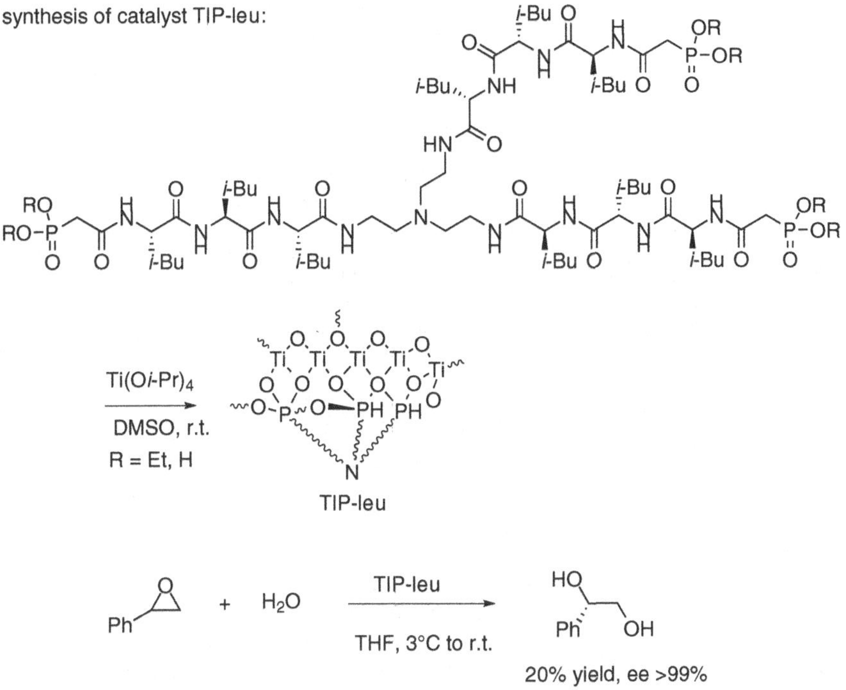

Ti(O*i*-Pr)₄
———————————→
DMSO, r.t.
R = Et, H

TIP-leu

20% yield, ee >99%

Scheme 8.7. Hydration of styrene oxide with a chiral tripodal peptidic titanium phosphonate catalyst

et al. developed the first example of a BINOL-derived titanium chiral catalyst in the highly enantioselective ring-opening reactions of *meso*-aziridines which employed this type of nucleophiles.[17] As shown in Scheme 8.8, the reaction was performed with titanium tetraisopropylate and a chiral tridentate ligand **7** or **8** at 10 mol% of catalyst loading in the presence of anhydrous MgSO₄, which was demonstrated to improve considerably the selectivity of the process. It provided the corresponding chiral *trans*-1,2-diamines in good to high yields and enantioselectivities of up to 95% ee.

These reactions were also independently investigated by Schneider *et al.* by using (*R*)-BINOL as titanium chiral ligand.[18] In this case, the active catalyst was *in situ* generated from this ligand and Ti(O*t*-Bu)₄, allowing the enantioselective ring-opening of a range of cyclic *meso*-aziridines with anilines to be achieved in remarkable enantioselectivities of up

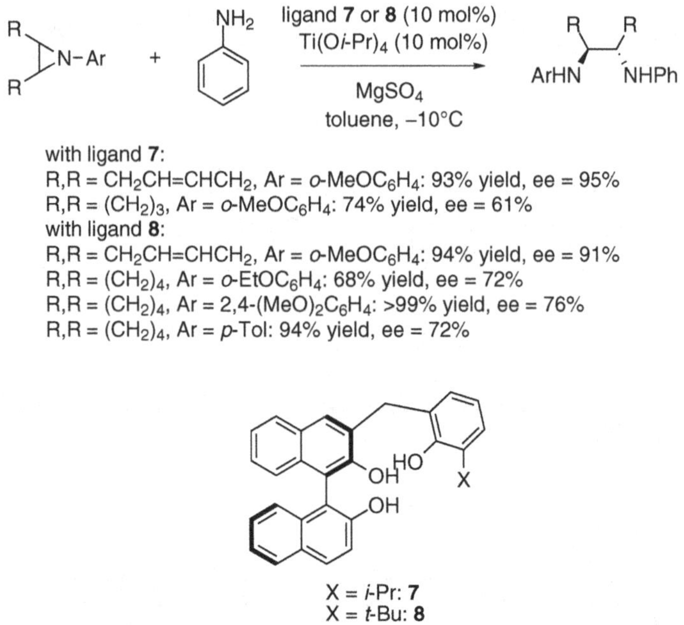
ligand 7 or 8 (10 mol%)
Ti(Oi-Pr)4 (10 mol%)
MgSO4
toluene, –10°C
N–Ar
NH2
ArHN
NHPh
with ligand 7:
R,R = CH2CH=CHCH2, Ar = o-MeOC6H4: 93% yield, ee = 95%
R,R = (CH2)3, Ar = o-MeOC6H4: 74% yield, ee = 61%
with ligand 8:
R,R = CH2CH=CHCH2, Ar = o-MeOC6H4: 94% yield, ee = 91%
R,R = (CH2)4, Ar = o-EtOC6H4: 68% yield, ee = 72%
R,R = (CH2)4, Ar = 2,4-(MeO)2C6H4: >99% yield, ee = 76%
R,R = (CH2)4, Ar = p-Tol: 94% yield, ee = 72%
OH
HO
OH
X
X = i-Pr: 7
X = t-Bu: 8

Scheme 8.8. Ring-opening of *meso*-aziridines with anilines in the presence of chiral tridentate ligands

to 99% ee. The process could also be applied to acyclic *meso* aziridines as was explicitly shown for *N*-phenyl-*cis*-2,3-dimethylaziridine which, upon reaction with aniline, gave rise to the corresponding product in 87% yield and 90% ee. Only in the case of aziridine with a furyl backbone was a low enantioselectivity (17% ee) observed, whereas aziridine attached to a seven-membered ring with an acetal moiety was ring-opened in poor yield (20%) albeit excellent enantioselectivity (90% ee), as shown in Scheme 8.9. Later, the same authors also promoted these reactions with the catalyst *in situ* generated from $Ti(Ot\text{-}Bu)_4$ and chiral bis-BINOL employed at 11 mol% of catalyst loading.[18b] This catalytic system in comparison with that derived from simple (*R*)-BINOL afforded higher yields and/or enantioselectivities in only selected cases. Generally, the enantioselectivities were lower, and less than 87% ee.

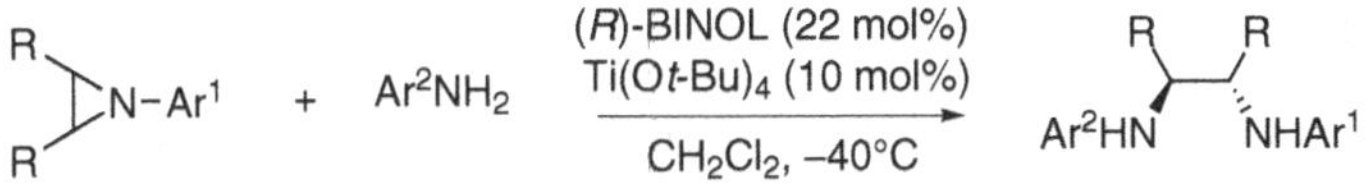

R,R = $(CH_2)_3$, Ar^1 = Ar^2 = Ph: 90% yield, ee = 99%
R,R = $(CH_2)_3$, Ar^1 = Ph, Ar^2 = p-Tol: 83% yield, ee = 99%
R,R = $(CH_2)_3$, Ar^1 = Ph, Ar^2 = p-MeOC$_6$H$_4$: 77% yield, ee = 97%
R,R = $(CH_2)_3$, Ar^1 = Ph, Ar^2 = p-(SH)C$_6$H$_4$: 76% yield, ee = 99%
R,R = $(CH_2)_3$, Ar^1 = p-MeOC$_6$H$_4$, Ar^2 = Ph: 96% yield, ee = 99%
R,R = CH$_2$CH=CHCH$_2$, Ar^1 = Ph, Ar^2 = p-MeOC$_6$H$_4$: 75% yield, ee = 99%
R = Me, Ar^1 = Ar^2 = Ph: 87% yield, ee = 90%
R,R = CH$_2$OCH$_2$, Ar^1 = Ar^2 = Ph: 85% yield, ee = 17%
R,R = CH$_2$OC(Me)$_2$OCH$_2$, Ar^1 = Ar^2 = Ph: 20% yield, ee = 90%

Scheme 8.9. Ring-opening of *meso*-aziridines with anilines in the presence of (R)-BINOL as ligand

8.4. Conclusions

Since 2008, remarkable results have been reported in the context of enantioselective titanium-catalysed ring-opening of epoxides. Most of them came from Kureshy's group, who reported highly enantioselective ring-openings of *meso*-epoxides, such as *meso*-stilbene oxide, cyclohexene oxide, cyclooctene oxide, and *cis*-butene oxide, with anilines catalysed by titanium complexes of novel chiral Schiff base ligands, which provided high yields of chiral β-amino alcohols with excellent enantioselectivities of up to >99% ee. At the same time, Wang and Ding also reported the same level of enantioselectivities in the enantioselective aminolysis of 4,4-dimethyl-3,5,8-trioxabicyclo[5.1.0]octane with a range of aliphatic amines using only 1 mol% of (R)-BINOL as ligand. Furthermore, Kureshy described the first report on asymmetric ring-opening of *meso*-epoxides using a polymeric titanium(IV) salen complex as catalyst. In the context of enantioselective titanium-catalysed ring-opening of *meso*-aziridines with anilines, good results have been obtained by Kobayashi, using chiral tridentate ligands. These reactions were also independently investigated by Schneider in the presence of (R)-BINOL as titanium chiral ligand, providing remarkable enantioselectivities of up to 99% ee.

References

(1) Y. Yu, K. Ding, G. Chen, in *Acid Catalysis in Modern Organic Synthesis*, H. Yamamoto, K. Ishihara, eds., Wiley–VCH, Weinheim, **2008**, Chapter 14, pp. 721–823.

(2) (a) M. Pineschi, *Eur. J. Org. Chem.*, **2006**, *22*, 4979–4988. (b) I. M. Pastor, M. Yus, *Curr. Org. Chem.*, **2005**, *9*, 1–29. (c) E. N. Jacobsen, *Acc. Chem. Res.*, **2000**, *33*, 421–431.

(3) (a) I. Paterson, D. J. Berrisford, *Angew. Chem., Int. Ed. Engl.*, **1992**, *31*, 1179–1180. (b) E. N. Jacobsen, *Acc. Chem. Res.*, **2000**, *33*, 421–431. (c) I. P. C. Nielsen, E. N. Jacobsen, in *Aziridines and Epoxides in Organic Synthesis*, A. K. Yudin, Ed., Wiley–VCH, Weinheim, **2006**, pp. 229–269.

(4) (a) P.-A. Wang, *Belstein J. Org. Chem.*, **2013**, *9*, 1677–1695. (b) R. Chawla, A. K. Singh, L. D. S. Yadav, *RSC Advances*, **2013**, *3*, 11385–11403. (c) S. Matsunaga, in *Comprehensive Chirality*, **2012**, *5*, 534–580, E. M. Carreira, H. Yamamoto, eds., Elsevier, Amsterdam. (d) C. Schneider, *Synthesis*, **2006**, *23*, 3919–3944.

(5) M. Hayashi, K. Kohmura, N. Oguni, *Synlett*, **1991**, 774–776.

(6) D. J. Ager, I. Prakash, D. R. Schaad, *Chem. Rev.*, **1996**, *96*, 835–875.

(7) (a) R. I. Kureshy, S. Singh, N. H. Khan, S. H. R. Abdi, E. Suresh, R. V. Jasra, *Eur. J. Org. Chem.*, **2006**, 1303–1309. (b) R. I. Kureshy, K. J. Prathap, S. Agrawal, N. H. Khan, S. H. R. Abdi, R. V. Jasra, *Eur. J. Org. Chem.*, **2008**, 3118–3128. (c) R. I. Kureshy, M. Kumar, S. Agrawal, N. H. Khan, B. Dangi, S. H. R. Abdi, H. C. Bajaj, *Chirality*, **2011**, *23*, 76–83.

(8) R. I. Kureshy, K. J. Prathap, S. Agrawal, N.-U. H. Khan, S. H. R. Abdi, R. V. Jasra, *Eur. J. Org. Chem.*, **2008**, 3118–3128.

(9) H. Bao, J. Zhou, Z. Wang, Y. Guo, T. You, K. Ding, *J. Am. Chem. Soc.*, **2008**, *130*, 10116–10127.

(10) R. I. Kureshy, M. Kumar, S. Agrawal, N.-U. H. Khan, B. Dangi, S. H. R. Abdi, H. C. Bajaj, *Chirality*, **2011**, *23*, 76–83.

(11) M. Kumar, R. I. Kureshy, D. Ghosh, N.-U. H. Khan, S. H. R. Abdi, H. C. Bajaj, *ChemCatChem*, **2013**, *5*, 1–8.

(12) J. Wu, X. Hou, L. Dai, L. Xia, M. tang, *Tetrahedron: Asymmetry*, **1998**, *9*, 3431–3436.

(13) J. Sun, F. Yuan, M. Yang, Y. Pan, C. Zhu, *Tetrahedron Lett.*, **2009**, *50*, 548–551.

(14) Y. N. Belokon, D. Chusov, A. S. Peregudov, L. V. Yashkina, G. I. Timofeeva, V. I. Maleev, M. North, H. B. Kagan, *Adv. Synth. Catal.*, **2009**, *351*, 3157–3167.

(15) A. Milo, R. Neumann, *Adv. Synth. Catal.*, **2010**, *352*, 2159–2165.

(16) (a) C. Schneider, *Angew. Chem., Int. Ed.*, **2009**, *48*, 2082–2084. (b) M. Pineschi, *Eur. J. Org. Chem.*, **2006**, 4979–4988. (c) X. E. Hu, *Tetrahedron*, **2004**, *60*, 2701–2743.

(17) R. Yu, Y. Yamashita, S. Kobayashi, *Adv. Synth. Catal.*, **2009**, *351*, 147–152.

(18) (a) S. Peruncheralathan, H. Teller, C. Schneider, *Angew. Chem., Int. Ed.*, **2009**, *48*, 4849–4852. (b) S. Peruncheralathan, S. Aurich, H. Teller, C. Schneider, *Org. Biomol. Chem.*, **2013**, *11*, 2787–2803.

Chapter 9

Enantioselective Titanium-Catalysed Domino and Tandem Reactions

This chapter covers efforts of the chemical community to develop novel enantioselective titanium-catalysed domino and tandem reactions. It looks at developments since the beginning of 2008, when the field was reviewed by Yu and co-workers in a book chapter dealing with titanium Lewis acids.[1]

An increasing number of enantioselective metal-catalysed domino processes have been developed since 2008. A domino reaction has been defined by Tietze as a reaction which involves two or more bond-forming transformations, taking place under the same reaction conditions, without adding additional reagents and catalysts; and in which the subsequent reactions occur as a consequence of the functionality caused by bond formation or fragmentation in the previous step.[2] It must be recognised that a relatively narrow distinction exists between domino and consecutive cascade or tandem reactions. From the point of view of an operator, the only difference between the two lies in the point along the sequence at which one or more catalysts or reagents had to be added to effect either the initiation of a sequence (that is, domino reaction) or propagation to the next step (that is, consecutive reaction). Thus the descriptors domino, cascade, and tandem are often used indistinguishably from one another in the literature,[3] and a variety of opinions exist on how such reactions should be classified.

According to Tietze, a domino reaction is strictly defined as a process in which two or more bond-forming transformations occur based on functionalities formed in the previous step and, moreover, no additional reagents, catalysts or additives can be added to the reaction vessel, nor can reaction conditions be changed.[2] Denmark further posited, however, that most domino reactions, as defined by Tietze, fell under the broader category of tandem processes.[4] Other tandem reactions which are not cascades involve the isolation of intermediates, a change in reaction conditions, or the addition of reagents or coupling partners. Other authors classify domino reactions with even stricter conditions.[5,6]

The quality and importance of a domino reaction can be correlated to the number of bonds generated in such a process, and the increase in molecular complexity. Its goal is to resemble nature in its highly selective sequential transformations. The domino reactions can be performed as single-, two-, and multi-component transformations.[7] The use of one-, two-, and multi-component domino reactions in organic synthesis is increasing constantly, since they allow the synthesis of a wide range of complex molecules, including natural products and biologically active compounds in an economically favourable way, by using processes that avoid the use of costly and time-consuming protection–deprotection processes, as well as purification procedures of intermediates.[8] Indeed, decreasing the number of laboratory operations required and the quantities of chemicals and solvents used have made domino and multicomponent reactions vital processes.[9] The proliferation of these reactions is evidenced by the number of recent reviews covering the literature through 1992.[2,5,7c,10]

Although asymmetric synthesis is sometimes viewed as a subdiscipline of organic chemistry, actually this topical field transcends any narrow classification and pervades essentially all chemistry.[11] Of the methods available for preparing chiral compounds, catalytic asymmetric synthesis has attracted most attention. The economical interest in combinations of asymmetric catalytic processes with domino reactions is obvious. In spite of the explosive growth of organocatalysis since 2008,[10l–o,12,13] and its application in the synthesis of a number of chiral products,[14] the catalysis of organic reactions by metals still constitutes one of the most useful and powerful tools in organic synthesis.[11,15] In this period the first examples of enantioselective titanium-catalysed domino reactions have been reported.

For example, Shibata *et al.* have developed enantioselective titanium-catalysed domino condensation/cyclisation reactions based on the use of chiral cinchona alkaloid ligands.[16] As shown in Scheme 9.1, the reaction between cyclic functionalised enamines and ethyl trifluoropyruvate led to the corresponding trifluoromethylated γ-lactams bearing a stereogenic centre through condensation followed by cyclisation. When the process

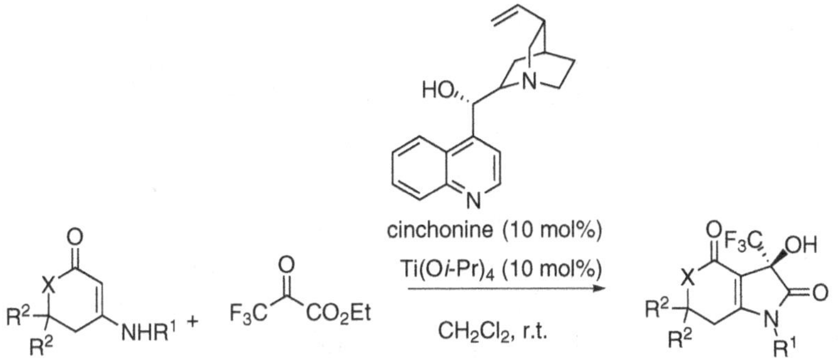

cinchonine (10 mol%)
Ti(O*i*-Pr)$_4$ (10 mol%)
CH$_2$Cl$_2$, r.t.

R^1 = CHPh$_2$, R^2 = Me, X = CH$_2$: 86% yield, ee = 88%
R^1 = CH$_2$-10-anthracenyl, R^2,R^2 = (CH$_2$)$_4$, X = CH$_2$: 93% yield, ee = 89%
R^1 = CH$_2$-10-anthracenyl, R^2 = Me, X = O: 58% yield, ee = 83%
R^1 = CHPh$_2$, R^2,R^2 = (CH$_2$)$_4$, X = CH$_2$: 99% yield, ee = 87%
R^1 = CHPh$_2$, R^2 = Me, X = O: 82% yield, ee = 84%

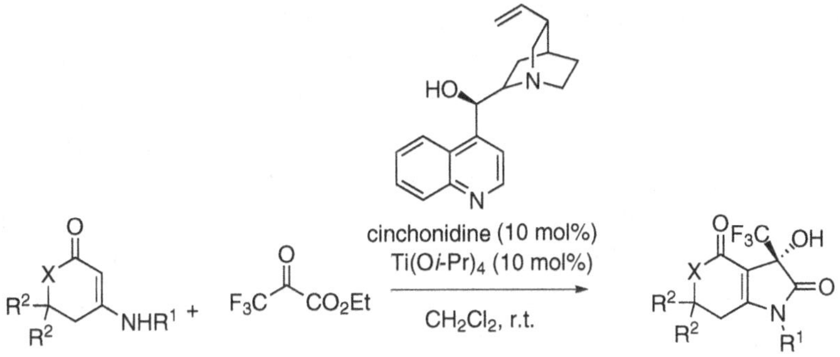

cinchonidine (10 mol%)
Ti(O*i*-Pr)$_4$ (10 mol%)
CH$_2$Cl$_2$, r.t.

R^1 = CHPh$_2$, R^2 = Me, X = CH$_2$: 90% yield, ee = 82%
R^1 = 9-fluorenyl, R^2 = Me, X = CH$_2$: 96% yield, ee = 93%
R^1 = CH$_2$-10-anthracenyl, R^2,R^2 = (CH$_2$)$_4$, X = CH$_2$: 99% yield, ee = 82%

Scheme 9.1. Domino condensation/cyclisation reactions with chiral cinchona alkaloid ligands was performed in the presence of 10 mol% of cinchonine as chiral ligand, it yielded the (*R*)-enantiomers of the domino products in high yields of up to 99% combined with good to high enantioselectivities of up to 89% ee. Also, the use of cinchonidine as chiral ligand allowed the formation of domino products with the (*S*)-configuration to be achieved in comparable yields and enantioselectivities of up to 93% ee, as shown in Scheme 9.1. This process constituted the first example of an asymmetric enamine-pyruvate domino condensation/cyclisation reaction catalysed by a titanium chiral catalyst. It offered the advantage of reaching both enantiomers of the multifunctionalised heterocycles by employing suitable pseudoenantiomeric cinchona alkaloids as titanium ligands. Furthermore, this type of products is attractive as medicinal chemistry templates.

In 2011, chiral 2-substituted-1,5-benzodiazepine derivatives were synthesised by Feng and Liu for the first time from an enantioselective domino reaction involving *o*-phenylenediamine and 2'-hydroxychalcones.[17] The process was induced by a titanium complex, formed from chiral ligand **1** derived from (*S*)-BINOL and L-prolineamide and Ti(O*i*-Pr)$_4$. It led to the formation of a range of 2-aryl-1,5-benzodiazepine derivatives in good yields, and enantioselectivities of up to 82% ee, as shown in Scheme 9.2. The 2'-hydroxy group of the α,β-unsaturated ketones was demonstrated to be critical for both the reactivity and stereoinduction of the process. In order to explain these results, the authors proposed that the formed α,β-unsaturated ketimine intermediate **A** might be stabilised by an intramolecular hydrogen bond of the hydroxyl group. In the presence of the chiral titanium complex, intermediate **A** could coordinate to the metal with the oxygen atom of the hydroxyl group and the nitrogen atom of the imine moiety to generate intermediate **B**. In intermediate **B**, the amino group preferred to attack the β-*Si* face of the C=C bond due to the fact that there was less steric hindrance between the aniline moiety of the ligand. Thus, the (*S*)-domino products were afforded through a final aza-Michael addition and the chiral titanium complex was released.

The field of *N*-heterocyclic carbene (NHC) catalysis has undergone an explosive growth since 2008.[18] In this context, Scheidt *et al.* have developed an enantioselective dimerisation of enals, catalysed by a combination of chiral NHC **2** and titanium tetraisopropylate.[19] The reaction occurred between two equivalents of enals, providing the corresponding

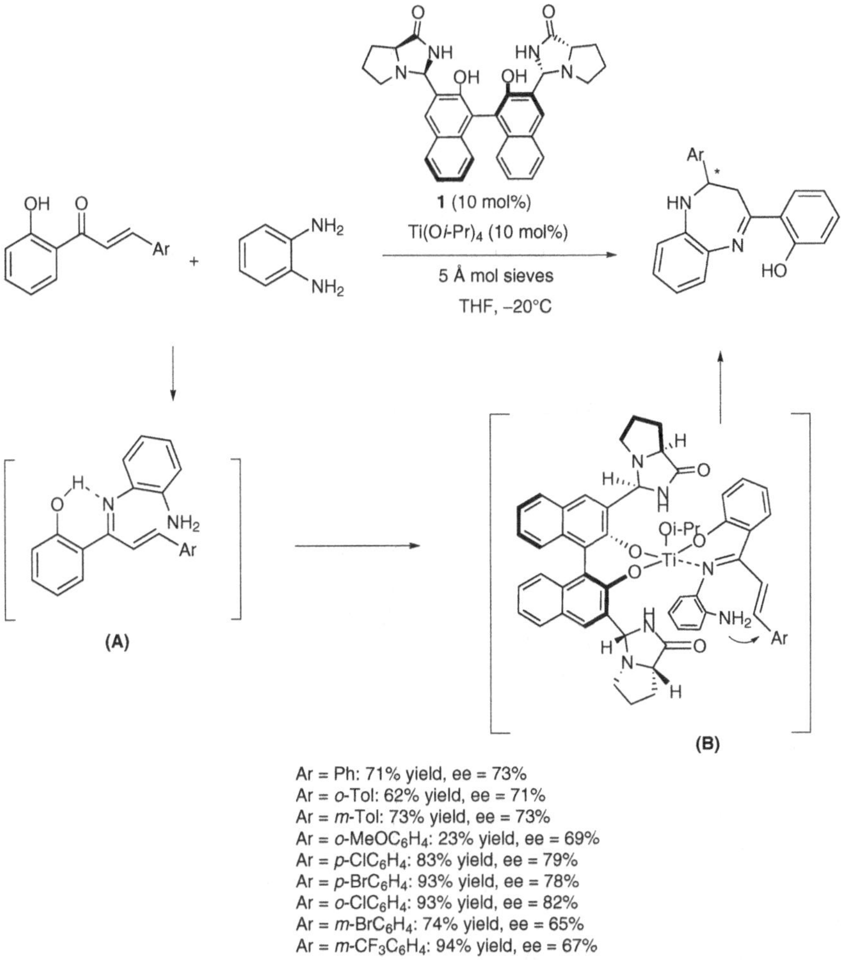
1 (10 mol%)
Ti(O*i*-Pr)4 (10 mol%)
5 Å mol sieves
THF, –20°C
(A)
(B)
Ar = Ph: 71% yield, ee = 73%
Ar = *o*-Tol: 62% yield, ee = 71%
Ar = *m*-Tol: 73% yield, ee = 73%
Ar = *o*-MeOC6H4: 23% yield, ee = 69%
Ar = *p*-ClC6H4: 83% yield, ee = 79%
Ar = *p*-BrC6H4: 93% yield, ee = 78%
Ar = *o*-ClC6H4: 93% yield, ee = 82%
Ar = *m*-BrC6H4: 74% yield, ee = 65%
Ar = *m*-CF3C6H4: 94% yield, ee = 67%

Scheme 9.2. Domino imine formation/aza-Michael addition reaction with a chiral (*S*)-BINOL ligand derived from L-prolineamide

chiral cyclopentenes in good yields and enantioselectivities of up to 90% ee, as summarised in Scheme 9.3. The synthetic utility of these products was demonstrated by the functionalisation of their alkene function. The authors proved that the use of titanium tetraisopropylate was a key element for the reactivity and selectivity of the domino process. It was assumed to promote a selective conjugate addition versus the typical

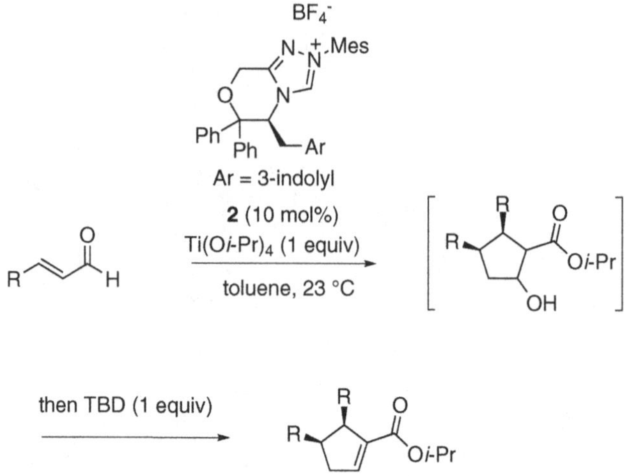

R = Ph: 61% yield, ee = 90%
R = *p*-BrC_6H_4: 61% yield, ee = 90%
R = *p*-ClC_6H_4: 59% yield, ee = 90%
R = *m*-ClC_6H_4: 58% yield, ee = 84%
R = *p*-$MeOC_6H_4$: 61% yield, ee = 90%
R = *p*-Tol: 61% yield, ee = 90%
R = *m*-Tol: 62% yield, ee = 85%
R = 1-Naph: 59% yield, ee = 85%
R = 2-Naph: 63% yield, ee = 86%

Scheme 9.3. Domino dimerisation of enals with a chiral NHC ligand

1,2-addition. The reaction finished with a dehydration of the domino alcohol intermediate which occurred after the subsequent addition of a base such as 1,5,7-triazabicyclo-[4.4.0]dec-5-ene (TBD).

In 2014, Lopp *et al.* reported an asymmetric synthesis of chiral tertiary 2-substituted 5-oxotetrahydrofuran-2-carboxylic acids, based on an enantioselective titanium-catalysed domino epoxidation/Baeyer–Villiger/acylation/hydrolysis reaction, using (+)-diethyl tartrate (DET) as chiral ligand.[20] As shown in Scheme 9.4, the reaction of 3-substituted 1,2-cyclopentanediones with *tert*-butyl hydroperoxide in the presence of catalytic amounts of titanium tetraisopropylate and (+)-DET afforded the corresponding chiral lactones in good yields, and high enantioselectivities of up to 94% ee. Different functional groups in the 3-alkyl substituent (R) of the substrate, such as hydroxyl, ether, Boc-protected amino, and ester groups, were tolerated. Boc-aminomethyl substituents led to β-amino acid analogues and Boc-aminoethyl substituents to γ-amino acid analogues, as well as

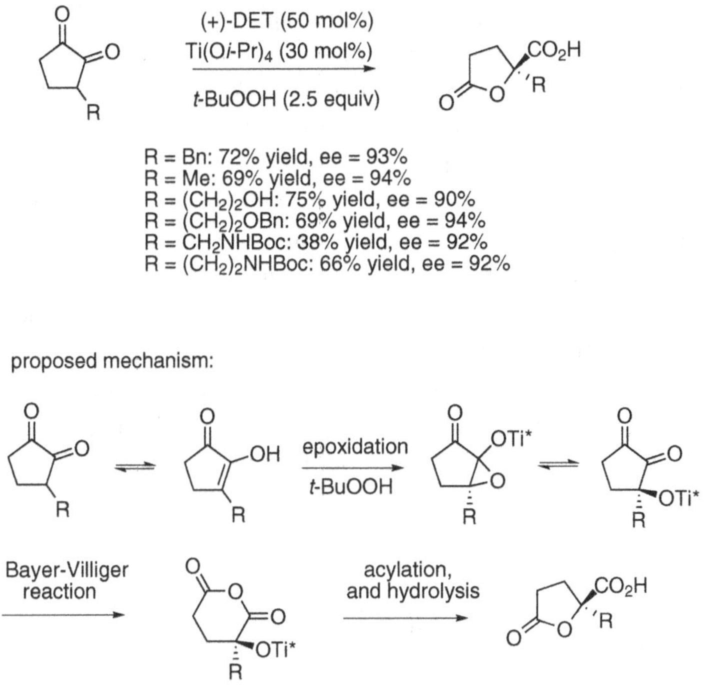
(+)-DET (50 mol%)
Ti(O*i*-Pr)$_4$ (30 mol%)
t-BuOOH (2.5 equiv)
R = Bn: 72% yield, ee = 93%
R = Me: 69% yield, ee = 94%
R = (CH$_2$)$_2$OH: 75% yield, ee = 90%
R = (CH$_2$)$_2$OBn: 69% yield, ee = 94%
R = CH$_2$NHBoc: 38% yield, ee = 92%
R = (CH$_2$)$_2$NHBoc: 66% yield, ee = 92%
proposed mechanism:
epoxidation
t-BuOOH
Bayer-Villiger reaction
acylation, and hydrolysis

Scheme 9.4. Domino epoxidation/Baeyer–Villiger/acylation/hydrolysis reaction with (+)-diethyl tartrate as ligand

spiro-lactone-lactams. This is clear for a chemist and, moreover all the steps are depicted in Scheme 9.4, in which the domino reaction starts with the epoxidation of the substrate, followed by a Baeyer–Villiger reaction of the resulted intermediate and then acylation and hydrolysis.

The enantioselective addition of allylmetal derivatives to ketones using chiral titanium complexes has been widely developed, especially with aldehydes as substrates. The first catalytic enantioselective process using ketones as substrates was published by Tagliavini in 1999.[21] In this work, enantioselectivities of up to 65% ee were obtained for the addition of tetraallyltin to various ketones in the presence of dichlorotitanium diisopropoxide and BINOL. In 2014, Pericas and Walsh reported the covalent immobilisation of (*R*)-BINOL to polystyrene, by employing a 1,2,3-triazole linker to heterogenise a titanium-BINOlate catalyst and its application in the enantioselective allylation of ketones.[22] Therefore, by using a simple synthetic route, enantiopure 6-ethynyl-BINOL was synthesised and anchored to an azidomethylpolystyrene resin through a copper-catalysed alkyne-azide cycloaddition reaction. The polystyrene-supported BINOL ligand **3** was prepared from 30 mol% of (*R*)-BINOL and Ti(O*i*-Pr)$_4$. It was then in situ converted into its diisopropoxytitanium derivative, and used as a heterogeneous catalyst in the asymmetric allylation of a variety of ketones with allyltributyl tin in CH$_2$Cl$_2$ at room temperature. This provided the corresponding chiral tertiary allylic alcohols in good to high yields of up to 98% and high enantioselectivities of up 95% ee. Furthermore, the authors demonstrated the reusability of the ligand, since both yields and enantioselectivities were preserved after three consecutive reaction cycles.

Importantly, this work constituted one of the very few examples of catalytic creation of quaternary centers involving the use of a heterogenised catalytic species. This methodology was applied to the development of two enantioselective tandem reactions. The first one consisted in a tandem allylation/Pauson–Khand reaction depicted in Scheme 9.5,

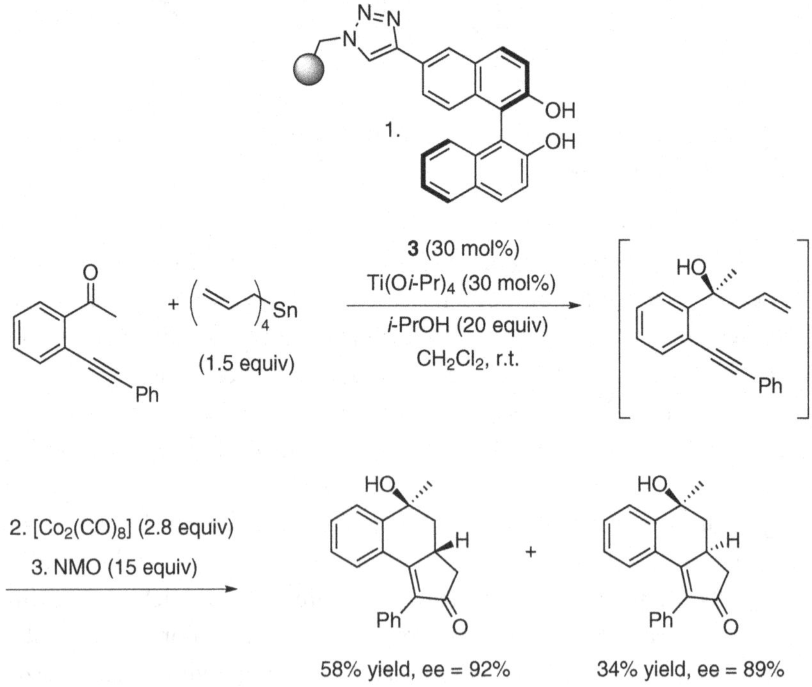

58% yield, ee = 92%   34% yield, ee = 89%

Scheme 9.5. Tandem allylation/Pauson–Khand reaction with a chiral polymer-supported titanium-BINOLate catalyst which afforded chiral tricyclic tandem products in 92% yield as a 63:37 mixture of *syn* and *anti* diastereomers respectively, obtained with enantioselectivities of 92% ee and 89% ee. While the first step of the sequence consisted in the enantioselective titanium-catalysed allylation reaction of an alkyne substrate with allyltributyl tin in the presence of polystyrene-supported BINOL ligand **3**, the second step delt with the Pauson–Khand reaction of the resulting allylic alcohol intermediate promoted by the addition of [Co$_2$(CO)$_8$] and *N*-methyl morpholine *N*-oxide (NMO) to the reaction mixture.

These authors developed a second highly efficient tandem reaction starting from the enantioselective titanium-catalysed allylation reaction of a cyclohexanone depicted in Scheme 9.6, with allyltributyl tin performed in the presence of polystyrene-supported BINOL ligand **3**.[22] The second step of the tandem reaction was the epoxidation of the intermediate allylic alcohol induced by the addition of *tert*-butyl hydroperoxide,

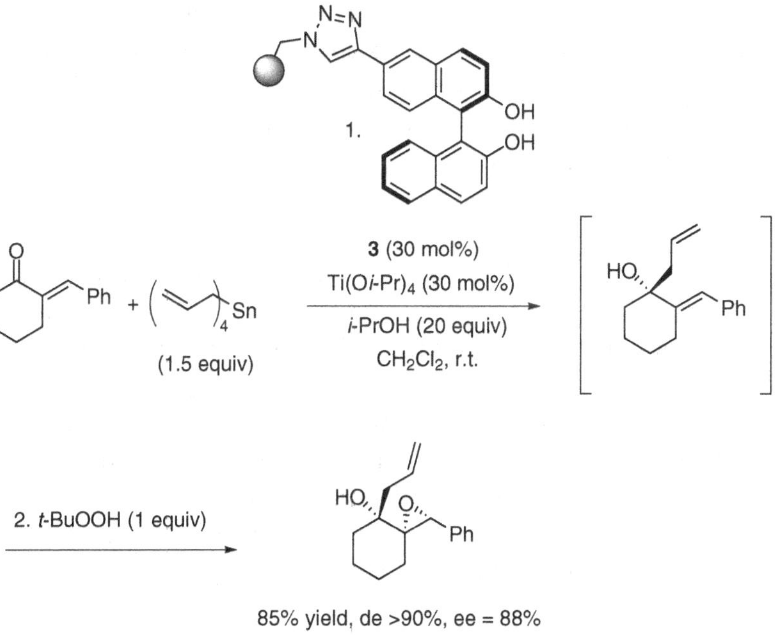

Scheme 9.6. Tandem allylation/epoxidation reaction with a chiral polymer-supported titanium-BINOLate catalyst

providing the corresponding functionalised chiral epoxide, bearing two adjacent quaternary centres as a single diastereomer in 85% yield and good enantioselectivity of 88% ee. This second contribution in addition to that depicted in the precedent Scheme well illustrate the utility of this novel polystyrene-supported BINOL ligand **3**.

Conclusions

This chapter illustrates how asymmetric titanium catalysis has contributed to the development of novel enantioselective domino (and tandem) reactions. It illustrates the power of these elegant one pot processes of two or more bond-forming reactions, evolving under identical conditions, in which the subsequent transformation takes place at the functionalities obtained in the former transformation, following the same principles that are found in biosynthesis from the nature. These fascinating reactions have rapidly become one of the most developing fields in organic chemistry. The years since 2008 have seen the first examples of enantioselective titanium-catalysed domino reactions, reported by several groups, using various chiral ligands, such as cinchona alkaloids, BINOL derivatives including polystyrene-supported ones, tartrates, and NHC, which all provided poly-functionalised mono- or poly-cyclic products, with excellent enantioselec-tivities in all cases of reactions. The economic interest in combinations of asymmetric titanium catalytic processes with the concept of domino and tandem reactions is obvious, and allows reaching higher molecular com-plexity and excellent levels of stereocontrol, with simple operational one pot procedures. There are also savings in terms of costs, solvents, time, and energy, by avoiding costly protecting groups and time-consuming purifi-cation procedures after each step. Undoubtedly, the future direction in this field is to continue expanding the scope of enantioselective domino and tandem reactions through the combination of these different types of reactions; the employment of novel chiral ligands; and the application of these powerful strategies to the synthesis of biologically interesting mole-cules, including natural products, novel chiral ligands, and functional materials.

References

(1) Y. Yu, K. Ding, G. Chen, in *Acid Catalysis in Modern Organic Synthesis*, H. Yamamoto, K. Ishihara, eds., Wiley–VCH, Weinheim, **2008**, Chapter 14, pp. 721–823.

(2) (a) L. F. Tietze, U. Beifuss, *Angew. Chem., Int. Ed.*, **1993**, *32*, 131–163. (b) L. F. Tietze, *Chem. Rev.*, **1996**, *96*, 115–136. (c) L. F. Tietze, G. Brasche, K. Gericke, *Domino Reactions in Organic Synthesis*, Wiley–VCH, Weinheim, **2006**.

(3) K. C. Nicolaou, D. J. Edmonds, P. G. Bulger, *Angew. Chem., Int. Ed.*, **2006**, *45*, 7134–7186.

(4) S. E. Denmark, A. Thorarensen, *Chem. Rev.*, **1996**, *96*, 137–165.

(5) C. J. Chapman, C. G. Frost, *Synthesis*, **2007**, 1–21.

(6) D. E. Fogg, E. N. dos Santos, *Coord. Chem. Rev.*, **2004**, *248*, 2365–2379.

(7) (a) L. F. Tietze, A. Modi, *Med. Res. Rev.*, **2000**, *20*, 304–322. (b) *Multicomponent Reactions*, J. Zhu, H. Bienaymé, eds., Wiley–VCH, Weinheim, **2005**. (c) D. J. Ramon, M. Yus, *Angew. Chem., Int. Ed.*, **2005**, *44*, 1602–1634. (d) *Synthesis of Heterocycles via Multicomponent Reactions*, R. V. A. Orru, E. Ruijter, eds., *Topics in Heterocyclic Chemistry*, Vols. I and II, Springer, Berlin, **2010**.

(8) (a) C. Grondal, M. Jeanty, D. Enders, *Nature Chemistry*, **2010**, *2*, 167–178. (b) K. C. Nicolaou, J. S. Chen, *Chem. Soc. Rev.*, **2009**, *38*, 2993–3009. (c) B. B. Touré, D. G. Hall, *Chem. Rev.*, **2009**, *109*, 4439–4486. (d) M. Colombo, I. Peretto, *Drug Discovery Today*, **2008**, *13*, 677–684. (e) A. Padwa, S. K. Bur, *Tetrahedron*, **2007**, *63*, 5341–5378. (f) C. Hulme, V. Gore, *Curr. Med. Chem.*, **2003**, *10*, 51–80.

(9) *Chem. Soc. Rev.*, *38(11)*, **2009**, Special Issue on Rapid Formation of Molecular Complexity in Organic Synthesis.

(10) (a) G. H. Posner, *Chem. Rev.*, **1986**, *86*, 831–844. (b) T.-L. Ho, in *Tandem Organic Reactions*, Wiley: New York, **1992**. (c) H. Waldmann, *Nachr. Chem. Tech. Lab.*, **1992**, *40*, 1133–1140. (d) K. Fukumoto, *Synth. Org. Chem. Jpn.*, **1994**, *52*, 2–18. (e) R. A. Bunce, *Tetrahedron*, **1995**, *51*, 13103–13159. (f) P. J. Parsons, C. S. Penkett, A. J. Shell, *Chem. Rev.*, **1996**, *96*, 195–206. (g) P. C. F. Balaure, P. I. A. Filip, *Rev. Roum. Chim.*, **2002**, *46(8)*, 809–833. (h) E. Capdevila, J. Rayo, F. Carrion, I. Jové, J. I. Borrell, J. Teixido, *Afinidad*, **2003**, *506*, 317–337. (i) L. F. Tietze, N. Rackelmann, *Pure Appl. Chem.*, **2004**, *76*, 1967–1983. (j) P. I. Dalko, L. Moisan, *Angew. Chem., Int. Ed.*, **2004**, *43*, 5138–5175. (k) H. Pellissier, *Tetrahedron*, **2006**, *62*, 2143–2173. (l) H. Pellissier, *Tetrahedron*, **2006**, *62*, 1619–1665. (m) D. Enders, C. Grondal, M. R. M. Hüttl, *Angew. Chem., Int. Ed.*, **2007**, *46*, 1570–1581. (n) G. Guillena, D. J. Ramon, M. Yus, *Tetrahedron: Asymmetry*, **2007**, *18*, 693–700. (o) D. M. D'Souza, T. J. J. Müller, *Chem. Soc. Rev.*, **2007**, *36*, 1095–1108. (p) A.-N. Alba, X. Companyo, M. Viciano, R. Rios, *Curr. Org. Chem.*, **2009**, *13*, 1432–1474. (q) J. E. Biggs-Houck, A. Younai, J. T. Shaw, *Curr. Opin. Chem. Biol.*, **2010**, *14*, 371–382. (r) M. Ruiz, P. Lopez-Alvarado, G. Giorgi, J. C. Menéndez, *Chem. Soc. Rev.*, **2011**, *40*, 3445–3454. (s) L. Albrecht, H. Jiang, K. A. Jorgensen, *Angew. Chem., Int. Ed.*, **2011**, *50*, 8492–8509. (t) H. Pellissier, *Adv. Synth. Catal.*, **2012**, *354*, 237–294. (u) C. De Graaff, E. Ruijter, R. V. A. Orru, *Chem. Soc. Rev.*, **2012**,

41, 3969–4009. (v) H. Clavier, H. Pellissier, *Adv. Synth. Catal.*, **2012**, *18*, 3347–3403. (w) H. Pellissier, *Chem. Rev.*, **2013**, *113*, 442–524. (x) *Asymmetric Domino Reactions*, H. Pellissier, Ed., Royal Society of Chemistry: Cambridge, **2013**. (y) H. Pellissier, *Tetrahedron*, **2013**, *69*, 7171–7210. (z) *Domino Reactions-Concepts for Efficient Organic Synthesis*, L. F. Tietze, Ed., Wiley–VCH: Weinheim, **2014**. (aa) *Enantioselective Multi-catalysed Tandem Reactions*, H. Pellissier, Ed., Royal Society of Chemistry: Cambridge, **2014**.

(11) *Comprehensive Asymmetric Catalysis*, E. N. Jacobsen, A. Pfaltz, H. Yamamoto, eds., Springer, Berlin, **1999**.

(12) (a) E. Knoevenagel, *Chem. Ber.*, **1896**, *29*, 172–174. (b) U. Eder, G. Sauer, R. Wiechert, *Angew. Chem., Int. Ed. Engl.*, **1971**, *10*, 496–497. (c) Z. G. Hajos, D. R. Parrish, *J. Org. Chem.*, **1974**, *39*, 1615–1621. (d) K. A. Ahrendt, C. J. Borths, D. W. C. MacMillan, *J. Am. Chem. Soc.*, **2000**, *122*, 4243–4244. (e) B. List, R. A. Lerner, C. F. Barbas, *J. Am. Chem. Soc.*, **2000**, *122*, 2395–2396. (f) B. List, *Chem. Commun.*, **2006**, 819–824. (g) C. Palomo, A. Mielgo, *Angew. Chem., Int. Ed.*, **2006**, *45*, 7876–7880. (h) A. Erkkilä, I. Majander, P. M. Pihko, *Chem. Rev.*, **2007**, *107*, 5416–5470. (i) S. Mukherjee, J. W. Yang, S. Hoffmann, B. List, *Chem. Rev.*, **2007**, *107*, 5471–5569. (j) X. Yu, W. Wang, *Org. Biomol. Chem.*, **2008**, *6*, 2037–2046.

(13) (a) P. I. Dalko, L. Moisan, *Angew. Chem., Int. Ed.*, **2001**, *40*, 3726–3748. (b) A. Berkessel, H. Gröger, in *Asymmetric Organocatalysis–From Biomimetic Concepts to Powerful Methods for Asymmetric Synthesis*, Wiley–VCH: Weinheim, **2005**. (c) J. Seayad, B. List, *Org. Biomol. Chem.*, **2005**, *3*, 719–724. (d) M. S. Taylor, E. N. Jacobsen, *Angew. Chem., Int. Ed.*, **2006**, *45*, 1520–1543. (e) P. I. Dalko, in *Enantioselective Organocatalysis*, Wiley–VCH: Weinheim, **2007**. (f) P. I. Dalko, *Chimia* **2007**, *61*, 213–218. (g) H. Pellissier, *Tetrahedron*, **2007**, *63*, 9267–9331. (h) A. G. Doyle, E. N. Jacobsen, *Chem. Rev.*, **2007**, *107*, 5713–5743. (i) M. G. Gaunt, C. C. C. Johansson, A. McNally, N. C. Vo, *Drug Discovery Today*, **2007**, *2*, 8–27. (j) *Chem. Rev.*, **2007**, *107(12)*, 5413–5883, Special Issue on Organocatalysis B. List, ed., (k) D. W. C. MacMillan, *Nature*, **2008**, *455*, 304–308. (l) X. Yu, W. Wang, *Chem. Asian. J.*, **2008**, *3*, 516–532. (m) A. Dondoni, A. Massi, *Angew. Chem., Int. Ed.*, **2008**, *47*, 4638–4660. (n) P. Melchiorre, M. Marigo, A. Carlone, G. Bartoli, *Angew. Chem., Int. Ed.*, **2008**, *47*, 6138–6171. (o) F. Peng, Z. Shao, *J. Mol. Catal. A*, **2008**, *285*, 1–13. (p) C. F. Barbas, *Angew. Chem., Int. Ed.*, **2008**, *47*, 42–47. (q) C. Palomo, M. Oiarbide, R. Lopez, *Chem. Soc. Rev.*, **2009**, *38*, 632–653. (r) S. Bertelsen, K. A. Jorgensen, *Chem. Soc. Rev.*, **2009**, *38*, 2178–2189. (s) L.-W. Xu, J. Luo, Y. Lu, *Chem. Commun.*, **2009**, 1807–1821. (t) M. Bella, T. Gasperi, *Synthesis*, **2009**, 1583–1614. (u) L. Gong in *Special Topic: Asymmetric Organocatalysis*, in *Chin. Sci. Bull.*, **2010**, *55*, 1729–1731. (v) B. List in *Asymmetric Organocatalysis*, in *Top. Curr. Chem.*, **2010**, *291*. (w) H. Pellissier, in *Recent Developments in Asymmetric Organocatalysis*, Royal Society of Chemistry: Cambridge, **2010**, (x) *Enantioselective Organocatalyzed Reactions, Vols. I and II*, R. Mahrwald, Ed., Springer: Berlin, **2011**.

(14) (a) R. M. de Figueiredo, M. Christmann, *Eur. J. Org. Chem.*, **2007**, 2575–2600. (b) E. Marquès-López, R. P. Herrera, M. Christmann, *Nat. Prod. Rep.*, **2010**, *27*, 1138–1167.

(15) (a) *Asymmetric Catalysts in Organic Synthesis*, R. Noyori, ed., Wiley, New-York, **1994**. (b) *Catalytic Asymmetric Synthesis*, 2nd ed., I. Ojima, ed., Wiley–VCH, New-York, **2000**. (c) *Transition Metals for Organic Synthesis*, 2nd ed., M. Beller, C. Bolm, eds., Wiley–VCH. Weinheim, **2004**. (d) D. J. Ramon, M. Yus, *Chem. Rev.*, **2006**, *106*, 2126–2208.

(16) S. Ogawa, N. Iida, E. Tokunaga, M. Shiro, N. Shibata, *Chem. Eur. J.*, **2010**, *16*, 7090–7095.

(17) X. Fu, J. Feng, Z. Dong, L. Lin, X. Liu, X. Feng, *Eur. J. Org. Chem.*, **2011**, 5233–5236.

(18) D. Enders, O. Niemeier, O. Henseler, *Chem. Rev.*, **2007**, *107*, 5606–5655.

(19) D. T. Cohen, B. Cardinal-David, J. M. Roberts, A. A. Sarjeant, K. A. Scheidt, *Org. Lett.*, **2011**, *13*, 1068–1071.

(20) A. Paju, K. Oja, K. Matkevits, P. Lumi, I. Järving, T. Pehk, M. Lopp, *Heterocycles*, **2014**, *88*, 981–995.

(21) S. Casolari, D. D'addario, E. Tagliavini, *Org. Lett.*, **1999**, *1*, 1061–1063.

(22) J. Yadav, G. R. Stanton, X. Fan, J. R. Robinson, E. J. Schelter, P. J. Walsh, M. A. Pericas, *Chem. Eur. J.*, **2014**, *20*, 7122–7127.

Chapter 10

Enantioselective Titanium-Catalysed Miscellaneous Reactions

10.1. Introduction

This chapter covers efforts of the chemical community to develop novel enantioselective titanium-catalysed miscellaneous reactions. It looks at developments since the beginning of 2008, when the field was reviewed by Yu and co-workers in a book chapter dealing with titanium Lewis acids.[1] The chapter is divided into three parts. The first part deals with enantioselective titanium-catalysed Michael-type reactions; the second explores enantioselective titanium-catalysed Friedel–Crafts reactions; while the third part includes other enantioselective titanium-catalysed reactions.

10.2. Michael-Type Reactions

The nucleophilic 1,4-addition of stabilised carbon nucleophiles to electron-poor olefins, generally α,β-unsaturated carbonyl compounds, is known as the Michael addition, although it was first reported by Komnenos in 1883.[2] Michael-type reactions can be considered as one of the most powerful and reliable tools for the stereocontrolled formation of carbon to carbon (C–C) and carbon–heteroatom (C–heteroatom) bonds,[3] as has been demonstrated by the huge number of examples in which it has been applied as a key strategic transformation in total synthesis. Compared with the intensively studied asymmetric catalytic cyanation of carbonyl compounds and derivatives, the asymmetric catalytic conjugated cyanation of activated olefins is less explored.

In 2010, Feng *et al.* reported a nice enantioselective titanium-catalysed conjugate addition of ethyl cyanoformate to diethyl alkylidenemalonates.[4] This process was promoted by an interesting modular catalyst generated *in situ* from a chiral cinchona alkaloid (such as cinchonidine), an achiral 3,3'-disubstituted biphenol **1**, and titanium tetraisopropylate, under neat and mild reaction conditions. As shown in Scheme 10.1, a series of diethyl

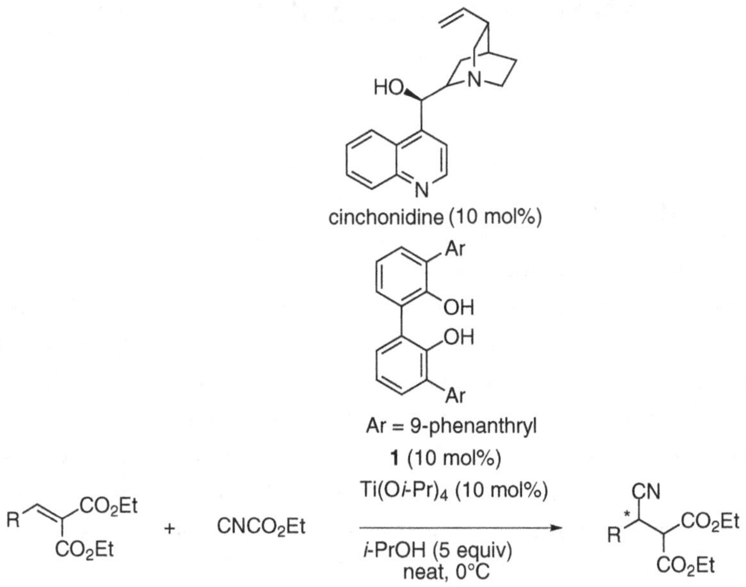

cinchonidine (10 mol%)

Ar = 9-phenanthryl
1 (10 mol%)

$$R\text{-alkylidenemalonate} + CNCO_2Et \xrightarrow[\text{neat, 0°C}]{\substack{Ti(Oi\text{-}Pr)_4 \ (10 \text{ mol\%}) \\ i\text{-PrOH (5 equiv)}}} \text{product}$$

R = Ph: 92% yield, ee = 93%
R = *p*-FC$_6$H$_4$: 97% yield, ee = 92%
R = *p*-ClC$_6$H$_4$: 87% yield, ee = 91%
R = *p*-BrC$_6$H$_4$: 90% yield, ee = 89%
R = *p*-Tol: 91% yield, ee = 92%
R = *p*-MeOC$_6$H$_4$: 92% yield, ee = 91%
R = *p*-PhC$_6$H$_4$: 99% yield, ee = 92%
R = *m*-Tol: 94% yield, ee = 94%
R = *m*-MeOC$_6$H$_4$: 84% yield, ee = 92%
R = 2-Naph: 93% yield, ee = 92%
R = Cy: 53% yield, ee = 88%
R = *i*-Pr: 68% yield, ee = 86%
R = *n*-Pent: 71% yield, ee = 74%

Scheme 10.1. Conjugated cyanation of diethyl alkylidenemalonates with a chiral cinchona alkaloid ligand alkylidenemalonates were tolerated, providing the corresponding 1,4-products in high yields and enantioselectivities of up to 94% ee in the case of aromatic substrates. For aliphatic substrates, both the yields and enantioselectivities were lower but still good (up to 71% yield and 88% ee, respectively). The utility of this highly efficient process was demonstrated with the conversion of the chiral products into many products of great utility and pharmaceutical importance, such as key precursors for the *gamma-aminobutyric acid* (GABA) synthesis. Moreover, the process presented the advantage of using a relatively cheap and less toxic cyanide source than usually employed trimethylsilyl cyanide (TMSCN), and substrates of low cost and easy preparation or availability.

Unfortunately, the preceding catalytic system was not compatible with the use of nitroalkenes as the substrates. Given this, Wang *et al.* have developed enantioselective cyanations of nitroolefins via silyl nitronate intermediates, achieved by using an *in situ* generated titanium catalyst of chiral salen ligand 2.[5] As shown in Scheme 10.2, this protocol was applied to a series of alkyl nitroolefins in good yields and with moderate to high enantioselectivities of up to 84% ee. The silyl nitronate intermediate pathway was confirmed through an *in situ* ^{1}H NMR investigation. The chiral

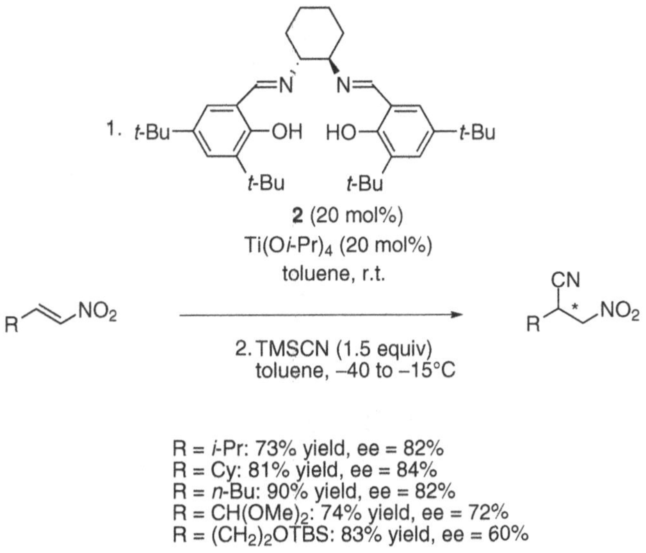

$$R = i\text{-Pr: } 73\% \text{ yield, ee} = 82\%$$
$$R = Cy: 81\% \text{ yield, ee} = 84\%$$
$$R = n\text{-Bu: } 90\% \text{ yield, ee} = 82\%$$
$$R = CH(OMe)_2: 74\% \text{ yield, ee} = 72\%$$
$$R = (CH_2)_2OTBS: 83\% \text{ yield, ee} = 60\%$$

Scheme 10.2. Conjugated cyanation of alkyl nitroalkenes with a chiral salen ligand products could be readily transformed into the corresponding β-amino acids, since β-peptides have received a growing attention for their potential pharmaceutical applications.

In 2013, North and Watson demonstrated that using preformed and structurally well-defined titanium salen complexes, made it possible to enhance the catalytic activity observed, by using *in situ* generated titanium salen complexes (Scheme 10.2) in these reactions.[6] As shown in Scheme 10.3, when the enantioselective conjugate addition of TMSCN to aliphatic nitroalkenes was performed with preformed chiral bis-titanium salen catalyst **3**, it yielded the corresponding chiral nitronitriles in higher yields and enantioselectivities of up to 93% and 89% ee, respectively, than those obtained by using the *in situ* generated titanium catalyst from salen ligand **2**. Importantly, only 2 mol% of catalyst loading of complex **3** was necessary to achieve these results, while the catalyst system derived from ligand **2** required 20 mol% of catalyst loading. Furthermore, the reactions catalysed by **3** were performed at 0°C, whereas those catalysed by the *in situ* generated catalyst from ligand **2** required –40°C rising to –15°C to be efficient.

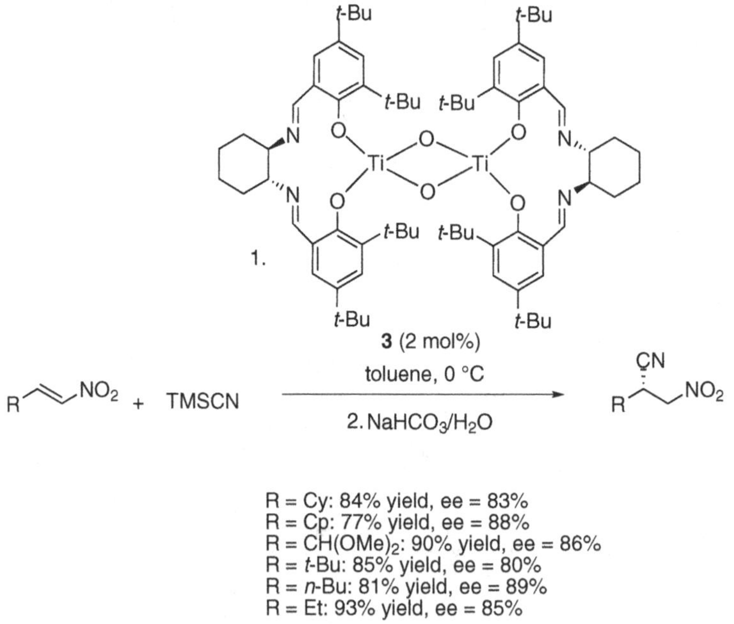

R = Cy: 84% yield, ee = 83%
R = Cp: 77% yield, ee = 88%
R = CH(OMe)$_2$: 90% yield, ee = 86%
R = *t*-Bu: 85% yield, ee = 80%
R = *n*-Bu: 81% yield, ee = 89%
R = Et: 93% yield, ee = 85%

Scheme 10.3. Conjugated cyanation of alkyl nitroalkenes with a preformed chiral *bis*-titanium salen catalyst

10.3. Friedel–Crafts Reactions

The Friedel–Crafts reaction is one of the most fundamental C–C bond-forming reactions in organic synthesis.[7] This reaction can be promoted by the presence of a Lewis acid, generally used in more than a stoichiometric amount to prevent the promoter deactivation by complexation to the ketone products. However, a wide number of catalytic versions of the Friedel–Crafts reaction have been developed using various catalysts, including asymmetric ones. This chemistry has been intensively explored since 2000, and interest in this field is still increasing.[8] Remarkable works have been recently reported by the group of Jurczak, using an *in situ* generated chiral titanium catalyst from BINOL-derived ligand **4**. Initially, these authors developed a highly enantioselective synthesis of 2-furanyl-hydroxyacetates on the basis of the Friedel–Crafts reaction of *n*-butyl glyoxylate with variously substituted furans, performed in the presence of titanium tetraisopropylate (2–5 mol%) and the same catalytic amount of 6,6′-dibromo-BINOL ligand **4**.[9] As shown in Scheme 10.4,

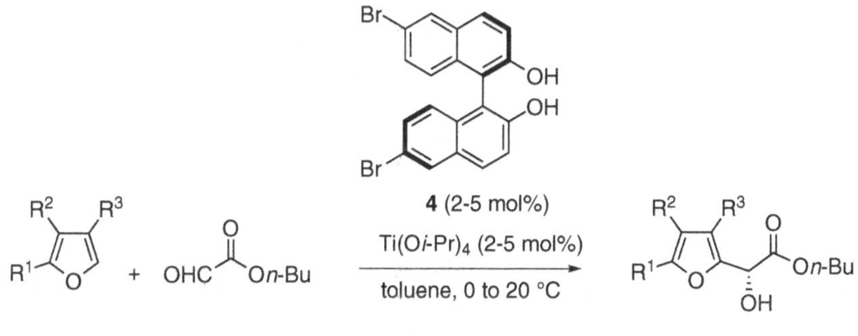

R^1 = Me, R^2 = R^3 = H: 92% yield, ee = 97%
R^1 = R^2 = R^3 = H: 45% yield, ee = 91%
R^1 = Et, R^2 = R^3 = H: 98% yield, ee = 94%
R^1 = *n*-Bu, R^2 = R^3 = H: 80% yield, ee = 90%
R^1 = Bn, R^2 = R^3 = H: 90% yield, ee = 95%
R^1 = CH_2OBn, R^2 = R^3 = H: 85% yield, ee = 95%
R^1 = Ph, R^2 = R^3 = H: 90% yield, ee = 90%
R^1 = *p*-BrC_6H_4, R^2 = R^3 = H: 96% yield, ee = 88%
R^1 = TMS, R^2 = R^3 = H: 76% yield, ee = 89%
R^1 = R^2 = Me, R^3 = H: 77% yield, ee = 96%
R^1 = R^3 = Me, R^2 = H: 95% yield, ee = 96%

Scheme 10.4. Friedel–Crafts reaction of furans with *n*-butyl glyoxylate with a chiral BINOL-derived ligand a range of chiral polyfunctionalised products were produced under simple conditions (no need for dry toluene or an inert atmosphere) in good to high yields and high enantioselectivities in the range of 90–97% ee in most examples.

These mild reaction conditions were also applied by the same authors to the reaction of thiophenes with *n*-butyl glyoxylate, which afforded the corresponding chiral hydroxyl(thiophene-2-yl)acetates in comparable yields but even higher enantioselectivities (92–98% ee) than those obtained with furans (Scheme 10.5).[10] This result constituted the first report on the efficient asymmetric Friedel–Crafts reaction of thiophenes with alkyl glyoxylates. Starting from simple thiophene and *n*-butyl glyoxylate, the authors demonstrated the formal synthesis of duloxetine.

In addition, the scope of this methodology was extended by these authors to the synthesis of the corresponding chiral pyrrole deriva-tives.[11] Indeed, the Friedel–Crafts reaction of pyrroles having one electron-withdrawing group at the α, β, or *N*-positions with *n*-butyl

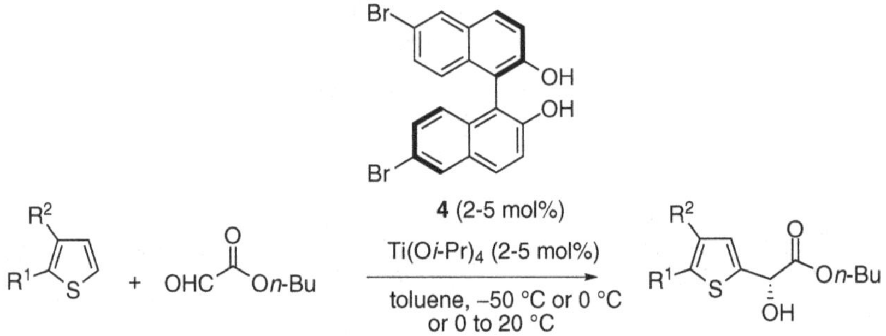

R^1 = Me, R^2 = H: 78% yield, ee = 94%
R^1 = Et, R^2 = H: 80% yield, ee = 95%
R^1 = *n*-Bu, R^2 = H: 80% yield, ee = 98%
R^1 = Bn, R^2 = H: 76% yield, ee = 96%
R^1 = OMe, R^2 = H: 92% yield, ee = 94%
R^1 = Ph, R^2 = H: 85% yield, ee = 94%
R^1 = TMS, R^2 = H: 65% yield, ee = 92%
R^1 = R^2 = H: 40% yield, ee = 92%
R^1 = R^2 = Me: 83% yield, ee = 93%

Scheme 10.5. Friedel–Crafts reaction of thiophenes with *n*-butyl glyoxylate with a chiral BINOL-derived ligand

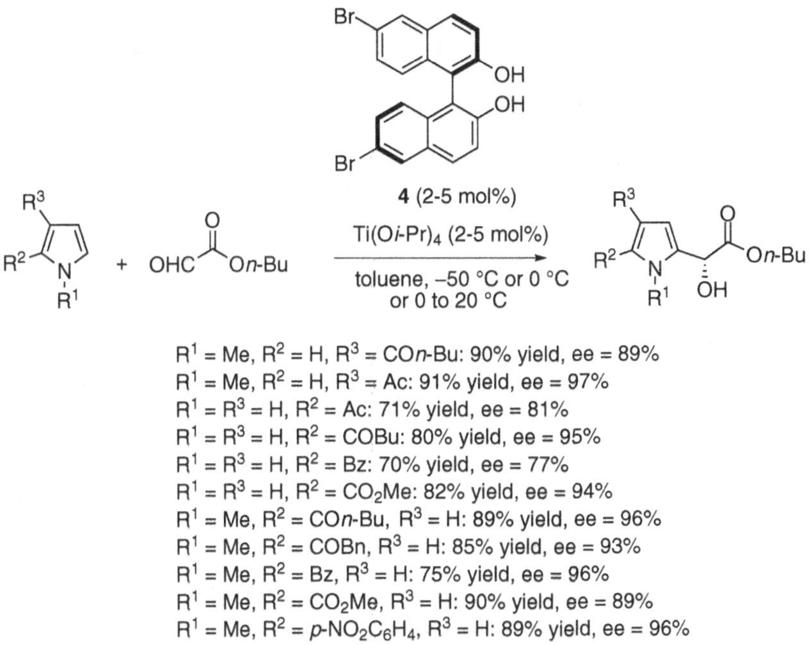
Br
OH
OH
Br
4 (2-5 mol%)
Ti(O*i*-Pr)4 (2-5 mol%)
toluene, –50 °C or 0 °C
or 0 to 20 °C
R3
R2
N
R1
OHC
O
O*n*-Bu
OH
R1 = Me, R2 = H, R3 = CO*n*-Bu: 90% yield, ee = 89%
R1 = Me, R2 = H, R3 = Ac: 91% yield, ee = 97%
R1 = R3 = H, R2 = Ac: 71% yield, ee = 81%
R1 = R3 = H, R2 = COBu: 80% yield, ee = 95%
R1 = R3 = H, R2 = Bz: 70% yield, ee = 77%
R1 = R3 = H, R2 = CO2Me: 82% yield, ee = 94%
R1 = Me, R2 = CO*n*-Bu, R3 = H: 89% yield, ee = 96%
R1 = Me, R2 = COBn, R3 = H: 85% yield, ee = 93%
R1 = Me, R2 = Bz, R3 = H: 75% yield, ee = 96%
R1 = Me, R2 = CO2Me, R3 = H: 90% yield, ee = 89%
R1 = Me, R2 = *p*-NO2C6H4, R3 = H: 89% yield, ee = 96%

Scheme 10.6. Friedel–Crafts reaction of pyrroles with *n*-butyl glyoxylate with a chiral BINOL-derived ligand

glyoxylate led, under the same reaction conditions, to the corresponding chiral pyrrole-hydroxyacetic acid derivatives in good to excellent yields (70–96%) and once again high enantioselectivities of up to 96% ee, as shown in Scheme 10.6. All *N*-methyl substituted pyrroles provided enantioselectivities ranging from 89–96% ee, whereas the lowest enantioselectivity (77% ee) was obtained in the case of an *N*-unsubstituted pyrrole. All these products provided an easy access to important synthons in chemical synthesis, which are commonly found in biologically active molecules, chiral catalysts, and receptors.

10.4. Other Reactions

In 2009, Zhang *et al.* developed a new family of achiral 3,3′,5,5′-tetrasubstituted-2,2′,6,6′-tetrahydroxy biphenyl ligand. The axial chirality of this ligand could be induced by the chelation of 2,2′,6,6′-tetrahydroxy groups

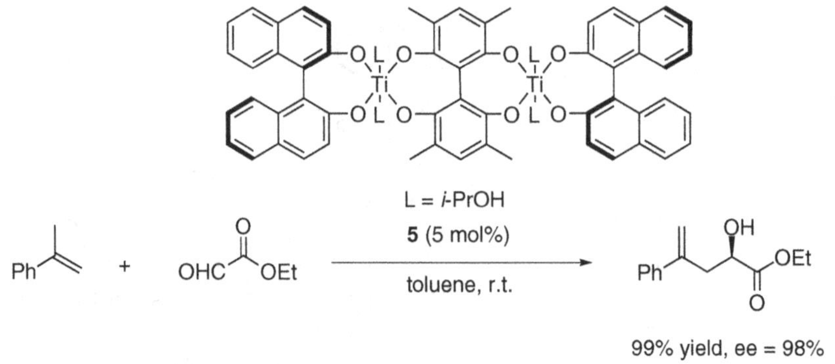

L = *i*-PrOH

5 (5 mol%)

toluene, r.t.

99% yield, ee = 98%

Scheme 10.7. Carbonyl-ene reaction of methylstyrene with ethyl glyoxylate with a chiral BINOL-derived bis-titanium catalyst

with (R)-BINOL-Ti(O*i*-Pr)$_2$ to form an axially chiral bimetallic titanium catalyst **5**.[12] This novel BINOL-derived bis-titanium catalyst **5** was shown to exhibit an excellent activity and enantioselectivity for the carbonyl-ene reaction of methylstyrene with ethyl glyoxylate, compared with (R)-BINOL-Ti(O*i*-Pr)$_2$ catalyst system. As shown in Scheme 10.7, the enantiopure product of the reaction, promoted by only 5 mol% of catalyst **5** at room temperature, was remarkably achieved in almost quantitative yield. The authors demonstrated that the presence of 3,3′,5,5′-tetrasubstituted groups on the catalyst had a remarkable effect on both the enantioselectivity and yield of the process. Additionally, it was shown that the bimetallic catalyst **5** showed better catalytic capability than the corresponding monometallic catalyst. Later, the same reaction was investigated by Yang *et al.* in the presence of a (R)-6,6′-Br$_2$-BINOL-derived titanium catalyst encapsulated in the nanocages of mesoporous silicas, affording the same carbonyl-ene product in 95% yield and 92% ee.[13]

In another context, a highly enantioselective synthesis of (+)- and (−)-fluvastatin was reported by Hayashi *et al.*, the key step of which was an enantioselective titanium-catalysed addition of diketene to a complex aldehyde.[14] As shown in Scheme 10.8, the corresponding chiral alcohol was achieved in 78% yield and 91% ee when the reaction was performed in the presence of 1.1 equivalent of titanium tetraisopropylate and chiral Schiff base ligand **6**. Notably, using only a catalytic amount of this catalytic system (0.5 mol%) provided both lower yield and enantioselectivity

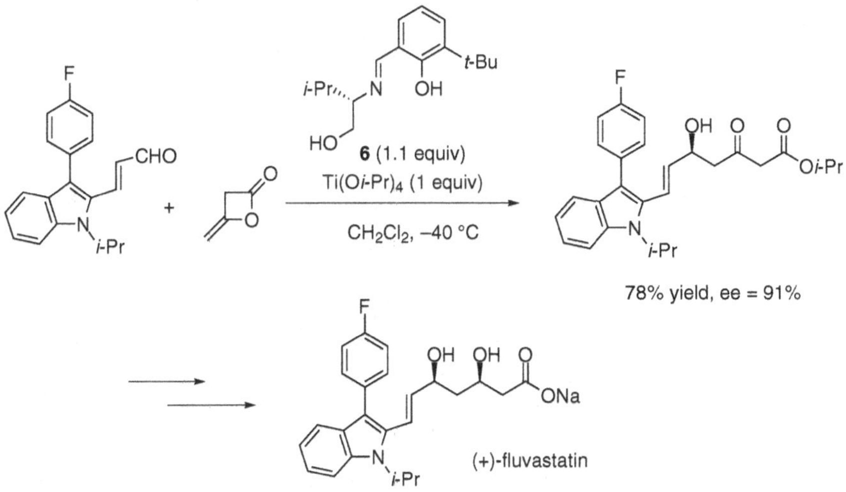

Scheme 10.8. Addition of diketene to an aldehyde with a chiral Schiff base ligand

of 73% and 83% ee, respectively. This key alcohol intermediate was subsequently transformed into fluvastatin, as expected.

The α-heterofunctionalisation of a carbonyl compound is a highly direct and strategically simple method for the synthesis of a large number of interesting molecules and synthetic building blocks, such as amino acids through amination, α-hydroxy acids through hydroxylation, and α-fluorinated products.[15] Asymmetric versions for fluorination, chlorination, and bromination of carbonyl compounds have all be developed since 2008, with fluorination attracting the greatest level of interest[16] due to the fact that fluoroorganic compounds have peculiar properties, rendering them interesting for a variety of specific applications.[17] Among these asymmetric versions, Togni *et al.* developed enantioselective titanium-catalysed fluorinations of β-ketoesters, using 1-(chloromethyl)-4-fluoro-1,4-diazoniabicyclo[2.2.2]octane bis(tetrafluoroborate) as the halogenating agent.[18] The reaction was performed at room temperature, and was promoted by 5 mol% of a chiral preformed titanium catalyst **7** derived from TADDOL. It afforded the corresponding fluorinated chiral products in moderate to high yields, and enantioselectivities of up to 91% ee, as shown in Scheme 10.9. Increasing the size of the benzyl ester group in β-ketoesters led to higher enantioselectivities. However, an equally

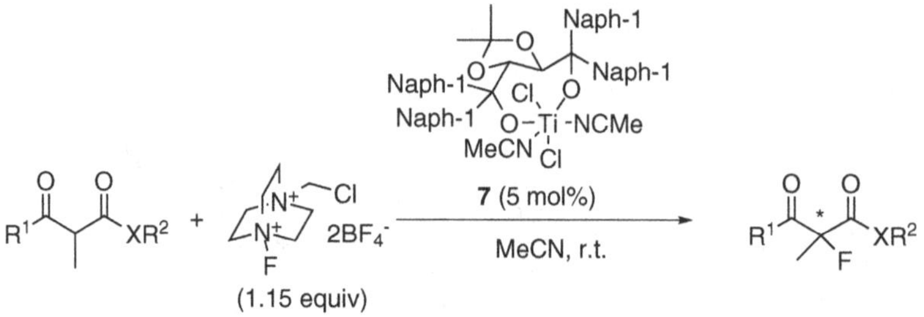

R^1 = Ph, R^2 = Bn, X = O: 85% yield, ee = 70%
R^1 = Ph, R^2 = F_5C_6, X = O: 44% yield, ee = 62%
R^1 = p-$MeOC_6H_4$, R^2 = Me, X = O: 94% yield, ee = 61%
R^1 = 2-Naph, R^2 = Et, X = O: 89% yield, ee = 61%
R^1 = Me, R^2 = 3,5-$(Me)_2C_6H_3$, X = O: 57% yield, ee = 74%
R^1 = Et, R^2 = 2,4,6-$(i\text{-}Pr)_3C_6H_2$, X = O: 89% yield, ee = 80–90%
R^1 = Et, R^2 = Ph, X = O: 50% yield, ee = 87-88%
R^1 = Et, R^2 = $CHPh_2$, X = O: 87% yield, ee = 81%
R^1 = Ph, R^2 = Bn, X = S: 39% yield, ee = 78%
R^1 = Et, R^2 = Ph, X = S: 78% yield, ee = 90–91%

Scheme 10.9. Fluorination of β-keto (thio)esters with a chiral titanium TADDOLate catalyst

high enantioselectivity was obtained with simple phenyl ester substrates. The scope of the methodology was extended to the fluorination of β-keto thioesters, with comparable results. Moreover, when the protocol was applied to the reaction of other activated carbonyl derivatives, such as β-ketoamides, it gave the corresponding products in moderate enantioselectivities (≤ 55% ee), while the use of 1,3-diketones as substrates afforded the corresponding fluorinated products in high yields, albeit with low enantioselectivities (≤ 27% ee). On the other hand, thioesters and dithioesters were not suitable substrates.

In 2013, Burns *et al.* described a novel catalytic enantioselective dibromination of allylic alcohols, constituting a new strategy for alkene dihalogenation.[19] This protocol employed a unique combination of reagents (such as ethyl dibromomalonate) as the bromonium source, and a titanium bromide species as the bromide source. The chirality was generated by using a chiral TADDOL ligand **8**, which allowed, when using in a stoichiometric quantity, the corresponding dibrominated products to be synthesised in moderate to good yields combined with high enantioselectivities of up to 90% ee. When this ligand was employed at 20 mol% of catalyst

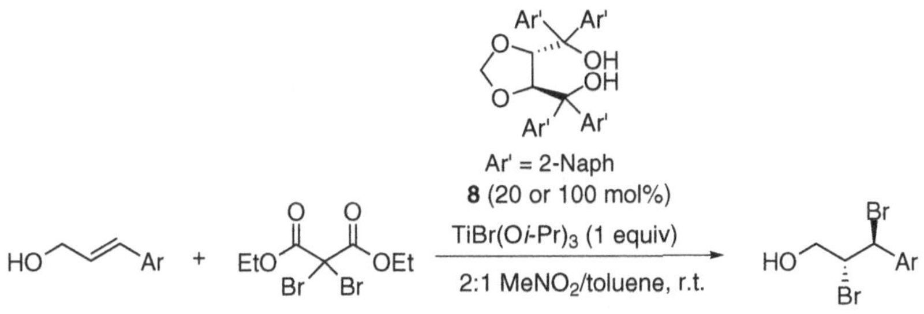

with 100 mol% of ligand:
Ar = Ph: 63% yield, ee = 86%
Ar = *p*-FC_6H_4: 65% yield, ee = 88%
Ar = *p*-ClC_6H_4: 64% yield, ee = 90%
Ar = *p*-BrC_6H_4: 64% yield, ee = 90%
Ar = *p*-$(CH{=}CH_2)C_6H_4$: 37% yield, ee = 80%
with 20 mol% of ligand:
Ar = Ph: 60% yield, ee = 76%
Ar = *p*-FC_6H_4: 47% yield, ee = 79%
Ar = *p*-ClC_6H_4: 60% yield, ee = 85%
Ar = *p*-BrC_6H_4: 51% yield, ee = 84%
Ar = *p*-$(CH{=}CH_2)C_6H_4$: 31% yield, ee = 71%

Scheme 10.10. Dibromination of allylic alcohols with a chiral TADDOL ligand

loading, both the yields and enantioselectivities were lower (≤ 85% ee), as shown in Scheme 10.10. The synthetic utility of the enantioenriched dibromides formed was demonstrated through their derivatisations into variously functionalised chiral monobromides.

Also in 2013, novel tetradentate bisoxazoline ligands derived from L-serine were reported by Wen *et al.* and further applied to promote an enantioselective titanium-catalysed pinacol coupling reaction of aromatic aldehydes.[20] The applications of these ligands revealed that the absolute configurations of the resulting pinacols were entirely dominated by the biphenyl component of the ligand, and that the bulky substituent adjacent to the hydroxyl group was favorable for increasing both the diastereo- and enantio-selectivities of the process. Among the ligands screened, ligand **9** exhibited the best asymmetric induction to furnish the pinacols with *dl*/*meso*-selectivity ratios of up to 99:1 and enantiomeric excesses of 63–89% ee, along with high yields (Scheme 10.11).

Finally, in 2014, Maruoka *et al.* described an enantioselective intramolecular cyclisation of *N*-aryl diazoamides, using an *in situ* generated

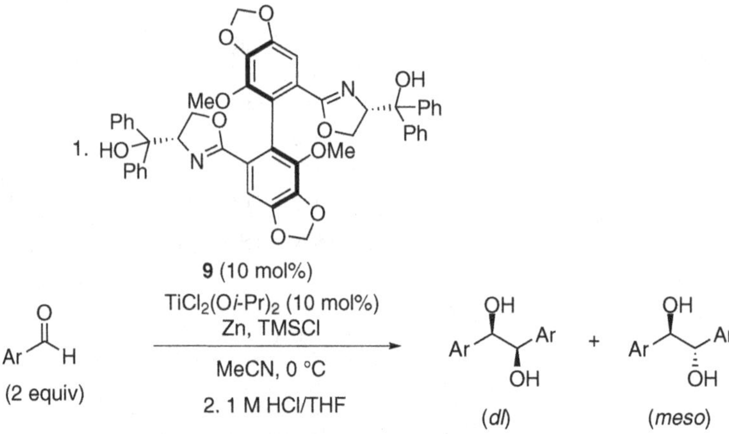
MeO
Ph
HO
Ph
N
O
OH
Ph
Ph
OMe
1.
9 (10 mol%)
TiCl2(Oi-Pr)2 (10 mol%)
Zn, TMSCl
MeCN, 0 °C
2. 1 M HCl/THF
Ar
H
(2 equiv)
(dl)
(meso)

Ar = Ph: 92% yield, *dl/meso* = 97:3, ee = 86%
Ar = *p*-Tol: 95% yield, *dl/meso* = 99:1, ee = 88%
Ar = *p*-MeOC_6H_4: 93% yield, *dl/meso* = 99:1, ee = 89%
Ar = *p*-ClC_6H_4: 91% yield, *dl/meso* = 91:9, ee = 81%
Ar = 2,4-$Cl_2C_6H_3$: 83% yield, *dl/meso* = 82:18, ee = 63%
Ar = 1-Naph: 90% yield, *dl/meso* = 98:2, ee = 87%

Scheme 10.11. Pinacol coupling reaction of aromatic aldehydes with a chiral bisoxazoline ligand

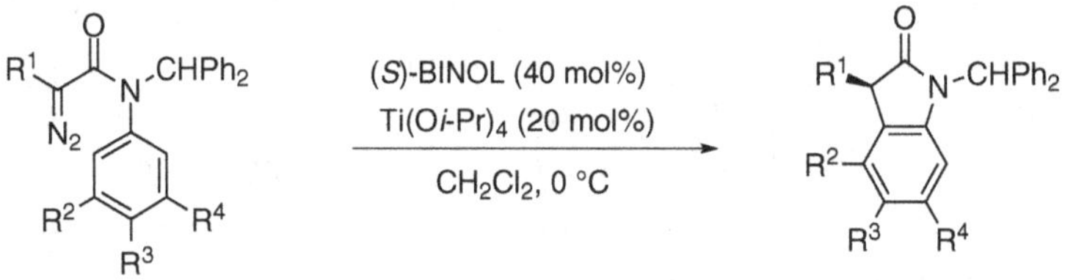
CHPh2
N2
(S)-BINOL (40 mol%)
Ti(Oi-Pr)4 (20 mol%)
CH2Cl2, 0 °C
N-CHPh2

$R^1 = R^3$ = Me, $R^2 = R^4$ = H: 76% yield, ee = 93%
R^1 = Me, $R^2 = R^4$ = H, R^3 = F: 59% yield, ee = 94%
R^1 = Me, $R^2 = R^4$ = H, R^3 = Cl: 58% yield, ee = 96%
R^1 = Me, $R^2 = R^4$ = H, R^3 = OMe: 64% yield, ee = 93%
R^1 = Me, $R^2 = R^4$ = OMe, R^3 = H: 92% yield, ee = 92%
R^1 = Me, $R^2 = R^3$ = H, R^4 = OMe: 64% yield, ee = 93%
R^1 = Me, R^2 = OMe, $R^3 = R^4$ = H: 9% yield, ee = 95%
R^1 = Et, $R^2 = R^3 = R^4$ = H: 61% yield, ee = 94%
R^1 = Bn, $R^2 = R^3 = R^4$ = H: 74% yield, ee = 85%
R^1 = $CH_2C(Me){=}CH_2$, $R^2 = R^3 = R^4$ = H: 63% yield, ee = 92%

Scheme 10.12. Intramolecular cyclisation of *N*-aryl diazoamides with (*S*)-BINOL ligand titanium catalyst from titanium tetraisopropylate and (S)-BINOL.[21] A formal asymmetric C(sp^2)–H insertion afforded the corresponding chiral oxindoles in low to high yields, with uniformly very good enantioselectivities of up to 96% ee, as shown in Scheme 10.12. The products could be further reduced with borane into the corresponding indoles without deterioration of the enantioselectivity.

10.5. Conclusions

Since 2008, very interesting new enantioselective titanium-catalysed reactions were successfully developed. In particular, high enantioselectivities of up to 94% ee were reported by Feng for an enantioselective titanium-catalysed conjugate addition of ethyl cyanoformate to diethyl alkylidenemalonates. In another context, remarkable results were achieved by Kureshy in titanium-catalysed Friedel–Crafts reactions of furans, thiophenes, and pyrroles with *n*-butyl glyoxylate using a chiral BINOL-derived ligand, with uniformly achieved excellent enantioselectivities of up to 97%, 98%, and 97% ee, respectively. Meanwhile, Zhang showed that a BINOL-derived bis-titanium catalyst exhibited an excellent activity and enantioselectivity (98% ee) for the carbonyl-ene reaction of methylstyrene with ethyl glyoxylate. In addition, a nice enantioselective intramolecular cyclisation of *N*-aryl diazoamides was reported by Maruoka, providing excellent enantioselectivities of up to 96% ee by using (S)-BINOL as titanium ligand. The ever-growing need for environmentally friendly catalytic processes has prompted organic chemists to focus on more abundant first-row transition metals such as titanium for developing new catalytic systems to perform reactions, such as C–C bond formations, C–heteroatoms bond formations, or C–H functionalisations. A bright future is undeniable for more sustainable novel and enantioselective titanium-catalysed transformations.

References

(1) Y. Yu, K. Ding, G. Chen, in *Acid Catalysis in Modern Organic Synthesis*, H. Yamamoto, K. Ishihara, eds., Wiley–VCH, Weinheim, **2008**, Chapter 14, pp. 721–823.

(2) T. Komnenos, *Justus Liebigs Ann. Chem.*, **1883**, *218*, 145–169.

(3) (a) P. Perlmutter, in *Conjugate Addition Reactions in Organic Synthesis*, Pergamon Press: Oxford, **1992**. (b) N. Krause, A. Hoffmann-Röder, *Synthesis*, **2001**, 171–196.

(c) M. P. Sibi, S. Manyem, *Tetrahedron,* **2000**, *56,* 8033–8061. (d) M. Kanai, M. Shibasaki, in *Catalytic Asymmetric Synthesis,* 2nd ed., Wiley, New York, **2000**, p. 569.

(4) J. Wang, W. Li, Y. Liu, Y. Chu, L. Lin, X. Liu, X. Feng, *Org. Lett.,* **2010**, *12,* 1280–1283.

(5) L. Lin, W. Yin, X. Fu, J. Zhang, X. Ma, R. Wang, *Org. Biomol. Chem.,* **2012**, *10,* 83–89.

(6) M. North, J. M. Watson, *ChemCatChem,* **2013**, *5,* 2405–2409.

(7) (a) *Friedel–Crafts Chemistry,* G. A. Olah, ed., Wiley, New York, **1973**. (b) G. A. OLah, R. Krishnamurti, G. K. S. Prakash, in *Comprehensive Organic Synthesis,* B. M. Trost, I. Fleming, eds., Pergamon Press, Oxford, **1991**, Vol. 3, pp. 293–339.

(8) (a) V. Terrasson, R. M. Figueiredo, R. M. Campagne, *Eur. J. Org. Chem.,* **2010**, 2635–2665. (b) S.-L. You, Q. Cai, M. Zeng, *Chem. Soc. Rev.,* **2009**, *38,* 2190–2201. (c) M. Bandini, A. Umani-Ronchi, *Catalytic Asymmetric Friedel–Crafts Alkylations,* Wiley, Weinheim, **2009**. (d) T. B. Poulsen, K. A. Jorgensen, *Chem. Rev.,* **2008**, *108,* 2903–2915.

(9) J. Majer, P. Kwiatkowski, J. Jurczak, *Org. Lett.,* **2008**, *10,* 2955–2958.

(10) J. Majer, P. Kwiatkowski, J. Jurczak, *Org. Lett.,* **2009**, *11,* 4636–4639.

(11) J. Majer, P. Kwiatkowski, J. Jurczak, *Org. Lett.,* **2011**, *13,* 5944–5947.

(12) F. Fang, F. Xie, H. Yu, H. Zhang, B. Yang, W. Zhang, *Tetrahedron Lett.,* **2009**, *50,* 6672–6675.

(13) X. Liu, S. Bai, Y. Yang, B. Li, B. Xiao, C. Li, Q. Yang, *Chem. Commun.,* **2012**, *48,* 3191–3193.

(14) J. T. Zacharia, T. Tanaka, M. Hayashi, *J. Org. Chem.,* **2010**, *75,* 7514–7518.

(15) A. M. R. Smith, K. K. Hii, *Chem. Rev.,* **2011**, *111,* 1637–1656.

(16) V. A. Brunet, D. O'Hagan, *Angew. Chem., Int. Ed.,* **2008**, *47,* 1179–1182.

(17) J. Ma, D. Cahard, *Chem. Rev.,* **2004**, *104,* 6119–6146.

(18) (a) L. Hintermann, M. Perseghini, A. Togni, *Belstein J. Org. Chem.,* **2011**, *7,* 1421–1435. (b) A. Bertogg, L. Hintermann, D. P. Huber, M. Perseghini, M. Sanna, A. Togni, *Helv. Chim. Acta,* **2012**, *95,* 353–403.

(19) D. X. Hu, G. M. Shibuya, N. Z. Burns, *J. Am. Chem. Soc.,* **2013**, *135,* 12960–12963.

(20) J. Wen, L. Liu, X. Zhou, R. Hu, Y. Xu, *Tetrahedron: Asymmetry,* **2013**, *24,* 860–865.

(21) T. Hashimoto, K. Yamamoto, K. Maruoka, *Chem. Commun.,* **2014**, *50,* 3220–3223.

General Conclusion

The enantioselective production of compounds is a central theme in current research. The broad utility of synthetic chiral molecules as single-enantiomer pharmaceuticals; in electronic and optical devices; as components in polymers with novel properties; and as probes of biological function, has made asymmetric synthesis a prominent area of investigation. Nearly all natural products are chiral, and their physiological or pharmacological properties depend upon their recognition by chiral receptors, which will interact only with molecules of the proper absolute configuration. Indeed, the use of chiral drugs in enantiopure form is now a standard requirement for virtually every new chemical entity, and the development of new synthetic methods to obtain enantiopure compounds has become a key goal for pharmaceutical companies. The search for new and efficient methods for the synthesis of optically pure compounds constitutes an active area of research in organic synthesis.

Of the methods available for preparing chiral compounds, catalytic asymmetric synthesis has attracted most attention. In particular, asymmetric transition-metal catalysis has emerged as a powerful tool to perform reactions in a highly enantioselective fashion over the past few decades. Efforts to develop new asymmetric transformations focused predominantly on the use of few metals, such as titanium, nickel, copper, ruthenium, rhodium, palladium, iridium, and more recently gold. However, by the very fact of the lower costs of titanium catalysts in comparison with other transition metals, and their nontoxicity which has permitted their use for medical purposes (prostheses), enantioselective titanium-promoted transformations have received ever-growing attention during the last decades, leading to exciting and fruitful researches.[1] This interest might also be related to the fact that titanium complexes are of high abundance, exhibit a remarkably diverse chemical reactivity, and constitute one of the most useful Lewis acids in asymmetric catalysis.

This book illustrates how much enantioselective titanium catalysis has contributed to the development of various types of enantioselective ecological and economical transformations. It updates the major progress in the field of enantioselective reactions promoted by chiral titanium catalysts, illustrating the power of these green catalysts of lower costs to provide all types of organic reactions — from basic ones, such as nucleophilic additions, cycloadditions, oxidations, and reductions; to completely novel methodologies, including, for example, domino reactions.

The excellent capability for the control of the stereochemistry demonstrated by titanium complexes in the catalysis can be attributed to their rich coordination chemistry and facile modification of titanium Lewis acid centre by structurally modular ligands. Despite the impressive number of results reported in the last seven years, however, many challenges remain, such as a better understanding of the mechanisms of the reactions, the role of the aggregation of titanium complexes, as well as the improving of the scope and mildness of the reaction conditions in some cases of transformations. The ever-growing need for environmentally friendly catalytic processes has prompted organic chemists to focus on more abundant first-row transition metals, such as titanium, to develop new catalytic systems that promote all types of organic reactions, such as C–C bond formations, C–heteroatoms bond formations, C–H functionalisations, oxidations, and reduction reactions. Therefore, a bright future is undeniable for more sustainable novel and enantioselective titanium-promoted transformations.

References

(1) (a) Y. Yu, K. Ding, G. Chen in *Acid Catalysis in Modern Organic Synthesis*, H. Yamamoto, K. Ishihara, eds., Wiley–VCH: Weinheim, **2008**, Chapter 14, p. 721. (b) D. J. Ramon, M. Yus, *Chem. Rev.* **2006**, 106, 2126. (c) F. Chen, X. Feng, Y. Jiang, *Arkivoc* **2003**, *ii*, 21. (d) J. Cossy, S. Bouz, F. Pradaux, C. Willis, V. Bellosta, *Synlett*

2002, 1595. (e) K. Mikami, M. Terada in *Lewis Acids in Organic Synthesis*, H. Yamamoto, ed., Wiley–VCH: Weinheim, **2000**, Vol. 2, Chapter 16, p 799. (f) K. Mikami, M. Terada in *Lewis Acid Reagents*, H. Yamamoto, ed., Wiley–VCH: Weinheim, **1999**, Vol. 1, Chapter 6, p. 93. (g) D. Ramon, M. Yus *Rec. Res. Dev. Org. Chem.* **1998**, *2*, 489. (h) A. H. Hoveyda, J. P. Morken, *Angew. Chem., Int. Ed. Engl.* **1996**, *35*, 1262. (i) B. M. Trost in *Stereocontrolled Organic Synthesis*, B. M. Trost, ed., Blackwell: Oxford, **1994**, p. 17.

Index

Printed in the United States
By Bookmasters